The Sociology of Farming

W0081292

This book provides a detailed and comprehensive introduction to the concepts and methods of the sociology of farming.

The sociology of farming focuses on co-production: the ongoing interaction and mutual transformation of the natural and the social (of 'human and living nature') which requires putting the farm labour process centre stage. While there are many books which discuss food and agriculture, this book is different: it delves into the methods and concepts used and presents a comprehensive conceptual framework and the associated methods for research to give students and research-ers of agriculture and rural studies a solid set of tools for unravelling the complex-ities of farming and rural life. Importantly, these tools also empower us to design new ways forward. A wide array of case studies, as wide-ranging as Brazil, Peru, China, the Netherlands, Italy and Guinea Bissau, help readers to grasp the com-monalities that underlie strongly diversified and divided rural worlds. The book lists over two hundred basic concepts and includes boxes that discuss the main methods of the sociology of farming.

This textbook is essential reading for students and scholars of food and agri-culture, agrarian studies, rural development, food and farming systems, peasant studies and environmental sociology.

Jan Douwe van der Ploeg is Emeritus Professor at Wageningen University, the Netherlands and an Adjunct Professor in the College of Humanities and Development Studies at the China Agricultural University in Beijing. He is the author of *The New Peasantries: Rural Development in Times of Globalization* (Routledge, 2018).

Earthscan Food and Agriculture

The Sociology of Farming
Concepts and Methods

Jan Douwe van der Ploeg

LONDON AND NEW YORK

Cover image: © Getty Images

First published 2023
by Routledge
4 Park Square, Milton Park, Abingdon, Oxon OX14 4RN

and by Routledge
605 Third Avenue, New York, NY 10158

Routledge is an imprint of the Taylor & Francis Group, an informa business

© 2023 Jan Douwe van der Ploeg

British Library Cataloguing-in-Publication Data
A catalogue record for this book is available from the British Library

Library of Congress Cataloging-in-Publication Data
Names: Ploeg, Jan Douwe van der, 1950- author.
Title: The sociology of farming : concepts and methods /
Jan Douwe van der Ploeg.
Description: New York, NY : Routledge, 2023. | Includes bibliographical references and index.
Identifiers: LCCN 2022014946 (print) | LCCN 2022014947 (ebook) |
ISBN 9781032321905 (hardback) | ISBN 9781032321875 (paperback) |
ISBN 9781003313274 (ebook)
Subjects: LCSH: Agriculture--Social aspects. | Sociology.
Classification: LCC S419 .P56 2023 (print) | LCC S419 (ebook) |
DDC 630.9--dc23/eng/20220722
LC record available at https://lccn.loc.gov/2022014946
LC ebook record available at https://lccn.loc.gov/2022014947

ISBN: 978-1-032-32190-5 (hbk)
ISBN: 978-1-032-32187-5 (pbk)
ISBN: 978-1-003-31327-4 (ebk)

DOI: 10.4324/9781003313274

Typeset in Goudy
by KnowledgeWorks Global Ltd.

To the memory of Ruffo Cárcamo,
Jaap Nieuwenhuize and Flaminia Ventura,
who all fought for the rural poor.

Contents

Preface

This book has grown out of many years of lecturing at agricultural universities in China, Brazil, the Netherlands, Italy and Spain. My courses, both introductory and graduate, were meant to offer students the instruments (i.e., concepts and methods) needed to come to grips with highly diverging, dynamic and complex agricultural realities to identify, in each of these, the main problems and to elaborate novel and hopefully effective ways forward.

The variety of students' backgrounds, experiences and (potential) jobs, as well as the diversity of empirical situations to be analyzed and discussed, led me to ponder how to bridge a series of formidable gaps – how to delineate and present concepts and methods that would be appropriate in both the Global North and the Global South and how these could embrace the social and economic, agronomic and political aspects of the topics being considered. Finally, the central concepts and methods had to simultaneously take into account the micro situations (of men and women working in agriculture, their problems and their struggles), the macro situations (of state apparatuses, markets, technology development, etc.) and their complex interactions.

I myself have been active in science but also in social movements and (sometimes) in the elaboration of policy proposals. This is also reflected in the book. It does not focus on theory *an sich*. Instead, it pays special attention to the interfaces between science, politics and social movements and the complex negotiations (and reversals of meaning) that take place at each of these interfaces.

It is important to clarify, right from the beginning, that this book is not a description and/or summary of world agriculture. It only offers the instruments to develop such a description (if indeed a single description is possible) and to elaborate critical diagnoses that allow for the search for alternatives. Neither does it go into length on each and every single issue. It searches, instead, for the interlinkages and contextual dependencies and tries to introduce students to the main debates. In so doing I hope to show that history and memory are concepts that should not be abandoned as irrelevant, for, apart from recent works, I also use scholarly work from previous times – simply because I think that it still is often very useful. Finally, I should explain why I do not use the usual *glossary* that one encounters in nearly all textbooks. Although at the beginning of each chapter I offer a list of main concepts that will be discussed, I refrain from giving a series of ready-made definitions of each of them. I think the contents, and especially the analytical and explanatory power, of the different concepts used reside in their interrelations. It is the web of interlinked concepts (i.e., theory) and the encounter with specific empirical situations that specify the meaning of each single concept. For a textbook such as this, it is more useful if students themselves explore the many ramifications of each single concept in order to obtain the definition they think to be applicable to, and useful for, their own situation.

When it comes to the critical application of concepts, the book offers a range of Text Boxes. These are accompanied by a series of Method Boxes that discuss research techniques. Together, these boxes contain a few storylines (that allow for a step-by-step deepening of the critical analysis of particular farming activities). These include rice production in tropical polders in West Africa; potato breeding and cultivation in Ecuador and Peru; dairy farming and cheese making in the Netherlands and Italy; the genesis and unfolding of agricultural landscapes in Italy, Mexico and the Netherlands; and the construction of radical changes in agricultural production and the associated development of new institutional patterns in France, Peru, China and Friesland (where I was raised).

Many people stimulated and helped me elaborate this textbook. This is particularly true of Ye Jingzhong and Sergio Schneider, who have kept insisting, for more than two decades, that it was my moral (if not socio-political or even scientific) obligation to work on this book and, more important, to finish it. Many thanks go also to the hundreds of international students (many of them from Africa and Asia) who have followed, over the last two decades or so, my course on the 'Sociology of Farming' at the College of Humanities and Development Studies of China Agricultural University. I trust that this book meets their need for critical instruments that they can use to probe into the rural realities they come from and where they are working now. I am also very grateful to Sabine de Rooij, who helped me avoid major lacunae, and to Rudolf van Broekhuizen, who discussed some strategic details with me. Credit also goes to Nick Parrott of TextualHealing.eu for having upgraded my sometimes clunky English. Hans Dijkstra of GAW in Wageningen was most helpful when it came to improving the artwork.

I have been working, throughout my academic career, with many gifted and highly motivated Ph.D. students. In this book I often refer to their work. Working with them and drawing now on their Ph.D. theses is, and has been, an enormous satisfaction and joy.

I dedicate this book to the millions of young men and women who engage in farming in order to change it and to the many students, researchers and activists who assist them in doing so whilst simultaneously documenting their heroic endeavours.

Jan Douwe van der Ploeg
Wageningen

1 The specificity of farming

Overview: Main concepts discussed in Chapter 1

   Co-production
   Craft
   Objects of labour
   Instruments
   Labour force
   Reproduction
   Social relations of production
   Cohesion
   Coordination
   Heterogeneity
   Modes of ordering
   Micro-macro interactions
   Interface
   Differential development patterns
   Autonomy
   Integration
   Scale of production
   Intensity of production
   Style of farming
   Domains of farming
   Transfer of meaning
   Equilibration
   Substitutability
   Polyvalence of agriculture
   The third class
   Non-capitalist segment
   Peasant agriculture
   Entrepreneurial agriculture
   Capitalist agriculture

DOI: 10.4324/9781003313274-1

In this introductory chapter I aim to consider the main characteristics of farming with an emphasis on delineating the specificity of agriculture as compared to other economic sectors. In doing so, I am not pursuing some kind of exceptionalism.[1] My aim is to simply assess some of the main conceptual and methodological features of the sociology of farming, for these theoretical features are derived from the specific nature and dynamics of agriculture, just as the former are helpful in uncovering the latter. Discussing these main characteristics allows me, at the same time, to present a brief overview of the main themes that will be discussed in this book.

Farming is co-production, albeit to a variable degree

A first, and probably decisive, characteristic of farming is that it implies the conversion of living nature into food, drinks and other agricultural products. Agriculture is *co-production*: it is the ongoing encounter, interaction and mutual transformation of man and living nature. Whereas industry mostly involves the conversion of dead material (iron, plastics, minerals, wood, scarce metals, etc.) into useful products (cars, bottles, chemical products, furniture, chips, etc.),[2] farming centres on the use and transformation of *living* nature into a very wide range of products that are essential to human survival. The natural resources used in farming (and which at the same time characterize it) not only entail living nature – they are also expressions of it. This applies to land, soil, soil biology and water; to animals, plants, seeds, trees, fruit trees, vines; or, more generally, to the genetic material used. It also applies to the eco-systems within which farming takes place and whose services are used (sunlight, temperature and rainfall, and their diurnal and seasonal changes and fluctuations) as well as 'wilderness' (i.e., the surrounding woods, rivers, lakes, prairies, etc.).

Living nature is never completely predictable. It repeatedly confronts us with surprises: sometimes pleasant ones, sometimes sour, if not dangerous, ones. Living nature can also not be completely standardized (although there are, and have been, a lot of efforts to do so). Living nature is capricious: its behaviour cannot be planned completely. All of this means that it is impossible to fully industrialize it. Consequently, farming is a craft that requires constant observation of the ways in which living nature behaves and reacts. Working with living nature implies that human labour has to continually adapt to the rhythms, setbacks and surprises of the former. Farming requires *care*: one needs to make sure that the best is obtained from the natural resources without harming them, for they will be used in the following cycles. Living nature needs to be given the possibility to recover and revitalize itself.

Living nature is embodied in the *objects of labour* that are central to agriculture: in the land, cattle, crops, fruit trees, vineyards, olive groves, fishponds, etc. Farming, as an activity, converts these objects of labour into grain, meat, milk, olive oil, wine, etc., thereby generating value. This conversion is at the heart of the agricultural labour process: it is labour that converts the objects of labour into food, drinks and other products. The workers might be well equipped (or not): using *instruments* (a spade, hoe, machete, plough, oxen for traction, tractors or whatever) to facilitate the conversion of natural resources into useful products.

Objects of labour, instruments and the labour force are the basic elements of every process of labour and production. Yet, agriculture has the specific characteristic of working with objects of labour that belong to (or 'contain') living nature. As already said, this implies a specific structuration of the labour process: living nature needs *craft*. Craft is the capacity to understand, and constantly adapt to, the changing objects of labour in order to intervene in such a way that the best possible results are obtained.

Living nature, as entailed in the natural resources used in agriculture, cannot be understood as an endless reservoir to be exploited without any further consideration. On the contrary: it needs to be reproduced, recovered, restored, improved, strengthened, regenerated and developed. If not, then agriculture becomes an activity that 'eats its own tail'. Agriculture cannot simply be extractivist: it needs to ensure the conditions for its own future existence.

I use the terms *agriculture* and *farming* as synonyms, although I prefer the term farming because it refers to an *activity* (the agricultural labour process) and the active and knowledgeable engagement of human actors in it.[3] In farming there are two fundamental actors: living nature, as embodied in the different natural resources used, and human labour, as embodied in the different people engaged in the agricultural process of labour. In line with this, Victor Toledo (1990) defined agriculture as implying two intertwined processes of exchange: ecological exchange and economic (or social) exchange (see Figure 1.1). The former involves interactions with the surrounding eco-system and the natural resources used; the latter refers to the supply of food and other useful products to society.

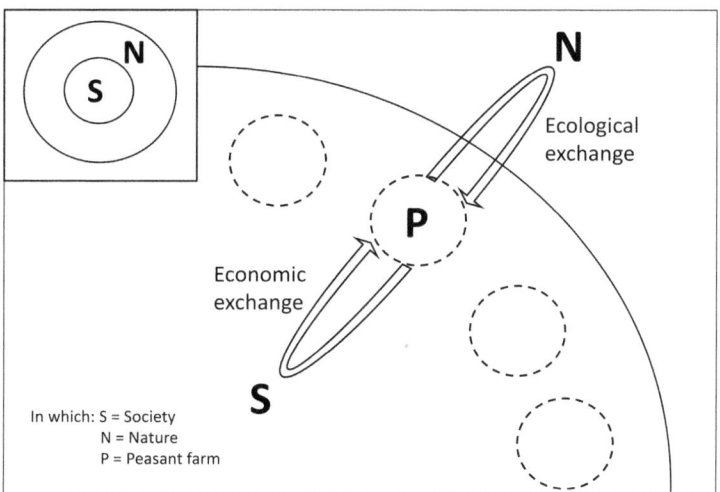

Figure 1.1 Agriculture as co-production (Toledo, 1990).

Both processes of exchange are two-sided. Ecological exchange not only entails the use of natural resources – it also includes many activities aimed at *restoring* the capacity of the natural resources to deliver the services required.[4] The same applies for the social exchange: there needs to be an equilibrium between the food supplied and the remuneration the farmers obtain.

The intertwinement of ecological and economic exchange makes farming into co-production – an ongoing encounter, interaction and mutual transformation of man and living nature.[5] Co-production implies the presence, combination, use and reproduction of both natural and social resources. The latter include the labour force, knowledge, networks, technologies, capital goods, etc. In short, co-production implies the permanent interplay of both natural and social resources.

Natural and social resources (or, more specifically: objects of labour, instruments and the labour force) cannot be combined and used in arbitrary ways. Their combination and use critically depend on the *social relations of production*. The social relations of production specify the relations between the people involved in production,[6] which, in turn, also specify the relations between people and 'things' (for instance, the land, the machines, etc.). These social relations of production underlie the processes of labour and production; they induce and sustain specific modalities for the organization and development of agriculture. At the same time, they regulate the distribution of the wealth produced.[7] When one talks about the social relations of production in agriculture (especially when it comes to family farming, which is not constituted, nor governed, by the capital-wage relation),[8] this implies embarking on a programme of empirical research. In some locations the social relations of production might be strongly shaped by gender relations; in others, the combination of market relations and the power of food empires may be more decisive. In yet other places it might be communal land ownership and patron-client relations between chiefs and followers. One never knows beforehand. In any situation, careful empirical research is needed to assess what functions as social relations of production.

From the beginning of their trajectory through life, the actors involved in farming encounter a given set of social relations of production. However, during their lives they not only reproduce these relations but they may also start to change them. These changes mostly occur in slow, stubborn and barely visible ways, but they may also occur abruptly.

The interplay of natural and social resources (i.e., co-production) induces specific features into the sociology of farming. It is not, as sociology generally is, only or mainly about the social. Nor is it, like agronomy and other agricultural disciplines, only or mainly about the natural. The sociology of farming (or agriculture) is about the combination, interaction and mutual transformation of *both* the social and the natural. Co-production shapes and continuously reshapes both man and nature and ties them together in specific land-labour institutions. The latter exist only insofar as they include both the natural and the human. Thus, the sociology of farming essentially is about soils, cows, farms and ways (or styles) of producing *as shaped, reshaped and developed by the people involved in farming* (in short: by farmers).[9] It is equally about farmers and their families, generations and institutions (such as the family farm, the rural community, cooperatives, etc.) *and how they are shaped by (and shape) ongoing interactions with living nature.* In general sociology it became clear, from the 1980s onwards, that the cultural and the natural, or the social and material, cannot be strictly separated into independent entities (let alone that the study of the social can be conducted independent of any knowledge and consideration of the material). However, the sociology of farming has, right from the beginning, focussed and been grounded on a thorough knowledge of both social and natural resources and, especially, their interactions.[10] The sociology of farming understands agriculture as socially constructed, and this applies in the broadest possible sense of the word. Because agriculture is a socially constructed reality, this implies that the actor-oriented approach is very important when doing research into it (and for elaborating theories about it). Methods Box #1 gives a short explanation of this approach.

Methods Box #1   The actor-oriented approach

In the actor-oriented approach, as elaborated by Norman Long (2001), the meaning that different actors attach to their own situation is central. In actor-oriented analysis, this situation is not represented and understood with schemes and concepts elaborated in, and by, science. Rather, it is about the situation as reflected in, described by and understood with folk concepts. These are the concepts used by actors to perceive, describe and understand themselves, their own activities, and the context in which these activities are embedded. In scientific analysis, a borehole and pump might represent a drinking water facility that provides good drinking water. If, however, the taste of the water does not meet the social definition of the actors involved (that is, does not align with local notions on taste, place, comfort, colour, etc.), it will *not be used* as drinking water. Local actors may well distinguish between 'drinking water', 'water to be used in the kitchen', 'water for washing your body', 'water to wash clothes', etc. Each category may have its own criteria, and if a recently constructed borehole and pump, intended to supply drinking water, do not conform to local criteria for 'drinking water', then the facility will not be used for this purpose (although maybe for other purposes). A borehole and pump do not necessarily constitute a drinking water facility but only *become* a drinking water facility if they are *used* as such. In Western Africa, some 80% of the facilities intended to supply drinking water (often constructed in the framework of international cooperation) are not used as the *only* source for drinking water. There is no need to say that this represents a considerable problem and possibly a gigantic waste of resources.

Actor-oriented research probes into the social definitions (the folk concepts) used by actors to describe their own situation, to attach specific meanings to it, to define the most appropriate course of action and also to negotiate meaning wherever this is needed.

Key assumptions in the actor-oriented approach are that (1) all people (all social actors) are *knowledgeable* of, and about, their own situation (although there are, of course, boundaries, just as the amount, syntax and nature of knowledge will vary), (2) social actors mostly behave in goal-oriented ways (the actions are purposeful in terms of their own views, notions and experiences), (3) much of this actors' knowledge is a compound elaboration of personal experiences, (4) often many of the activities come with explicit and unambiguous wording (see, e.g., Jean-Paul Darré, 1985), and (5) there are ongoing negotiations at the different interfaces between actors and others.

Participant observation is an excellent technique to get to know, in a step-by-step way, the folk concepts that influence a community's, society's or individual's actions and, especially, the way in which they compose a more or less coherent worldview. During such research, the researcher needs to frequently check whether he or she has correctly grasped the meaning of the notions being used. Interestingly, being a relative outsider often is an advantage for doing this type of research because it allows you to ask, 'What do you mean by ...' (insiders are assumed to know).

Farming styles research (to be discussed later) can be seen as a specific form of actor-oriented research that puts the knowledge repertoires and strategies of the actors involved centre stage.

It is labour (i.e., the knowledgeable and goal-oriented action of the actors involved in the labour process) that sets the instruments and objects of labour 'into motion'. Through labour the natural resources are converted into new values: food and other agricultural products that have use values and (at least in a market economy) exchange values. The same labour process also reproduces the instruments, labour force and the objects of labour – at least partly (this will be discussed further in Chapter 2). Finally, it is within, and through, the labour process that a particular *way of producing* (a 'style of farming') is produced (this will be outlined further in Chapter 4).

Consequently, the sociology of farming first and foremost understands the number of products produced, their quality, the continuity or volatility of their supply, the environmental effects of their production, and the levels of employment and income generated as (co-)resulting from the *social* organization of agricultural production and its interaction with the *politico-economic* context in which it is embedded. Economic and technological determinism are rejected. This does not mean that markets and technology are irrelevant; the point is that their effects are mediated and conditioned by the social organization of agricultural production (see Chapter 3).

The agricultural process of production needs cohesion. The natural and social resources contained within the farm need to be combined in a way that creates a well-equilibrated and well-functioning whole. If not, resources will get lost, part of the work will be in vain and production will be sub-optimal (and the farmer concerned will come to be known as someone who is not very good at his or her job). But it is more: this socially constructed whole also needs to be aligned with the main contextual elements. If not, there will be all kind of frictions and the performance will be less than satisfactory. Thus, every farm needs a permanent and complex process of *coordination*: the many different internal elements as well as the many relations with the relevant context need to be turned into a coherent, consistent and well-functioning constellation. This coordination needs to be ongoing and is an intrinsic and strategic part of the labour process. By carefully tuning the farm's resources and combining them into a well-equilibrated whole, the farm and its operation are brought, as much as possible, in line with the specific interests and experiences of the farming family, including developing a pathway for developing future (often trans-generational) prospects. Moulding the natural resources into 'promises for the future' (making a piece of land into a fertile and well-producing field, breeding more productive cows, selecting seeds that will give more abundant harvests, etc.) is a strategic part of this process. In short: *an evolving and actively developed co-production* is at the heart of the farm and central to the labour done by the farming family.

A diametrically opposed view (i.e., the mirror image of the concept of co-production) is encountered in, and central to, contemporary agronomy (or 'production ecology' as it prefers to describe itself). Agricultural sciences today understand agriculture as the bio-physical transformation of inputs as nutrients and energy into outputs of food and other agricultural products. 'Agriculture is applied ecology. Subjected to the laws of physics, chemistry and biology' (Zadoks, 1985:378).[11] The social aspects are typically left out of the equation.

Today's agronomy has a strong focus on calculating the theoretically possible yield levels. Yet, it is weak in understanding farming as it manifests itself in, and through, all of its empirical manifestations. The best it can do is to conclude that the latter are far from 'optimal'. They represent a *yield gap* when compared to the theoretically possible levels. Thus, different degrees of sub-optimality are put centre stage and they are to be remedied, typically, through different bio-physical interventions (such as increased levels of fertilizer application, the introduction of genetically modified varieties, etc.).

Currently, this essentially technocratic approach comes strongly to the fore in debates about total world food production, its potentials and its (natural) limits (see, e.g., Koning et al., 2008). The main parameters in the models used are the amounts of water, nutrients, solar energy and agricultural land (ranked in different classes that reflect levels of fertility). Together these physical elements determine (according to the algorithms used) the total amount of food that can be produced. What is suspiciously absent in such models is human labour: the capacity to cleverly increase and regulate the amount of nutrients, improve soil quality, construct irrigation systems, select new varieties, etc. Yet, it is labour that has moved co-production ahead through the ages. Of course, the limits of the natural (of living nature) impose, at least in the short term, tenacious boundaries. But it is equally true that, through man's labour, these same boundaries have been moved to allow for ever growing levels of production and productivity. While there is, indeed, often a considerable gap (between the calculated optimum yield levels and the empirical ones), it is equally true that steady, widespread and con-sistent improvements in the practice of farming allow for ongoing growth and development. It is telling, however, that the technocratic versions of agronomy do not explore this space. Strategic questions such as *where* production is to be increased, *how* this is to be done and by *whom* are basically neglected (even if the largest gains in production growth can be reached if the focus is on poor peasant producers in the Global South and on agroecological methods of production[12]).

Farming is highly heterogeneous

Most agricultural sectors are highly decentralized, with production occurring in large numbers of relatively small units. These operate according to the decisions taken within these units: there is no single centre of command that directs the mass of single units; there is no centralized prescription and sanctioning. At the same time, all of these single units of production have to deal with ecosystems, rural communities, markets, industries, banks, state apparatuses, regulatory schemes, etc. These 'contextual elements' obviously have a strong impact upon the socially constructed process of agricultural production. Nonetheless, they do not determine one single solution whilst excluding all others.

Typically, in regions characterized by a context that is one and the same for all farms (all farms have to deal with the same ecosystem, face the same set of markets and institutions, and have access to the same array of technologies), there is – regardless of the homogeneous context –considerable heterogeneity in the way the processes of production are organized and farming is practised. Sometimes these differences become more pronounced over time: creating mutually contrasting developmental trajectories. That is: co-production unfolds in different ways. It is important to note that such (growing) differences are not articulated on just one single dimension – the differences are many, and they are multi-dimensional.

Figure 1.2 offers an illustration of this. It shows a constant data set of 113 Friesian dairy farms – farms that all face a context that is largely the same. The figure shows

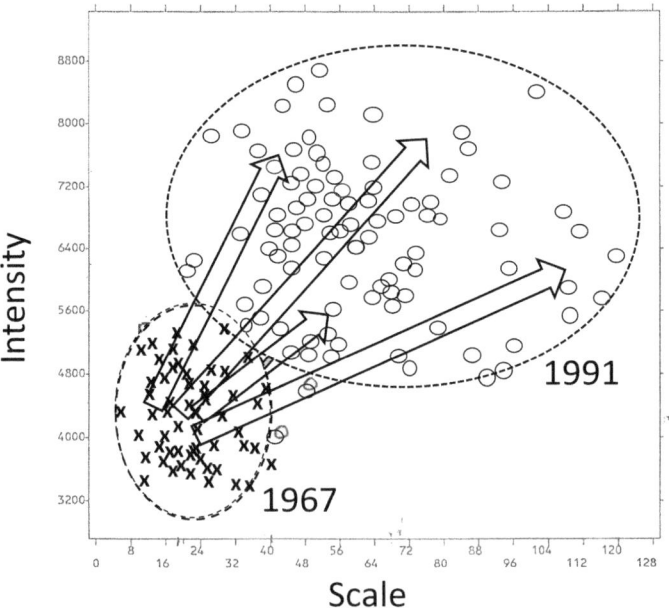

Figure 1.2 Growing heterogeneity in the Dutch dairy sector 1967–1991 (derived from Bruin, 1997).

their distribution within the space defined by two axes: the number of cows per unit of labour force (a proxy for scale) and the production realized per cow (a proxy for intensity). This is done for two years: 1967 and 1991. In the first year there was already considerable variety. In some farms the scale was three times higher than in others, whilst the range of intensity went from 3,500 (kilograms of milk per year per dairy cow) to some 5,500. Twenty-four years later, this diversity had increased even further. More detailed analysis (synthesized in the four arrows) shows the differential development trajectories. Some farms mainly intensified, others mainly increased their scale of farming, whilst yet others combined intensification and scale enlargement. In all of these trajectories, some made 'big jumps', whilst others proceed in a 'step-by-step' way. Examples like these show that co-production is moulded in different ways and over time unfolds along different trajectories, creating a widening range of expressions that increasingly differ from each other. Underlying all of this are contrasting modes of ordering. I will discuss the main ones later.

The heterogeneity that results from differently ordered processes of co-production is reflected in the sociology of farming in three ways. First, the discipline pays considerable attention to *micro-macro interactions*. The trends observed at the macro level and those occurring at the micro level do not necessarily run parallel to each other, nor can the former be simply understood as determining the trends or being indicative of what happens at micro level. The micro level always merits careful research (and theorization) in its own right. Methods Box #2 explores this point in more detail.

Methods Box #2   On avoiding determinism

A frequent mistake in the analysis of agriculture is to use specific macro conditions to directly explain specific sets of macro outcomes. The macro conditions are assumed to operate as causes that directly bring, and therefore explain, specific results at the macro level (i.e., macro outcomes). For example, a specific (and assumed low) price level for food products is presented as causing a stagnation of food production as a whole at, for example, the national level. Or the lack of new technologies – for example, for potato production – is assumed to explain the low level of average yields in a specific potato producing region. What is critically lacking in these examples of (economic and technological) determinism is the *micro level*. How do farmers perceive the prices on offer, and how do they translate this into their practices? And why and how do farmers implement a new technology (if available and if no other obstacles hinder its application)? In agriculture, the explanation of whatever general feature, trend or problem needs to always take into account the micro level. It is at this level that the macro conditions are effectively translated, mostly in differential ways, into (elements that compose) the macro outcomes. The linkage between macro causes and macro outcomes is always mediated by the micro level. Analyses and explanations cannot ignore the decisive mediating role of the micro level.

That said, it follows, in the first place, that heterogeneity at the micro level needs to be taken very seriously into account, even, and maybe especially, when seeking an explanation of the macro situation. Secondly, that the 'fallacy of the wrong level' should be absolutely avoided. This methodological failure consists in ascribing a trend that is observed at one level (e.g., the macro level) to another level (e.g., the micro level). For instance, at a national level we may note an ongoing tendency towards larger farm sizes and a decrease in the number of small farms. To generalize this undeniable macro tendency to the micro level ('all farms are being enlarged – if not, they will disappear') is a fallacy of the wrong level.

Secondly, *interface analysis* comes to the fore as decisive. The concept of interface refers to the points at (and the fields within) which the micro and the macro meet (Long, 1989, 2004). At these interfaces, general or macro tendencies and relations (as entailed in, e.g., food markets and/or agrarian policies) are perceived and actively interpreted and translated into specific courses of action to be applied in the micro situations. These perceptions, interpretations and translations are, of course, highly actor dependent. How do farmers 'read' their economic and institutional environment? How do they 'translate' changes in this environment into guidelines for the organization and development of their own farms? Such questions are necessarily at the core of the sociology of farming. Without appropriate answers, an adequate understanding of farming and agriculture is impossible (just as appropriate policies will be impossible).

Thirdly comes the need to conceptualize, understand, explain and correctly represent the phenomenon of *differential* development patterns. Agricultural development occurs through differentiated processes. The overall trend is *just the average of contrasting developmental patterns* (see Figure 1.2). General trends (at the macro level) are actively translated (at the different interfaces) into contrasting courses of action at the micro level. But, contrary to conventional wisdom, one cause (or one set of 'structural factors') will trigger different processes (just as a differential in a car produces contrasting velocities in different wheels).

At this point, some additional observations are needed. The translation of *macro* conditions into a coherent course of action at the *micro* level (implying a remoulding of co-production to fit with changing circumstances) mostly occurs through processes of *negotiation* in which extension officers, agronomists, bank officials, agents of the ministry of agriculture, etc., deal with farmers, farmers' leaders and, in the background, farmers' women, the youth, those who have migrated, etc. (Long and Long, 1992). It also applies that the outcomes of these processes of negotiation become a strategic input in the dynamics at the interfaces. They will feed perceptions, interpretations and translations in new, supplementary or sometimes, even, overpowering ways that may show that particular and hitherto neglected ways of farming are unexpectedly superior (this especially occurs in times of crisis). In this way, differentiation and selection may go together in unexpected ways. Finally, there is what is known as the *problem of aggregation*. The sum of all agricultural practices often does not inform us at all about the single practices. The underlying diversity might be so large that the calculated average is a meaningless 'fact'. This applies, for instance, to agricultural sectors characterized by a *latifundia-minifundia* complex (see also Figure 5.2) in which, on the one hand, there are a few very large-scale extensive farms (often run as capitalist enterprises) whilst, on the other hand, there is a multitude of small-scale, very intensive farms. The contrasting trends occurring in such a constellation may very well neutralize each other. The problem of aggregation applies wherever there are differentiated developmental trajectories. The net sum may appear to suggest stagnation, but stagnation noted at the macro level does not inform us about the (opposing) dynamics at the micro level. Despite this, such failures of aggregation abound in policy documents, national statistics and consultancy reports.

Methods Box #3   The comparative method

Given the large heterogeneity of farming and the many, mutually contrasting, developmental trajectories, the comparative method is of central importance in the sociology of farming. This method carefully brings together contrasting constellations and then assesses, in the first place, the commonalities between the different constellations. In this vein, one might bring together, for example, a set of Dutch family farms with sets of family farms from, for example, Africa and Latin America. The search for commonalities might help to define what is decisive to family farming; that is, assess the central features of family farming that are present everywhere (i.e., regardless of the many differences in context). This search might be extended by purposively looking for other constellations that lack some of the features that were identified, in the previous step, as being decisive and indispensable.

Secondly, the same method also allows, after having assessed the common elements, identifying the dissimilarities. These help to specify the *uniqueness* of, for example, the Dutch family farm as compared to its African and/or Latin American antipode. The same can be done for the African family farm, etc.

In such comparisons it is important to take into account as many dimensions as possible.

It is also important to take into account that the differences between various contrasting constellations do not always reside in different 'scores' on different variables (e.g., Dutch agriculture having higher levels of production per hectare than African farms) but in the *interrelations* between the different variables (e.g., the intensity in African farming being closely related to the available family labour whilst the intensity in Dutch family farms is strongly related to the technologies applied) and/or on the 'gravitational centre' of each constellation (for instance, decision making in the African farm will be located very much inside the *extended farming family*, whilst in the case of the Netherlands it will probably be located in the professional support structure – cooperatives, banks, agro-industries – linked to the farm). These different 'gravitational centres' consequently impact differently on the organization and development of the farm.

For further reading, the following publications are recommended: Glaser and Strauss (1967); Hofstee (1982) and Ragin (2007).

Heterogeneity is the outcome of different modes of ordering

Co-production can be moulded in different, mutually contrasting, ways. This leads to the construction of different styles of farming and the agricultural sector unfolding along different, mutually contrasting, trajectories. This can be due to external factors, but it might also be the outcome of internal drivers (although, mostly, the two go together in specific, self-affirming constellations).

When it comes to modes of ordering, *care* and *control* are two keywords. They describe different ways of relating to living nature. Care, a word that is often central in the cultural repertoire of peasant societies, refers to work being purposely done in such a way that the productive results per object of labour (i.e., per hectare, per dairy cow, etc.) are optimized, whilst the same object of labour is reproduced in a way that carries promises for the future (i.e., soil fertility is maintained or even improved, the dairy cow gives strong and beautiful calves).[13] Care refers to a specific structuration of the process of labour and an ordering of the farm as a whole that allows for sufficient labour time per object of labour. Those doing the work can dedicate enough time to each animal, to the different fields, thus allowing for care. Care expresses as craftmanship (for a general discussion, see Sennett, 2008), and it evidently assumes knowledge as well: one has to know the fields, the crops, the animals and their development during the productive cycle. One has to understand the different signs coming from living nature in order to intervene in appropriate ways. Hence, specific requirements apply when it comes to both the quantity and the quality of labour. To be able to put care centre stage in the labour process, highly skilled labour is needed, and it needs to be sufficiently (if not abundantly) available. Only when the quantitative relation between available labour force, on the one hand, and the amount of labour objects, on the other, is sufficient can the work be done with care, and only then are timely interventions possible (the need for them usually comes at unforeseen moments).

This relation (between labour input and the objects of labour) is internationally referred to as the *scale of production*. Scale cannot simply be equated to magnitude. Scale is the quantitative ratio of the amount of objects of labour and labour input. It expresses, for example, the number of hectares or cows per unit of labour force (as in Figure 1.2).[14] The scale of production depends on a number of factors, including the instruments (the technical means) that are used within the process of production. The level of production per object of labour is referred to as the *intensity of production*. If the input of both labour and technical means is relatively high, and if the work is done with care, the intensity of production will be relatively high. Working with care and dedication, and thus realizing relatively high intensity levels, is highly esteemed in most peasant societies and communities. According to Graham Brade-Birks (1950:XVI): 'Good farming means farming so carried out as to produce the maximum economic output from the land' whilst 'the practice of using the minimum amounts of labour, cultivation and manure [represents] a low standard of farming'.

However, working with care is not the only possibility. The polar opposite approach centres on *control*. The notion of control mostly refers to the amount of labour objects that can be held and managed by the owner of the farm. Control is aligned, in a way, with scale. The larger the farm, the more the owner is seen as somebody who is able to control a lot of land or many animals (or both).

The differences between care and control are subtle, but they are, nonetheless, of strategic importance. If care represents the feminine side of the equation, control represents the masculine side. Care and control represent two different attitudes towards living nature, two different modalities of co-production. Control involves an intent to impose discipline upon nature; it aims to submit living nature directly to human will. Control assumes that living nature can be commanded and forced to evolve in a particular way. Care, on the other hand, is about understanding nature, nurturing it and awaiting its gifts. Care requires patience, whilst control is about hurrying up. Care and control also have an impact on the actors involved, in highly contrasting ways. Actors applying care act as stewards; those who exercise control are more like managers.

Conceptually, care and control might seem to be opposites – and in a way, they are. In real life, though, they combine, albeit in highly different degrees: they may overlap and they may equally follow each other in time, one laying the foundations for the other and vice versa. Nonetheless, in real life, the differences between the different balances of the two can also be very real. There is a well-known dictum in sociology that things are real if their consequences are real. That is exactly the case here. For while the concepts may seem subtle and intangible, they materialize in often characteristic and mutually contrasting ways of farming: in treating the animals differently, different designs for the barns, fields that are worked in divergent ways and relations with co-workers that strongly differ. And once fields and buildings are shaped in a particular way and the cows and horses accustomed to a particular treatment, work has to adapt to these particularities. Cows that need care need to be dealt with in a caring way – if not, the productive results will be disappointing.

All of this points to yet another methodological issue that is central to the sociology of farming. The *explanans* (i.e., that which explains) does not reside solely in the attitudes of the actors involved. Nor does it reside solely, or even mainly, in the materiality of things. It is, instead, the ongoing interaction of the two, the implied co-constitution as well as their co-evolution through time, that results in certain phenomena being very real. Concepts such as care and control manifest at the *interfaces* between the natural and the social. It is at these interfaces that particular modes of ordering (centring on care, or on control, or a combination of the two) are constructed, negotiated, defended and/or changed. Here the specific mode of ordering constitutes the *explanans*. However, the different modes of ordering cannot be directly observed. That is why the sociology of farming tries to study practices (i.e., the ways in which material things and relations are dealt with) and cultural repertoires (which are expressions of the best possible way to do so) as a *unity* in which each informs and reinforces the other.

Care puts co-production centre stage. Care assumes that dealing gently with living nature will make the best out of co-production.[15] The notion of control is the quintessence of co-production (namely, that by dealing well with nature, the productive results will be obtained). Control, by contrast, mainly focusses on the *quantity* of labour objects: the more the better. Thus, the desire for control tends to push the scale of farming upwards, whilst the intensity of production (the production per labour object) receives less attention. Care fosters opposite relations: restricting scale whilst looking to steadily increase the levels of intensity.[16]

These contrasting modes of ordering might be difficult to grasp directly, but this does not exclude the involved actors (farmers, but also extensionists, accountants, traders) from being quite familiar with these (and similar) expressions. Thus, care and control (and autonomy and integration, discussed next) clearly figure in cultural repertoires. They are reflected in novels, farmers' journals, scripts for folk theatre, poems, paintings and photographs (see Figure 1.3). They are even reflected in contemporary commercial propaganda that seeks to address farmers and their basic feelings in a direct way.

While care and control shape the ecological exchanges contained in farming, the socio-economic exchanges (see Figure 1.1) are governed by another pair of ordering modes: *autonomy* and *integration*. The socio-economic realities in which farming occurs often manifest themselves in awkward and hostile ways to those doing the agricultural work, who frequently seek protection through an actively constructed autonomy. There are many forms of autonomy. Being autonomous

Figure 1.3 Care and control expressed in everyday life images (GAW/Hans Dijkstra).

differs from time to time and from place to place. There are also many ways to actively construct it. But, however it materializes, it always implies lessening one's dependence on others for the reproduction of one's own livelihood. Having a certain degree of autonomy means that one (at least partially) provides oneself with the means to organize one's own livelihood. Self-provisioning[17] brings and supports autonomy, just as autonomy helps to increase self-provisioning (this will be further explained in Chapter 3).

Socio-economic realities can sometimes also offer opportunities and/or protection which makes entering into, and becoming part of, those realities attractive. This spurs integration. However, more often than not, integration is forced upon those involved in farming. Farmers are forced to plant cash crops in order to pay heavy taxes, or into labour migration to be able to maintain the family and earn the money to buy a pump and diesel for irrigating their own fields or, to use a contemporary example from the Global North, to invest in expensive milk tanks and cooling facilities and a large parking lot because the dairy industry decided to not collect milk churns anymore.

Autonomy and integration have many similarities with care and control. They are opposite notions that are difficult to precisely define. Nonetheless, in real life it is often very clear how each materializes. It is also clear that each exists because of the presence of the other. The longing for autonomy may be spurred when integration, and the subordination that comes with it, is dominant or when distrust in the temptations of the dominant socio-economic order becomes widespread and manifest. But the opposite may occur, for example, when the enticements of following the exhortations of the powerful seem to make the search for autonomy a futile form of self-denial.

Exploring the threats and enticements that come with integration requires a politico-economic analysis that not only explores unequal exchange, the dynamics of capital accumulation and the main mechanisms of socio-political power but which also probes into the degree to which farmers can distantiate themselves from the circuits directly controlled by capital, the state and ruling elites and establish a degree of autonomy. This, critically, involves paying attention to the micro-macro relations and their interfaces and the ways in which farmers, especially when operating collectively, might counterbalance the hegemony of the powerful and (re-)shape their labour and production processes in ways that align better with their own interests and prospects.

The sociology of farming takes the practicalities of real life very seriously. If autonomy and integration are represented as clouds in the sky, it will be very difficult to grasp them – even to the degree of considering them to be meaningless abstractions. If, on the other hand, the difference between, for example, feeding your cattle with feed and fodder from your own meadows and fields (i.e., self-provisioning) versus the need to acquire the alimentation for the animals in the market (i.e., dependency) is taken into account, then autonomy and integration come to the fore as socio-material facts that are very real – the more so when one includes the accompanying impacts in the analysis.

Farming includes the construction and further unfolding of specific, internally consistent and mutually contrasting, styles

Different farming systems contain considerable differences in terms of both the scale and intensity of production. Such differences were extensively documented by Hayami and Ruttan in their well-known 1985 publication that represented, at that time, a milestone in the international comparison of agricultural systems. However, *within* the many different farming systems, wherever located in time and space, there also nearly always is a considerable diversity in the scale and intensity of farming (that is, the national data used by Hayami and Ruttan are mathematical averages that both synthesize and hide the diversity internal to the agricultural sectors of different countries). Figure 1.2 gave an illustration of this phenomenon. It showed that even in a homogeneous setting (where the ecosystem, the markets and the available technologies are comparable for all producers) there is considerable heterogeneity when it comes to the way of farming. In each place, wherever located, at any given time, there is hardly ever just one way of farming – there are always alternatives which are being actively explored and used. Hence, among the many small units of production, a wide variety of ways of farming are practised. This empirical diversity indicates de facto that it would be ludicrous to assume any determinism (see also Methods Box #2). Of course, the markets, the relative factor prices (that is the relative prices of the factors of production: land, capital and labour), the available technologies, the reigning agrarian policies, etc., are important parameters – but their importance is not the same for all producers. What is decisive here are the *interfaces* and the processes of interpretation and translation (of the macro parameters) into specific courses of action (at the micro level) that occur at these interfaces. Equally important is the actual dependency on the different markets (i.e., the degree of market integration).

A style of farming might be (theoretically) defined as a multi-level phenomenon. The first level regards cultural repertoire. This includes a set of strategic notions that specify how farming should be organized and developed. It builds on experience and collective memory and often entails pathways that specify how progress might be wrought. The second level regards the practice of farming. This is informed by the first level, but the combination of cultural repertoire (that specifies how things ought to be) and the many practicalities may very well contain tensions and frictions. Third is the level of interrelations between farms and markets.[18] Markets are not neutral – nor are farmers' relations with them. Involvement with markets implies transaction costs (the costs related to using the market) as opposed to governing costs (the costs related to running and managing one's own farm). The organization of these relations with the markets needs to be in line with the specific practice of farming (as specified at level 2). It is also to be in line with cultural repertoire (level 1). Fourth, a style of farming often comes to the fore as a (socio-political) response to the state – especially when the latter explicitly prescribes (through 'carrots' or 'sticks') how farming should be organized. Doing it differently thus becomes a distinctive statement.

The definition of farming style presented above is comprehensive and theoretically the most satisfactory one. Empirical studies, though, mostly focus on levels 1 and 2. Patterns of cohesion between the many characteristics of the on-farm process of production are carefully explored and, wherever possible, related to the strategies that inform them.

Figure 1.4 summarizes a wide range of empirical enquiries into farming styles.[19] It shows their distribution in the space defined by the scale and intensity of farming. It also shows a hypothetical link with the underlying modes of ordering (care, control, autonomy and integration). The top left illustrates the style of farming intensively. Here, care is the decisive mode of ordering. Labour focuses on the individual labour objects with the main aim being to obtain high and sustainable levels of production per object of labour. Consequently, fine-tuning the many aspects of, and relations within, the process of production is central. Labour here is characterized by skills and craftmanship. In contrast to this position, we find (in the bottom right corner) the style of farming that aims at the highest possible labour productivity: to produce as much as possible per unit of labour force. Here, control is central. This results in large-scale, relatively extensive farming. Interestingly, both styles can render the same incomes.

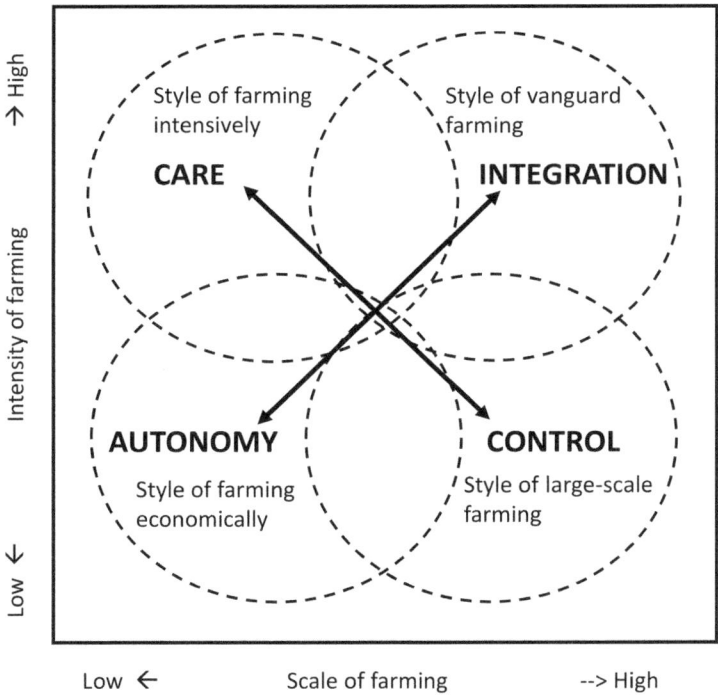

Figure 1.4 Farming styles and the main modes of ordering underlying them (author's own elaboration).

The second diagonal in Figure 1.4 runs from bottom left to top right. The bottom left shows the style of farming economically. The search for *autonomy* is the main mode of ordering here. It implies keeping most markets (for factors of production and inputs) at arm's length. These farmers try to minimize their integration in the markets for land, capital, labour, machine services, feed and fodder, nutrients, animals, etc., in order to keep monetary costs low. This does not imply running a poorly equipped farm; rather, it is a case of seeking to provide as many of the needed inputs from one's own strongly developed autonomous and self-governed resource base. Opposed to this style, we find (in the top right position) the style of vanguard farming (this wording is derived from hegemonic discourse that strongly favours this type of farming). Intensity levels here are, like the scale of farming, a function of the applied technologies. In historical data sets (prior to the 1960s and 1970s), it is difficult or nearly impossible to find this style. But modernization policies (as they are called in the Global North) and Green Revolution policies (in the Global South) drastically changed this panorama. In many places, though not everywhere, they resulted in the making of this new style of farming. Integration into hegemonic discourse as well as integration with the markets for factors of production (including new technologies and new animal breeds) and inputs (seeds, chemical fertilizers, pesticides, herbicides, growth hormones, high-energy fodder, etc.) is the main mode of ordering here. Despite the very high volume of production, the incomes derived from this strategy may be more or less identical to those of other styles (especially if the price ratio of output and input is unfavourable). In general, though, the highly intensive and large-scale vanguard farms (*les grands intensifs*, as the French say) generate the highest incomes.

Farming embraces several domains which require coordination

Farming covers four domains (or fields of activity), just as it always occurs in and through the interaction of these domains. The most visible one is the domain of production – the domain that can be properly described, analyzed and represented in terms of farming style (as indicated in the previous section). Alongside this domain there are the domains of reproduction, of family and household and, finally, of market relations. Figure 1.5 gives a simple synthesis and also shows the overlaps.

The domain of family and household[20] includes the people linked to the farm (through ownership relations, labour input and sharing in the benefits generated by the farm) as well as their mutual relations (intergenerational, gender and, often, patriarchal relations). These relations may involve the extended family, but they may well be limited to the household nucleus. The family and household are also related to the village and/or the wider community through all kinds of social networks. Normally the family shares the cultural repertoire and reciprocity relations of the wider community, but is also possible that there are significant differences within a community. Sometimes groups of farming households distantiate themselves from the community and the reigning repertoire, and this allows them to follow other roads when it comes to farming. The size of the household

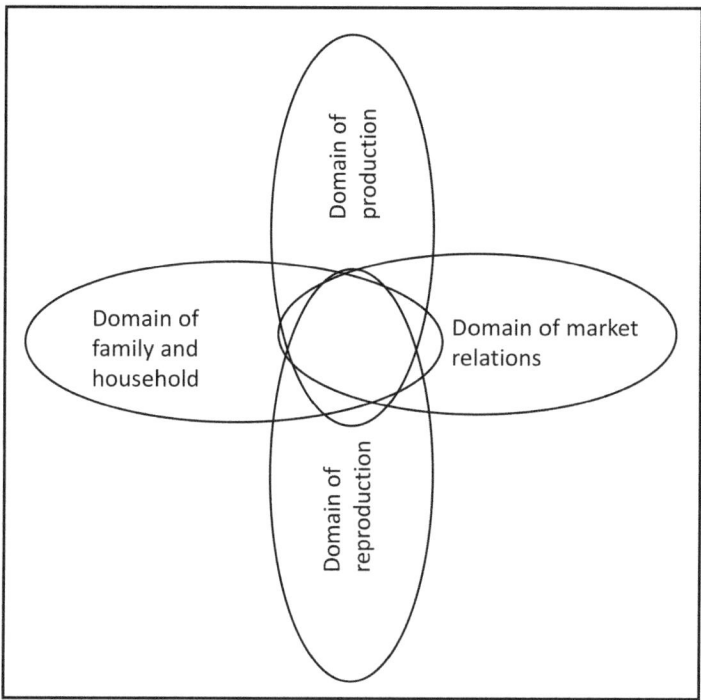

Figure 1.5 The domains of farming (author's own elaboration).

and its internal composition have a strong impact on the organizational plan and the developmental trajectory of the farm.[21] The Russian agrarian economist Chayanov (1925/1966) began outlining these interrelations that 'balance' the composition of the household and the organization of the farm. Since then, the sociology of farming has uncovered many more such balanced interrelations. The processes of socialization that turn the youngsters born and raised on the farm into the farmers and farmers' women of tomorrow play an important role here. The same applies to gender relations that mould young women (and men) into providing the female (and male) inputs needed in the farm and farming household as well as the corresponding identities.

Markets constitute another important domain in which farmers need to be active. Products need to be sold, money to be acquired, the services of others to be contracted and missing inputs, machinery and tools to be bought. If needed, farmers engage in credit relations. They might also get involved in the labour market, whether through contracting workers, selling their own labour, or both. Operating in these different markets always implies an engagement in institutional relations, for the markets are not anonymous: for the involved farmers they manifest as the dairy cooperative, the trader, the banker, the commercial house that represents agribusiness, the officials who control compliance with regulatory schemes of the state, etc.

Although the relations reigning in this domain might be highly skewed (they often are), they never are completely one-sided (that is, with one side imposing its full and unconditioned will on the other side). They are two-sided: they can be mediated, negotiated, but also sabotaged and even ended from both sides. There always is, in short, room for manoeuvre. Throughout this book I will offer examples. Analytically important is that this room for manoeuvre can be smaller or larger, partly because farmers will actively try to increase it because they experience this room for manoeuvre as autonomy – or, as Bernhard Slicher van Bath (1978) argued, as *farmers' freedom*.

The domain of reproduction covers the renewal of labour objects, instruments and the labour force and evidently overlaps with the other domains – firstly, because production and reproduction are often tightly tied together: no potato seedlings without potato production, no milk without calves, no cuttings and sprigs without fully grown fruit trees. But seedlings, calves, etc., can also be acquired in the relevant markets. The reproduction of the labour force occurs in the household, although much of the socialization (learning the best way to deal with the many practicalities included) is located in the domain of production. In reality, the many interrelations here are highly mouldable. Following Anne Lacroix (1981), reproduction might be basically *ecology* dependent (deriving new objects of labour and instruments, or replacing existing ones, from the surrounding ecosystems). But it can become *farmer* dependent as well. In this case, reproduction is located within the farm and governed by the farmers' insights, experiences and plans for the future. Finally, reproduction can be *market* dependent, with objects of labour, instruments and/or workers that need replacing being obtained through the relevant markets. These different modalities not only compose a historical sequence – they also co-exist in nearly all agricultural systems of the world (albeit in highly variable ratios).

Together the four domains of farming constitute a complex whole that can only exist (and evolve over time) through careful and complex coordination. The relations, parameters and actions within each domain need to be brought in line with those in the other domains. Such coordination occurs through different mechanisms. I will briefly discuss three of them here.

A first mechanism is the *transfer of meaning*. This implies that parameters and/or relations governing in one domain are actively transferred to other domains. Take, for example, a farmer who wants to actively take part in local community life. This requires time and must allow for his or her absence from the farm from time to time. Consequently, the domain of production (and probably that of re-production, too) will be structured in a way that allows both for a lower labour input and for his or her occasional absence. Or take the farmer's wife. She might feel very isolated from social contacts (or she wants to reapply previously gained experience in, e.g., care activities). This might translate into a farm shop or in care activities on the farm (for children or disabled people), and this will require a considerable reshuffling of nearly all relations within the domain of production. But meaning might also be transferred from other domains to that of production. A well-known example is the many product specifications from market agencies: only products that meet specified requirements (which may include quality, outer aspects, amounts, time and method of delivery, etc.) will be accepted and paid for. In order to meet such criteria, these specifications become internalized within the domain of production as a set of protocols that (increasingly) govern the concrete organization and development of the farm. Bruno Benvenuti (1982; Benvenuti et al., 1988) analyzed this specific transfer in terms of TATE: a technico-administrative task environment, which increasingly interferes in the different domains of farming.

Interestingly, the transfer of meaning sometimes occurs through the labelling of money as well. Independent rural women in West Africa who are not able to do the heavy preparation of the soil in their own *bolanhas* (tropical rice polders) may raise and fatten pigs, sell them and use the money obtained exclusively for paying a working group of young men to plough and prepare their polders. The money obtained is not all-purpose money. It is earmarked – it is, in other words, socially targeted: to be used solely for paying the working group. Such examples abound: Small dairy farms in Emilia Romagna in northern Italy that had to deal with a lack of fields to produce enough fodder (mainly *luzerne*) normally dedicated a part of their fields to the production of tomatoes. The money thus obtained (tomatoes render a high income per unit of land) was exclusively earmarked for the acquisition of fodder on the market (and de facto kept apart from other flows of money). Thus, the needed fodder appeared in the barn as 'already payed for' (thus allowing for *autonomy* in cattle feeding).

Equilibration offers a second mechanism through which coordination occurs. It also helps in the construction of actual autonomy. Figure 1.6 shows the multiplicity

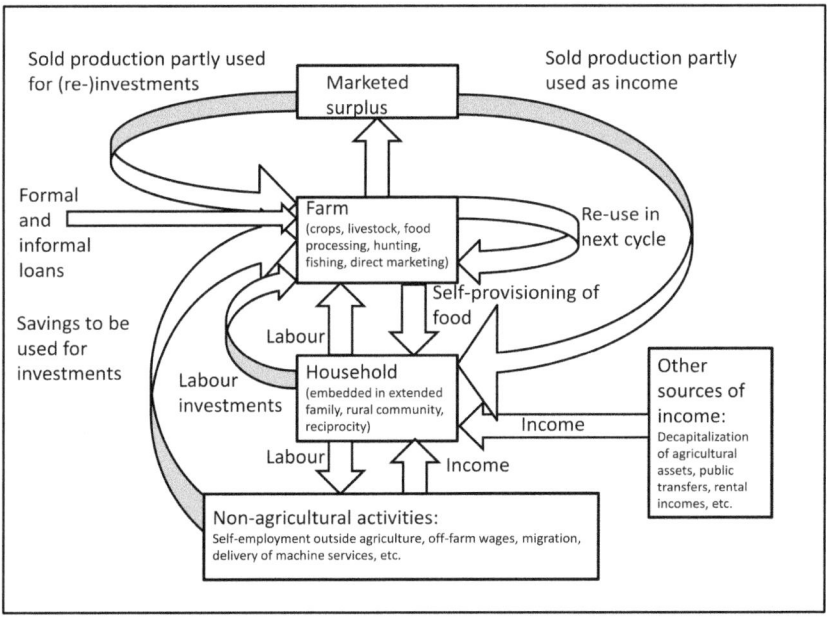

Figure 1.6 The multiplicity of flows and the possibility to reset some balances (HLPE, 2013).

of flows entailed in the typical peasant farm: it shows the flows of food, labour, income and investments within and between the farm, the household, the market and the non-agricultural activities in which (some of) the actors might be engaged.

Figure 1.6 shows that labour can flow in different directions, from the household to the farm to non-agricultural activities, and/or can be used to construct farm buildings, an irrigation system, fences, etc. These are called *labour investments*. Food from the farm may be consumed by the household, but it will also be sold in the market (the 'marketed surplus'). Investments can have many different sources, again originating different flows (summarized on the left in Figure 1.6). The same applies for the flows of income (the self-provisioning of food included). These can potentially be derived from different sources and go to several possible destinations. Thus, income flows may take different pathways and forms (summarized on the right-hand side of Figure 1.6). It is important to note that these flows can complement each other and that substitution is also possible. In specific situations they may also mutually exclude each other. In each farming family the balances between the different flows need to be assessed and equilibrated in such a way that they align with the interests and prospects of the farming family as well as with the potentials and limitations of their farm. Together these different flows and the possibility to bring them into an optimal balance constitute an important part of creating room for manoeuvre.

The strategic importance of the mouldability of flows is that it allows the farm household to bypass barriers that threaten to block the farm's development. If, for instance, banks are unwilling to provide a badly needed loan (or if the conditions coming with it are unacceptable for the farming family), there are alternatives, or possible ways out of the *cul de sac* created by the bank. Some family members might engage in non-agricultural activities (including labour migration) to raise the needed money. Or maybe direct investment of labour is possible, or the sale of some animals or a piece of land. Admittedly, all of these alternatives come with a price, but they also allow for bypassing the obstacle created by the bank. Thus, *the re-setting of some of the many balances* entailed in the unity composed by household and farm offers new ways forward.

A third mechanism for coordination relates to the capacity to actively engage in network building. Whilst the previous mechanism mainly involves changes that are *internal* to the farm and the farm household, this third one involves changes that are triggered from within but are basically located *outside* the realm of farm and household. Through the construction of new networks that link with consumers in novel ways (thus reshuffling the domain of market relations), a range of effects might be realized in the other domains. This creates *substitutability*. To return to a previous example, if engagement in non-agricultural activities and labour investments is not feasible, the networks existing in the extended family, the caste and/or the neighbourhood might be used to mobilize informal loans whose use is governed by reciprocity rules. Most Chinese peasant farms (which number 200 million) rely solely on such informal loans. In the Global North this underlying principle is now being rediscovered and applied through such innovations as CSA (community-supported agriculture).[22]

The relevance of farming extends far beyond agriculture

Farming involves a rich tapestry of links to wider society. Food represents, of course, a first linkage, and it is of utmost importance (whether or not such importance is recognized). Without the delivery of sufficient food, of good nutritional value, in line with the prevailing culinary traditions, and for acceptable prices, agriculture would lose its raison d'être, whilst society as a whole would face a huge problem that most probably would include destabilization, disintegration and insurrection. It is notable that many constitutions and international treaties include and specify a non-negotiable right to food. Many people understand this as part of a wider set of human rights that includes the right to farm (and, more specifically: the right to access land). The 'UN Declaration on the Rights of Peasants and Other People Working in Rural Areas' (UNDROP; endorsed by the United Nations on 17 December 2018)[23] can be understood as a synthesis and further specification of these rights (United Nations, 2019). The principle of *food sovereignty* (see also Text Box 9.4) does the same insofar as the relation between people generally and (national) agricultural systems is concerned. But there are far more linkages – and their nature, impact and the way they are balanced all differ according to time and place.[24]

Agriculture is often an important economic sector within the national economy as a whole. It is linked through input-output relations with other economic sectors and sometimes contributes considerably to export earnings. Agriculture also provides employment opportunities to many people, work that is (at least sometimes) attractive, possibilities (albeit often small) to improve one's own livelihood by means of one's own efforts, and incomes that (hopefully) are satisfactory.

If the employment generated by farming is considerable and incomes are good (that is, they allow, at the least, for an improvement of rural livelihoods), agriculture may provide an important 'internal market' (for both means of production and consumption goods) that will spur the development of national industry and thus contribute to overall economic development. However, the possible contribution to overall economic development is often grounded on extractivist mechanisms: heavy taxation, unequal exchange (between industrial and agricultural products and services), coerced deliveries of grains to cities, forced labour, etc. What these mechanisms have in common is that they 'squeeze value' out of agriculture in order to allow other sectors to flourish. Such a squeeze is not necessarily limited to poor countries in an assumed 'take-off stage' (as Rostow, 1960, would have it) or to 'primitive accumulation' used to feed the beginning of industrial capitalism or the industrialization of young socialist countries – it is omnipresent in today's 'modern' world as well (where it is associated with the dominance of food empires that exploit farmers 'to the bones').

The polyvalence of agriculture is not just limited to economic values, for agriculture equally shapes, contributes to and further unfolds the many scenic landscapes that are part of the heritage of mankind. The same applies for natural values and biodiversity. Farming might increase them – but it is equally possible that farming eats into and in the end will destroy them. It all depends on the way farming is structured and how it relates to nature. Increasingly important nowadays is the way in which farming deals with energy. The ways in which fossil fuel energy and non–fossil fuel energy (as represented by solar energy, human labour, soil biology, etc.) are *combined* and the efficiency with which they are *converted* to output differ greatly, between and within countries. The mouldability of energy use is inherent to co-production, just as the empirical diversity of energy use is increasingly becoming a decisive feature within the framework of agricultural heterogeneity. To say it in a somewhat polemic way: some styles of farming contribute greatly to climate change, whilst others help to cool the planet.

There is yet another important linkage that needs to be mentioned here. The possibility to engage in farming and to develop it further has been, and is, in many places all over the world a kind of last resort for the marginalized, the downtrodden and those who are persecuted. In a world where the availability of formal and informal jobs is limited and where no free lunch is offered (to say it politely), farming emerges as one of the few possibilities that offers the opportunity to forge a livelihood, gain some dignity and forge progress. Thus, agriculture comes to the fore as the space that offers the poor and downtrodden the possibility to (at least partly) realize their emancipatory aspirations. The problem, though, is that agriculture is increasingly fenced off by other interests – thus contributing to the further marginalization of those who most urgently need some space.

Most farmers, but especially peasants, belong to, and co-constitute, the 'third class'

Within the framework of the wider capitalist economy, the typical family farm can be seen as an intriguing exception: it represents (together with urban artisans) *a non-capitalist segment* of the capitalist economy. Farmers (and artisans) are neither capitalists nor wage workers. They compose *a third class* (Thiemann, 2014, 2022). And this is far from insignificant.

Co-production requires an organic unity of mental and manual labour if it is to flourish. If the two were separated, the needed cycles of observation, interpretation, experimentation, adaptation and re-organization would involve too many frictions and shortcuts. To unfold well, co-production also needs an organic unity of ownership and labour. Only when the fruits of dedicated and careful labour go to those doing the work (that is, if the ones doing the work actually own the farm) will co-production prosper and be actively developed. In this context, it is important that the scale of farming does not exceed specific (historically variable) limits in order to allow the labour process to be tuned to the (highly variable) specificities of fields, animals, crops, local eco-systems and communities and allows for the development of the required detailed and place-specific knowledge on social and natural resources (I will spell this out further in Chapter 2).

To this reasoning we may add a complementary argument. The exchange of genetic material (seeds, sires, bloodlines), knowledge, labour, instruments and fields is only really feasible when farmers perceive each other as colleagues (instead of being competitors). Reciprocity is only possible if the pursuit of profits is kept at arm's length.

Together these arguments strongly suggest that the family farm is the best possible place for co-production to flourish. It probably is also the best possible institutional form for dealing with, and surviving, the vagaries of the surrounding societal formations. This applies especially to the peasant farm (later on I will spell out the commonalities and differences between the concepts of a family farm and a peasant farm).

In family farms, no wages are paid. Consequently, the notion of profit becomes irrelevant. What matters is the total income that is left after the monetary expenses are deducted from the monetary benefits obtained by selling the harvested products. This is the income that remunerates the labour of those involved in production. It is, indeed, the 'labour income' (as Chayanov [1925/1966] would have it), or 'the clean part', as it is colloquially called in many parts of the world: it is what remains for the farming household. Nowadays the concept used is *value added* (VA).

As simple and straightforward as this observation might be ('if there are no wages paid, one can neither distinguish a profit'), it comes with two far-reaching and highly relevant corollaries. First, if there are no profits, farm development is driven by aims and mechanisms (such as the balances I mentioned earlier) other than the *maximization* of profit.[25] Second if no profits can be calculated, there is equally no *capital* (in the Marxist sense of the word).[26] Certainly there are material resources that often will be referred to as *capital goods*. A poor farming couple in the Andean mountains may proudly look to their potato seedlings that they just selected from their potato harvest and say, '*Mírame, de la buena cosecha viene nuestro capital*' ['Look, the good harvest renders us our capital']. The seedlings are a capital good because they allow them to produce new and more value (a new potato harvest) in the next cycle. But they definitely are not a means that is to be combined with wage labour in order to produce a profit. Neither are they a means for the farming couple to exploit themselves. More generally, *capital goods* (such as the mentioned seedlings, but also a pair of oxen or, for that matter, a tractor and previously constructed improvements, such as terraces to facilitate potato growing on steep hills, storehouses to avoid post-harvest losses, hedgerows to stop cold winds from freezing the growing plants or, even more poetic: the manure collected to increase soil fertility) are not necessarily capital as understood in Marxist analysis. Capital (in the Marxist sense) combines with the labour force of wage workers (those who do not own capital) in order to produce surplus value (or 'profit', as it is colloquially called) that is subsequently invested as capital in order to produce more surplus value. Capital is central to processes of accumulation. It assumes a series of specific relations: relations between the owners of the means of production, relations between the production obtained and both workers and owners, relations between the obtained results and their use, etc. Such relations are definitely absent in the family farm.

Again: there might be a plenitude of capital goods in the family farm just as there might be many people involved in 'setting them into motion'. However, this does not imply, by itself, that it is a *capitalist farm*. It is the outcome of what Philip Huang et al. (2012) coined, in a convincing analysis of agriculture in the rich Yang Tze delta, as 'capitalization without proletarianization'. Through their stubborn work, farmers build a set of capital goods that help them to work even better *as farmers*. In this process, they definitely are *not* turning themselves into wage workers. There is no proletarianization, and the many capital goods do not constitute capital.

In short: an agricultural sector composed of family farms represents a non-capitalist segment of the wider capitalist economy (Chayanov, 1924/1966). This does not exclude, however, that this non-capitalist segment is subordinated to, and exploited by, capital groups located outside of the agricultural sector. Nor does it exclude capitalist relations from penetrating into the agricultural sector. Equally, there are ways and means of resisting this subordination and penetration (see Chapters 6 and 9).

Through mechanisms of *unequal exchange*, the agrarian processes of labour and production might be (and often are) exploited by capital groups that operate in and through the trading and processing of agricultural products and/or in and through the supply of required means.[27] Through such unequal exchange, growing parts of the value produced in primary production are appropriated by the capital groups located in agri-business, food industries, trading companies and large retail. Heavy *taxation* is another mechanism for value shift, whilst the banking circuit composes yet another one: through the payment of *interest*, a part of the total value is channelled away from the direct producers. *Contract farming* (implying the payment for obligatory 'services') equally brings a value shift.

Two of the modes of ordering discussed before are evidently related to the subordination to, and exploitation by, external capital groups. These are system integration and autonomy. By systematically integrating their farms in the web of commodity relations (and the opportunities they undoubtedly offer), some farmers try to 'make the best' out of their subordination. Others, though, try to distantiate themselves and their farming practices through striving for autonomy. The mechanics of both positions will be discussed in Chapters 2, 3, 6 and 7.

Capitalist relations may also penetrate into the agricultural sector itself. Imagine a highly indebted farm. Capital (in the Marxist sense) is present here *within* the farm itself. It is *embodied* in the newly built barn, the acquired machinery, the bought cattle, etc. They are all financed with loans. Therefore, the barn, machinery and cattle need to be used in such a way that repayment of the credit (according to the deadlines specified by the bank) plus payment of the interest rate is secured. Failure to do so would imply the loss of the farm. Thus, the barn, machinery and cattle become part of a *capital relation*: a relation between bank and farm that penetrates into the core of the latter; that is, into the labour process itself. This sometimes brings far-reaching and radical changes even in the bio-physics of production. An example of this is that yields are no longer optimized per animal but per available place (per 'stall post') in the stable, which in its turn brings a shortened longevity and accelerated replacement of animals.[28]

In this example, the barn, machinery and cattle represent capital within the farm. By means of, and through, these 'things', capital starts to (re-)constitute the processes of labour and production; it also starts to redefine the distribution of the wealth produced: which part is to go to the bank and which part remains in the farm.

The same occurs when sophisticated machinery (e.g., robotized milking equipment) or artificialized nature (e.g., genetically modified seeds) comes with obligatory maintenance and repair contracts (or, in the case of seedlings, obligatory technical assistance and prescribed inputs and outlet schemes): they induce an external prescription and sanctioning of on-farm decisions and establish a permanent flow of money from the farm to outside agencies. This penetration of capital (as relation) into the heart of the farm is especially evident in entrepreneurial farming (to be introduced in the next paragraph).

The penetration of capital relations into farming can occur in other ways as well. Taking over the resource base of farms (partly or completely) is one modality. This is currently known as *land grabbing* (see, e.g., Edelman et al., 2015). Introducing regulatory schemes (mostly by the state) to align farms with the interests of capital is another. Cognitive schemes (such as, e.g., farm accountancy grounded on neo-classical views or classification schemes that present small farms as inevitably doomed to fail) represent a third modality.

Methods Box #4 Going beyond the immediacy of 'things'

One of the important methodological challenges of, and for, the sociology of farming is that it has to go beyond outer appearances. An extended machine pool inside a farm might look impressive and suggest a successful and rich farmer. It could, however, imply many different and even highly contrasting meanings. The powerful tractor and the implements that come with it might be an expression of the autonomy that the farmer has de facto constructed through arduous work (and probably cooperation with others). By having this self-owned (and self-governed) machine pool, he (or she) is no longer dependent on third parties (contractors) for delivering machine services. Nor is there any need to lease machinery. But a similar machine pool might, on another farm, be the expression of severely restrictive dependency relations. The machinery here could represent indebtedness and probably dependency on dealers (including for maintenance and repair). In yet another farm, it might reflect cooperation with other farmers through which the machinery is jointly owned and operated by different farmers who also support each other through labour exchanges (and probably joint learning and experimentation as well). This latter case has been very well documented by Veronique Lucas (2018).

This means that it is always necessary to probe into, and reconstruct, the *social biography* of 'things' (when and how have they been acquired, under what conditions, who and what regulates their use, etc.) and the *social networks* within which they operate (the dependency relations that come with the presence and use of farm buildings, technologies, genetic material, etc.). The *relations* inherent to this biography and the surrounding networks shape the meaning and significance of 'things'. It definitely is not the outer appearance (the 'immediacy') of these things themselves that tells the story. More often than not, the exterior hides the true significance.

The subordination to, exploitation by and penetration of capital sometimes generate considerable resistance and socio-political struggles. It is not possible to assess a priori how these will unfold, the mechanisms that will be used and whether or not they will be successful. Such questions need to be the starting point for empirical inquiry. In Chapter 9 I will systematically discuss peasant resistances and struggles.

Peasant farmers perform in ways that differ distinctly from those of entrepreneurial and capitalist farmers

So far I have centred the exposition on family farms and the nature, internal structure and dynamics of farming located within them. Family farming relates in specific ways to nature; it equally relates in a specific way to society. The family farm is where the third class is located – at least a large part of it. There have been family farms throughout history, but that does not make it an ahistorical category. The family farm as we know it today has been 'perfected' through longstanding struggles. It is the outcome of parts of the third class fighting for its emancipation. All of this has made the family farm into a vehicle that bests corresponds with the interests and prospects of the farming population (including the possibility to leave the farm for those who choose to do so). Being a promising vehicle for emancipation also explains why so many non-farmers (rural workers, for instance, but also people with urban backgrounds) have been interested in building their own farms – today probably even more than ever before.

The family farm has been made into a very strong land-labour institution. It ties land and labour, the past and the future, work and income, living and working, drudgery and satisfaction together into a dogged entity that is passionately defended by those involved and which, in turn, helps them to move forward. It is also omnipresent: most estimations suggest that there are at least some 500 million family farms in the world. However, the family farm is not the only strongly present land-labour institution. Next to it there is the capitalist farm where those owning the land and obtaining the profits are not those who do the work (the wage workers). In the capitalist farm, land and labour are *separated*. At the same time, there is a widening (although not always very clear) gap *within* the family farming sector: alongside those producing in peasant-like ways, family farms operating in an entrepreneurial way have emerged. Here the *means* to work the land are increasingly controlled by others (even if the land and labour are formally in the same hands).

Today, peasant, entrepreneurial and capitalist farming represent the main socio-technical constellations in global agriculture. They are, to use a Marxist notion, different modes of production.[29] They represent different ways to relate to the different markets (see Table 1.1 for an overview). Consequently, their resource bases are structured differently.

Apart from domestic production (production directly oriented at consumption by the producers themselves),[30] all farming is oriented, partly or completely, to the market. The products are marketed and the monetary benefits thus obtained represent an important (though often far from only) income flow for the farming

Table 1.1 Modes of farming, relations with markets and commoditization

	Labour force	Means of production	Products produced
Domestic mode of production (DMP)	−	−	−
Peasant mode of production (PMP)	−	−	+
Entrepreneurial mode of production (EMP)	−	+	+
Capitalist mode of production (CMP)	+	+	+

−Not commoditized; +commoditized.

family. Next to this indispensable link with the 'output' market (also referred to as 'downstream' market) there are the 'input' markets (situated on the 'upstream' side of the farm). These are the markets for the means of production and labour force (or, in neo-classical language, for factors of production and non-factor inputs). Theoretically, the relations with these 'upstream markets' are highly variable (I will explore this issue further in Chapter 3). Empirical analysis shows that both historical and contemporary realities display a huge variability in this respect. Some farms depend very much on 'upstream' markets, and others are relatively independent. These differences are associated with different modes of production. Or, in different terms: different modes of production imply different sets of relations with the main markets. The relational patterns constitute specific modes of farming, whilst the latter result in and reproduce the specific relations with different markets. The relation with the market implies that the elements mobilized in that market (labour, means of production, etc.) appear as *commodities* within the farm labour process. The absence of a systematic relation with the relevant markets means that the same elements function, and are operated as, *non-commodities*. If labour is mobilized on the labour market, it enters the farm as a commodity. This means that wages are paid and that there is a hierarchy of workers and boss (or manager). If labour is mobilized through the family (i.e., family labour), the situation will clearly be different. As I will explain in Chapter 3, the penetration of commodity relations into the heart of the agricultural process of production – that is, into the farm labour process – has far-reaching consequences.

Table 1.1 indicates how different 'modes of farming' *relate* in distinctively different ways to the main markets. This overview is, of course, highly abstract but theoretically helpful.

Domestic production is situated outside of the markets: the resources used (land, labour, instruments, seeds, etc.) are reproduced with the farm (or obtained through socially regulated exchange) and the products produced are not commodities. They are consumed within the family or passed to others as gifts or through barter.

Peasant agriculture, in its turn, produces wholly or partly for the (output) market, and it does so without (or minimally) depending on upstream markets. Labour and the means of production (land, seeds, animals, buildings) are available in the farm and operated as non-commodities. Thus, the impact of market relations upon the farm labour process is minimized. The organization and development of production are, instead, regulated by a search for autonomy and care.

Text Box 1.1   Peasant agriculture, family farms and smallholdings

---

In everyday language, peasant farms are often equated with family farms and small-holdings. In itself this is not wrong per se. Analytically, though, these concepts refer to *different theoretical dimensions*. The notion of a family farm refers to relations of ownership. In the family farm, the ownership of the farm (of the means of production) typically resides within the family, whilst the required labour force is supplied by the same family. As said before, there is an organic unity between 'capital' and labour. The guiding question here is 'who *owns* the land?' If those who work the land also own it, we talk about family farms. The notion of a smallholding refers to the relative magnitude of the farm. It takes into account the distribution of all available means of production over the total number of farms. The central question here is 'how does the *amount* of land compare to others?' Finally, the concept of a peasant farm refers to the concrete use of the land; that is, the organization of the processes of labour and production within the farm. Here one needs to ask, 'What is actually *done* with the land?'

Mostly, these three dimensions overlap. But there are important exceptions. Peasant agriculture (i.e., a particular way of farming) can be practised in large cooperatives of rural workers (instead of being limited just to family farms). Family farms are not necessarily identical to smallholdings. In some parts of the world, the state may own the land (as occurs in China), whilst farming is typically peasant-like. But property may reside also in the peasant community (as occurs in many Andean countries) or be allocated by traditional chiefs who control the commons (as occurs in several parts of Africa), etc.

In capitalist agriculture (see Table 1.1), both labour and the means of production are commoditized. The means of production are operated as capital (in the Marxist sense). When combined with, and set in motion by, the wage workers, they aim to render profit, and this impacts the organization of the farm labour process in particular ways. Because the capitalist mode of farming is characterized by the separation of labour and land, control (of both living nature and labour) is a decisive mode of ordering. This can lead to variable degrees of system integration.

Entrepreneurial agriculture is yet another constellation. The means of production are mobilized in different markets and show up in the farm labour process as commodities. This is the main contrast with farms structured in a peasant-like way. There is equally a basic difference from capitalist farms: in entrepreneurial farms, the labour force is not a commodity: it is provided by the family who own the farm. Entrepreneurial farms are strongly geared towards system integration, which involves the use of means of production designed, re-modelled, constructed and supplied by agencies and institutions that are external to agriculture. Integration also involves the processing of 'raw materials' produced on entrepreneurial farms by food industries and their subsequent distribution through large retail organizations. In this model, farming follows the script of external agencies (on both the input and output sides of the farm). Thus, labour and living nature are controlled through the externally provided means of production. This represents, as I will argue later, an interesting form of control at a distance.

One significant difference here is that the performance of peasant agriculture is distinctively different from that of capitalist and entrepreneurial agriculture. This applies to the processes of production (including polemic issues such as the yield levels realized) but also applies to their wider impact (that is, to the type, nature and reach of their polyvalence). This introductory chapter is not the appropriate place to discuss these different performances. What is important here is that the sociology of farming continuously and systematically documents the huge variety in outcomes (whatever their specific nature) and subsequently analyzes and explains how these different outcomes are rooted in differently moulded processes of labour and production.

Knowledge relates in complex and sometimes twisted ways to the practice of farming

Different knowledge systems, which relate in different ways to farming, can be discerned in agriculture. Farmers' knowledge (discussed further in Chapter 2) is the tacit knowledge operated by those actors directly involved in agricultural production. It is built on their experience, comparison and mutual exchange. It is about knowing 'how to do the job'. It is what the French call *savoir faire*, which intimately reflects the conditions under which it has been generated. It is partly embodied in specific breeds, varieties, buildings, irrigation systems and the networks that link the producers with traders and consumers, etc. It is very much interwoven with local cultural repertoires. In this respect, it is also strongly value laden: it defines how the job is to be done, what a well-worked field should look like, etc. Farmers' knowledge is sometimes reflected in notebooks, local theatre plays, music, songs, etc. (which in turn can provide important sources for learning the local rules and insights). Farmers' knowledge is mostly informal, not codified, highly heterogeneous and dynamic. In the past, classical agronomy (at that time an important scientific discipline) accumulated and communicated farmers' knowledge. Classical agronomy tried to describe, catalogue, understand, explain and – if possible –improve farmers' knowledge.

Currently, alongside farmers' knowledge, we distinguish an agricultural knowledge system (AKS). This AKS systematically differs from farmers' knowledge as much as it differs from classical agronomy. Central to the AKS are the modern agricultural sciences (including theoretical agronomy or production ecology, as it is also known), but it also includes other types and sources of knowledge (such as the specific knowledge blocks developed in ministries of agriculture, extension services, banks and industries, etc.). Farmers' knowledge of today differs from that of the past, but what the two have in common is that they reflect *how the job is done*. Contrary to this, the knowledge developed and operated in the AKS *prescribes* how the farm (and consequently agriculture as a whole) ought to be and

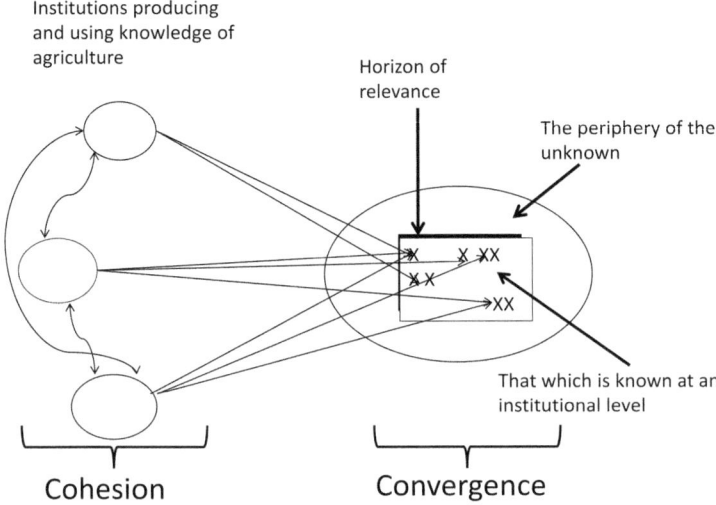

Figure 1.7 The AKS and the production of myopia (author's own elaboration).

what steps need to be taken to get to that point. Farmers' knowledge is about *today* and about the prospects entailed in today's practices. The AKS operates the other way around: it specifies a *future* and then prescribes how to get there (which very well might be a selective process that includes some farmers and excludes others). In this respect, an AKS is also very much value laden, but the value orientation might differ from that entailed in farmers' knowledge even though the AKS claims to be value free.

Another series of differences can be explained with the help of Figure 1.7, which synthesizes the internal structure and specific focus of the AKS and thus explains why AKS often represents a kind of institutionalized myopia.

The left side of Figure 1.7 shows different institutions, such as agricultural universities, ministries, centres for applied research, agricultural banks, food industries, etc., which all produce, use and exchange knowledge. Mostly these institutions are tightly interwoven, partly because staff members in these different institutions are usually trained in the same institution(s) and often change positions. In order to move forward within, or between, these institutions, they quickly internalize the rules. This means that these institutions are often tied together through 'old boys' networks. In short: there often is considerable *cohesion* in the institutional pattern that sustains the AKS.[31]

These different knowledge institutions need an (implicit) agenda that defines what needs to be known, because not everything can be known. There are always boundaries between the known and the unknown, and with the expansion of knowledge, new areas of knowledge, as well as adjacent areas of ignorance, are created (Visser, 2010). The boundaries of the known are referred to as the *horizon of relevance*. This horizon specifies what needs to be known (because it is relevant to know) and what remains unknown (because it is considered irrelevant).

Thus, the horizon of relevance a priori delineates what is known from what is unknown. Here the issue of *convergence* comes to the fore. The different knowledge institutes converge in what is considered to be relevant and, consequently, what is irrelevant to know.

As an example, the current AKS considers it to be more or less irrelevant to study how small farms might render a good income. There is, one could say, a tacit agreement that (1) small farms cannot obtain adequate incomes ('they are too small'), (2) they will disappear anyway ('so why study them') and (3) even if there are some small farms rendering a good income, they are 'black swans' – or 'refrigerator anomalies', as Herman Koningsveld (1976) aptly characterized them,[32] thus making any further inquiry irrelevant. In this vein, there is a nearly endless list that ranges from manure and mountainous areas, from traditional breeds to young non-agricultural people wanting to farm, to alternative methods for pursuing sustainability. On the other hand, there is a range of issues (the x's in Figure 1.7) that represent themes considered as highly relevant and thus 'meriting' further research (and funding). But this is a narrowly delineated range, which is reduced by what is considered as being relevant by a *limited* set of *specific* institutions.

The danger of such convergence (which is further strengthened by the carefully governed *funding* of agricultural research) is twofold. In the first place, it generates a constellation that is highly *self-referential* (thus reproducing its own in-built biases). If a new opportunity emerges, it can easily be argued that it has no potential because 'there is no scientific proof for it' (and there is no such proof as the issue has been considered as not relevant and there has been no research to systematically probe into it). The second problem is that it is precisely in the *periphery of the unknown* (which empirically exists even if it is 'under the radar' of the AKS) that many new insights and practices (to be discussed later as *novelties*) germinate and are tested and explored further (see, e.g., Osti, 1991). However, due to the in-built bias (the horizon of relevance), such novelties are mostly neglected, which often turns the AKS into a conservative force (the more so because this often comes with the 'not-invented-here syndrome'[33]).

The AKS produces specific bodies of knowledge that, due to the relations discussed above, relate in complex, and often twisted, ways to the interests, prospects and knowledge of the direct producers, despite the same AKS nearly always claiming that the knowledge it produces is scientifically grounded, universally applicable, value free and beyond possible critique and contestation. This claim sits well with the need of state agencies and private capital groups to represent their policies and views as both unavoidable and in line with the general interest. This means that it is often useful to critically examine the information produced by the AKS (Sumberg, 2017).

The same biases inherent to the operation of the AKS and the knowledge it produces necessarily imply a research agenda for the sociology of farming – not for wanting to be critical per se but because that which is made invisible today (because it is located beyond 'the horizon of relevance') might be highly needed tomorrow.

Notes

1. That would be to assume that there is a radical gap between agriculture and other economic sectors. There is no such fundamental gap; rather, there are both dissimilarities and commonalities. Here I focus on the dissimilarities which, together, create the specificity of agriculture.
2. I am aware of the implications of the current attempts and interest to create a bio-based economy. These carry the possibility that the differences between agriculture and industry will become blurred – at least partially.
3. This may initially seem to be a completely trivial statement. However, it is not. It is related to an important paradigmatic struggle over the 'proper' definition of agriculture. I will return to this issue later on in this chapter, when discussing the historically evolving definition of agriculture.
4. This reflects a purposefully constructed reciprocity between man and living nature. Such reciprocity characterizes many of the typical knowledge repertoires used in peasant agriculture. More details are provided in Text Box 2.2.
5. I am using here the historically delivered, classical wording. 'Man' does not refer here to male actors only but to mankind generally. It refers to the social actors engaged in farming.
6. Here, two of the four questions formulated by Bernstein (2010) in his synthesis of politico-economic analysis in agriculture apply: 'who does what' and 'who owns what'.
7. Here the second two of Bernstein's four questions come to the fore: 'who gets what' and 'what is done with the surpluses'.
8. The wage relation is the immediate expression of the capitalist mode of production, which is widespread in the urban economy but mostly absent, or only indirectly present, in agriculture.
9. I use the notion of a farmer here as generic concept: a farmer is someone who is involved in farming. Later I will delineate more specific concepts, such as peasants, agricultural entrepreneurs, capitalist farmers, domestic producers, etc., who practise farming in specific and mutually contrasting ways.
10. Outstanding examples can be found in, to name but a few, Weber (1891/2008), Hofstee (1946), Bourdieu (1958) and Mendras (1967; see, in particular, chapter 1).
11. This view represents what was, at that time, a clear demarcation between classical agronomy and social agronomy. In the latter tradition, agriculture was defined, more or less, at farmers' practices; that is, what farmers were doing. The turning point (from classical to technocratic agronomy) can be located in the 1930s (see, e.g., Minderhoud, 1954).
12. See, for example, Tittonell (2015).
13. A recent and vivid description based on herding sheep can be found in Rebanks (2016).
14. With sophisticated statistical methods, it is possible to standardize labour objects. Thus, the total amount of all labour objects utilized on a farm (whatever their specific nature) can be related to labour input.
15. The words chosen here ('gently dealing with living nature') are inspired by the thesis of Zuiderwijk (1998), titled *Farming Gently, Farming Fast*. See also Bergh (1989) and Steenhuijsen Piters (1995).
16. The complex relations between scale and intensity have been, and are, the object of heated debates. The 'inverse productivity hypothesis' (large scale but low productivity) is a point in case (see, e.g., Larson et al., 2012). In industrialized agriculture, other interrelations may be found.
17. The notion of self-provisioning is not to be confused with self-sufficiency or subsistence. Self-provisioning is defined at the level of the production unit. It involves the provisioning of the resources needed for farming (see Chapter 3). Self-sufficiency is

defined at the level of the farming household (see, e.g., Galeski, 1972) and is, at most, only a tiny part of self-provisioning. Self-sufficiency and the associated concept of subsistence production relate to the food needed to feed those who are working and living in the farm.

18. And the technologies that are made accessible through these markets.
19. This distribution of farming styles and the modes of ordering underlying them typically reflects Western European agriculture of the late 20th and early 21st centuries. Politico-economic conditions in Europe and the wilfulness of European farmers allowed for a clear crystallization of different farming styles. The most outspoken expressions each represent a single mode of ordering. In other epochs and other parts of the world, such articulation was often not possible. The reigning politico-economic and power relations often implied an amalgamation of the search for autonomy and the centrality of care. This made for the typical peasant farm of the past. System integration was impossible in many of these politico-economic constellations, simply because the 'system' had nothing to offer. Currently, though, the search for control and system integration are increasingly fused and becoming the engine driving agriculture towards mega-farms. On the other hand, today's agroecological movement is once again inducing the actively constructed combination of autonomy and care, which is strengthening the reality of 'new peasant farms'.
20. Enrique Mayer (2002:1–42) offers a very good discussion of different theoretical approaches to this domain, as does Robert Netting (1993:58–101). Mayer focusses on the Andean region; Netting's discussion is more general, with special emphasis on Africa and Asia.
21. The organic unity of farm and farming family is beautifully summarized in the Latin concept of *domus*. See, for example, Le Roy Ladurie (1980).
22. The list of possible alternatives is far longer. Depending on the specificities of time and space, there can be arrangements as *ayni* (one farmer bringing in land and labour, the other capital), *tontines* (saving clubs in Africa), shared investments and joint use of machinery, etc.
23. See https://www.eurovia.org/wp-content/uploads/2016/03/UN-Declaration-on-the-Rights-of-Peasants.pdf
24. A historically important linkage was the possibility to mobilize large masses of peasants as, or coerce them to become, soldiers fighting for the nation. Soldiers were often remunerated, after active service, with a piece of land. This is still reflected in crofter agriculture in Scotland; the failure to remunerate former soldiers in this way provoked, at least partly, the rise of fascism in Italy in the 1920s and 1930s.
25. It is easy to show that the search for a maximum income gives results that differ markedly from the pursuit of maximum profits (see, e.g., Ploeg, 2013:102–103).
26. The same conclusion also follows from the organic unity of ownership and labour supply mentioned earlier.
27. I limit myself here to current forms of value shifts. In feudal times, in colonial regimes and in the 'empirically existing socialist societies', other forms could be found (such as forced labour, to mention just one).
28. A detailed discussion of this chain of effects is provided in Ploeg (2018:79). See also Tony Weis (2010).
29. A mode of production is a specific, internally consistent, combination of productive forces (land, labour, technologies, technologies embodied in capital goods and inputs) and social relations of production. It assumes and also specifies a specific social division of labour and thus brings about a specific patterning of the relations between productive units and markets.
30. There are many different forms of domestic production found all over the world. These range from *datsjas* in Russia (numbering in the millions), to vegetable gardens in Western Europe (some countries, provinces and/or municipalities even recognize

a legal right to have access to such vegetable gardens), to the uncountable number of Chinese peasant farms that produce partly for consumption within the household, partly for the market. In Africa and Latin America, one encounters other, place-specific, expressions of DMP (domestic mode of production).

31. This is not always the case. In some countries, there are several agricultural universities (or faculties), each having the need to 'get ahead' of the others. This may generate different institutionalized points of view. So, in empirical reality there are *degrees* of cohesion which are sometimes stronger and at other times weaker.

32. Meaning that they represent deviant information that, at best, is 'put in the refrigerator'. For the moment, it is not used because it is supposedly useless.

33. Meaning new insights, practices, artefacts (etc.) are only deemed to be valid if they have been developed with the present AKS. This 'syndrome' helps very much to sustain the hegemony of the AKS – especially when agricultural and rural policies (as well as policies regulating food production and consumption) are exclusively built on advice emerging from the AKS. In many countries, AKS and policy (informed by the former) compose an 'unholy coalition'.

Bibliography

Benvenuti, B. (1982), 'De technologisch administratieve taakomgeving (TATE) van landbouwbedrijven', *Marquetalia* 5, pp. 111–136.

Benvenuti, B., S. Antonello, C. de Roest, E. Sauda, and J. D. van der Ploeg. (1988), *Produttore Agricolo e Potere; Modernizzazione delle relazioni sociali ed economiche e fattori determinanti dell'imprenditorialita agricola*, CNR/IPRA, Roma.

Bergh, P. J. M. van der. (1989), *La Tierra no da Así Nomás: Los Ritos Agrícolas en la Religión de los Aymara-Cristianos*, Ph.D. thesis, CEDLA, Amsterdam.

Bernstein, H. (2010), *Class Dynamics of Agrarian Change*, Fernwood, Halifax, NS, Canada.

Bourdieu, P. (1958), *Sociologie d'Algerie*, PUF, Paris.

Bruin, R. de. (1997), *Dynamiek en Duurzaamheid: Beschouwingen over bedrijfsstijlen, bestuur en beleid*, Ph.D. diss., Studies van Landbouw en Platteland 23, LUW, Wageningen, the Netherlands.

Chayanov, A. V. (1924/1966), 'On the theory of non-capitalist economic systems', in *The Theory of Peasant Economy*, edited by D. Thorner et al., 1–28, Manchester University Press, Manchester, UK.

Chayanov, A. V. (1925/1966), *The Theory of Peasant Economy*, edited by D. Thorner et al., Manchester University Press, Manchester, UK.

Darré, J. P. (1985), *La Parole et la Technique: L'Univers de Pensée du Ternois*, Editions L'Harmattan, Paris.

Edelman, M., C. Oya, and S. M. Borras Jr. (2015), *Global Land Grabs: History, Theory and Method*, Routledge, Oxon, UK.

Galeski, B. (1972), *Basic Concepts of Rural Sociology*, Manchester University Press, Manchester, UK.

Glaser, B. G., and A. L. Strauss. (1967), *The Discovery of Grounded Theory: Strategies for Qualitative Research*, Aldine, Chicago.

Graham Brade-Birks, S. (1950), *Modern Farming: A Practical Illustrated Guide*, Heinemann, London.

Hayami, Y., and V. Ruttan. (1985), *Agricultural Development: An International Perspective*, John Hopkins University Press, Baltimore, MD.

HLPE. (2013), 'Investing in smallholder agriculture for food security', A report by the High Level Panel of Experts on Food Security and Nutrition, Rome.

Hofstee, E. W. (1946), *Over de Oorzaken van de Verscheidenheid in de Nederlandsche Landbouwgebieden* (inaugurele rede), Landbouwhogeschool, Wageningen, the Netherlands.

Hofstee, E. W. (1982), *Differentiële Sociologie in Kort Bestek: Schets van de Differentiële Sociologie en haar Functie in het Concrete Sociaal-Wetenschappelijk Onderzoek*, Mededelingen van de Vakgroepen Sociologie, Landbouwhogeschool, Wageningen, the Netherlands.

Huang, P., G. Yuan, and Y. Peng. (2012), 'Capitalization without proletarianization in China's agricultural development', *Modern China* **42** (4), pp. 339–376.

Koning, N. B. J., M. K. van Ittersum, G. A. Becx, M. A. J. S. van Boekel, W. A Brandenburg, J. A. van den Broek, J. Goudriaan, et al. (2008), 'Long-term global availability of food: Continued abundance or new scarcity?' *NJAS - Wageningen Journal of Life Sciences* **55** (3), pp. 229–292.

Koningsveld, H. (1976), *Het Verschijnsel Wetenschap: Een Inleiding tot de Wetenschapsfilosofie*, Boom, Meppel, the Netherlands.

Lacroix, A. (1981), *Transformations du Proces de Travail Agricole, Incidences de l'Industriali-sation sur les Conditions de Travail Paysannes*, INRA, Grenoble, France.

Larson, D. F., K. Otsuka, T. Matsumoto, and T. Kilic. (2012), *Should African Rural Development Strategies Depend on Smallholder Farms? An Exploration of the Inverse Productivity Hypothesis*, World Bank, Washington, DC.

Le Roy Ladurie, E. (1980), *Montaillou: Cathars and Catholics in a French Village, 1294–1324*, Penguin, Harmondsworth, UK.

Long, N. (1989), *Encounters at the Interface: A Perspective on Social Discontinuities in Rural Development*, Wageningen Studies in Sociology No. 27, Agricultural University Wageningen, Wageningen, the Netherlands.

Long, N. (2001), *Development Sociology: Actor Perspectives*, Routledge, London.

Long, N. (2004), 'Actors, interfaces and development intervention: Meanings, purposes and powers', in *Development Intervention, Actor and Activity Perspectives*, edited by T. Kontinen, 14–36, University of Helsinki, Helsinki, Finland.

Long, N., and A. Long. (1992), *Battlefields of Knowledge: The Interlocking of Theory and Practice in Social Research and Development*, Routledge, London.

Lucas, V. (2018), *L'Agriculture en Commun: Gagner en autonomie grâce à la coopération de proximité. Expériences d'agriculteurs en CUMA à l'ère de l'agroécologie*, Ph.D. thesis, Université d'Angers, Angers, France.

Mayer, E. (2002), *The Articulated Peasant: Household Economies in the Andes*, Westview, Boulder, CO.

Mendras, H. (1967), *La Fin des Paysans – Innovations et Changement dans l'Agriculture Française*, Futuribles/SEDEIS, Paris.

Minderhoud, G. (1954), *Inleiding tot de landhuishoudkunde*, De erven F. Bohn N.V., Haarlem, the Netherlands.

Netting, R. McC. (1993), *Smallholders, Householders: Farm Families and the Ecology of Intensive, Sustainable Agriculture*, Stanford University Press, Stanford, CA.

Osti, G. (1991), *Gli Innovatori della Periferia, la Figura Sociale dell'Innovatore nell'Agricoltura di Montagna*, Reverdito Edizioni, Torino, Italy.

Ploeg, J. D. van der. (2013), *Peasants and the Art of Farming: A Chayanovian Manifesto*, Fernwood Publishing, Halifax, NS, Canada.

Ploeg, J. D. van der. (2018), *The New Peasantries: Rural Development in Times of Globalization*, 2nd ed., Routledge, London and New York.

Ragin, C. C. (2007), 'Comparative methods', in *The Sage Handbook of Social Science Methodology*, edited by W. Outhwaite and S. Turner, 67–81, SAGE, London.

Rebanks, J. (2016), *The Shepherd's Life: A Tale of the Lake District*, Penguin, London.

Rostow, W. W. (1960), *The Stages of Economic Growth: A Non-communist Manifesto*, Cambridge University Press, Cambridge.

Sennett, R. (2008), *The Craftsman*, Yale University Press, New Haven, CT.

Slicher van Bath, B. H. (1978), 'Over boerenvrijheid' (inaugurele rede Groningen, 1948), in *Geschiedenis van Maatschappij en Cultuur*, edited by B. H. Slicher van Bath and A. C. van Oss, 71–92, Basisboeken Ambo, Baarn, the Netherlands.

Steenhuijsen Piters, B. de. (1995), *Diversity of Fields and Farmers: Explaining Yield Variations in Northern Cameroon*, Agricultural University, Wageningen, the Netherlands.

Sumberg, J., ed. (2017), *Agronomy for Development: The Politics of Knowledge in Agricultural Research*, Earthscan, London.

Thiemann, L. (2014), *Artisans of the World, Unite: The 'Peasant Way' and Alliances for an Artisan Mode of Production*, Research Paper, ISS, The Hague, the Netherlands.

Thiemann, L. (2022), *The Third Class: Artisans of the World Unite?* Ph.D. diss., Erasmus University, Institute of Social Studies, The Hague, the Netherlands.

Tittonell, P. (2015), *Produce More in Western Countries?* http://www.pablotittonell.net/2015/06/whos-producing-our-food (accessed 16 May 2017).

Toledo, V. M. (1990), 'The ecological rationality of peasant production', in *Agroecology and Small Farm Development*, edited by M. Altieri and S. Hecht, 53–60, CRC Press, Ann Arbor, MI.

United Nations. (2019), *United Nations Declaration on the Rights of Peasants and Other People Working in Rural Areas*, Resolution adopted by the General Assembly on 17 December 2018, UN, New York.

Visser, J. (2010), *Down to Earth, a Historical-Sociological Analysis of the Rise and Fall of 'Industrial' Agriculture and the Prospects for the Re-rooting of Agriculture from the Factory to the Local Farmer and Ecology*, Ph.D. thesis, Wageningen University, Wageningen, the Netherlands.

Weber, M. (1891/2008), *Max Weber, Roman Agrarian History*, trans. R. I. Frank, Regina Books, Claremont, CA.

Weis, T. (2010), 'The accelerating biophysical contradictions of industrial capitalist agriculture', *Journal of Agrarian Change* **10** (3), pp. 315–341.

Zadoks, J. C. (1985), 'Landbouw tussen oecologie en economie', in *Inleiding tot de Oecologie*, edited by K. Bakker, 375–421, Bohn, Scheltema en Holkema, Utrecht/Antwerpen.

Zuiderwijk, A. (1998), *Farming Gently, Farming Fast: Migration, Incorporation and Agricultural Change in the Mandara Mountains of Northern Cameroon*, PhD thesis, CLM [Centrum voor Milieustudies], Rijksuniversiteit, Leiden, the Netherlands.

2 The farm labour process

The labour process in agriculture involves the ongoing and well-coordinated interaction of the labour force, the objects of labour and instruments (see Figure 2.1). The objects of labour are part of living nature: fields, crops, animals, trees, flowers, bees, water, seeds, seedlings, vines and so forth. In and through the labour process these objects of labour (or natural resources) are both reproduced and converted into useful products that contain value. The instruments facilitate and improve the labour process. The labour force, in its turn, sets the process of production in motion. The ways in which labour force, instruments and the objects of labour are combined and used depend on the social relations of production.

The amount of value produced per object of labour reflects the intensity of production. The intensity can be expressed both in monetary terms and in physical units (as, e.g., the amount of grain produced per hectare or the amount of

DOI: 10.4324/9781003313274-2

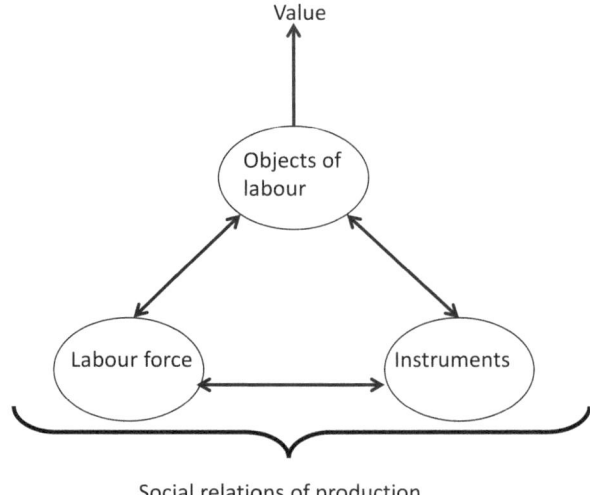

Figure 2.1 The elements constituting the labour process (author's own elaboration).

milk produced per dairy cow per year). The scale of production considers the quantitative relation between the objects of labour and the labour force. It can be represented as, for example, the number of hectares or the number of dairy cows per unit of labour force.

Together the scale and intensity of farming define the total production per unit of labour force. If GVP is the gross value of production (expressed in, e.g., euros), T is the amount of land (expressed in, e.g., hectares), L is the labour force (expressed in, e.g., full-time equivalents), then GVP/T is the intensity of farming and T/L is the scale of farming, whilst GVP/LU is the total production per unit of labour force (which is a more or less a satisfactory proxy for income per unit of labour force).[1]

Mathematically, the following equation can be derived[2]:

$$GPV/T * T/L = GVP/L.$$

This shows that incomes earned from farming depend on both scale and intensity. It also indicates that incomes can be improved by enlarging the intensity of farming, or the scale of farming, or through a combination of the two. Agricultural development may occur as ongoing intensification (often referred to as the 'Asian model') or as ongoing scale enlargement (the 'American model') or, again, as a combination of the two (the 'European model').

The farm labour process as a unity of different tasks

The labour process in agriculture embraces an extended range of different tasks (as shown in Figure 2.2). In dairy farming, for instance, the meadows need to be

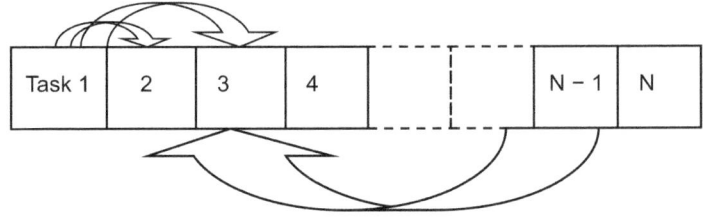

Permanent cycles of observation, comparisons, analysis,
evaluation and reorganization

Figure 2.2 The structure and dynamics of the labour process (author's own elaboration).

fertilized, the grass to be mown, converted into hay or silage and transported to the barn; if there is grazing, the animals have to be brought to the pastureland; if the animals remain permanently in the stable, they need to be fed with fresh grass. Then the milking has to be done, the milk cooled and stored until it is collected (or processed, in the farm itself, into butter, cheese, etc.); the cows need to be mated, the calves cared for and the replacement of the cows organized in a proper way; the manure needs to be collected, worked well and be brought to the fields for fertilization; and so on. Even in a relatively simple process such as grain production, some 100 tasks and sub-tasks can be distinguished, each implying a range of decisions. In more complex processes, such as dairy farming, the range is far wider, and the amount of daily, weekly, seasonal and yearly decisions that have to be taken is enormous.

The many different tasks and sub-tasks require careful coordination so that they flow together into a flawless whole. Let task 1 (in Figure 2.2) be the fertilization of the field (how much manure needs to be brought to each field, when and how it is to be spread, when and how to relate manure distribution to the rhythm of the rains, how to combine manure with the application of chemical fertilizer, etc.). Task 2 could be the mowing (when, how many times, etc.) and grazing (also: when, how, how to relate it to mowing). Task 3 could be the feeding of the animals and task 4 the milking. It is evident that these tasks need to be carefully coordinated: they need to be interlinked through feedback and feed-forward information. Cattle feeding (task 3) needs to be tuned to the expected performance (milk yield) of the dairy cows as shown during milking (task 4). But the relations between tasks 3 and 4 depend, in their turn, on the fertilization of the land (task 1), which impacts on the energy content, nitrogen level and taste of roughage, silage and hay, just as task 2 (mowing and grazing) will impact on the total amount of dry matter[3] obtained per hectare.

If the different tasks and sub-tasks are not well coordinated, a level of production that falls short of the optimal will be the result, with unnecessary losses and/or costs that are too high.

The required coordination occurs through, and is actively organized as a series of ongoing, mutually interdependent cycles of observation, interpretation, analysis, evaluation and re-organization (see Figure 2.2). Insofar as these cycles concern several fields (not a single one) and different dairy cows (from different bloodlines, combined with different sires), such cycles are comparative by nature. The different fields and the effects that different levels of fertilization have on them and on the amount of dry matter (DM) they produce are carefully observed and compared, as are the effects that the different quality grades of the produced hay, silage and fresh roughage have on the appetite of the animals, the composition of their milk and the smell and structure of the manure. All of these effects are carefully recorded, and this renders information that is processed, as it were, into tacit knowledge that, in its turn, offers insights into possible improvements.

These ongoing cycles of observation, interpretation, etc., allow for a fine-tuning of the process of production. They also generate, consolidate and increase knowledge of the experiential type (I will discuss this later in this chapter). The labour process that has these cycles at its heart and the type of knowledge that it generates are both artisanal. They crucially depend on the organic unity of mental and manual labour. This unity allows for observations that are quickly, but in a well-grounded way, translated into corrections. When a 'mistake' is observed (a cow not eating well, or bad-smelling manure, or a corner of a field does not render sufficient dry matter), immediate interventions are needed to correct the underlying 'mistake'. This explains why losses are generally far lower in artisanal production than in production organized in an industrialized way.

Methods Box #5   Doing research on the farm labour process

Mapping out a specific farm labour process can be done in different ways. Participation in the labour process itself can be very useful, but extensive interviewing is a good method as well (and a combination of participant observation and interviewing is probably the best approach). All stages (all tasks) need to be carefully discussed and the *why* question needs to be asked time and again. It is important to repeat this inquiry on different farms (preferably showing important differences in their styles). By doing so, one will surely encounter all kinds of small differences. These differences might be important; they can also be used as a tool in the interviews, by saying that one has met other farmers that perform this task in a (slightly) different way. This allows one to ask why such differences exist, the backgrounds of such different ways-of-doing, whether the respondents have experimented with other ways, and so on.

Depending on the objectives of the research, it might be useful to relate different ways of organizing the farm labour process to the yields obtained.

In research done in a potato-producing area in southern Peru (the community of Antapampa near Cuzco), farmers were asked how they performed a range of tasks (Ploeg, 1990, chapter 4). These involved the enhancement of soil fertility (the amount of land lying fallow, the amount under potatoes, the quantity of dung and fertilizer used, etc.) and the intensity of cultivation (labour input/*topo* for sowing, two rounds of cultivation and ridging up and weeding). All of these tasks were recoded as +1 if they positively compared to the average and as 0 if the performance was average or below. Thus, two scales were constructed and (using multiple regression) related to the yield estimates. This rendered the following model (in which 'labour structured as craftmanship' is the interaction of 'production of soil fertility' and 'intensity of cultivation'):

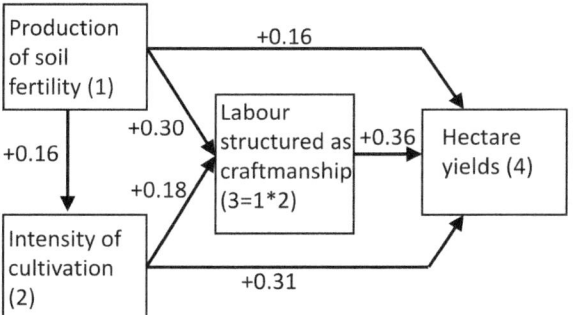

This approach is simple and straightforward (more sophisticated approaches are also possible, of course), and the result shown (the interaction having the strongest effect on yields, which is strengthened by additional effects of the independent variables) is not surprising at all. The relevance, though, is that once constructed, such a simple model can be used to check the impact of all other kinds of variables. In the next chapter, I will return to this model and show how it can be linked to contextual variables.

If task N − 1 (in Figure 2.2) represents cheese making (mostly a task belonging to farming women), there is active feedback from the cheese-making woman to her husband (who is involved in task 3): 'stop feeding them Brussels sprouts because the cheese has started stinking like hell' (or indicating that the hygiene of milking needs to be improved in order to avoid the loss of cheeses that start to ferment, for example). That is, the coordination of tasks and fine-tuning of production often translate into specific forms of cooperation as well.

Back to co-production

The barrel with staves (shown in the lower part of Figure 2.3) is a well-known *meme* out of the rich history of agricultural sciences. Originally developed by Justus von Liebig (in 1840) but still being used in today's curriculum, it refers to the many growth factors that together sustain and condition the processes of biological growth (the unfolding of living nature, as we could say, using the terms introduced in Chapter 1). The amount of nitrogen in the subsoil, for instance, is one such growth factor. Plants cannot grow without nitrogen – the question, though, is how much of it is needed (too much nitrogen, for instance, will even 'burn' – that is, kill – the germinating plants). But then once the nitrogen is encapsulated in the soil, it needs to be 'transported' to the roots of the growing plant. This makes *transportability* a second growth factor. *Absorbability* – that is,

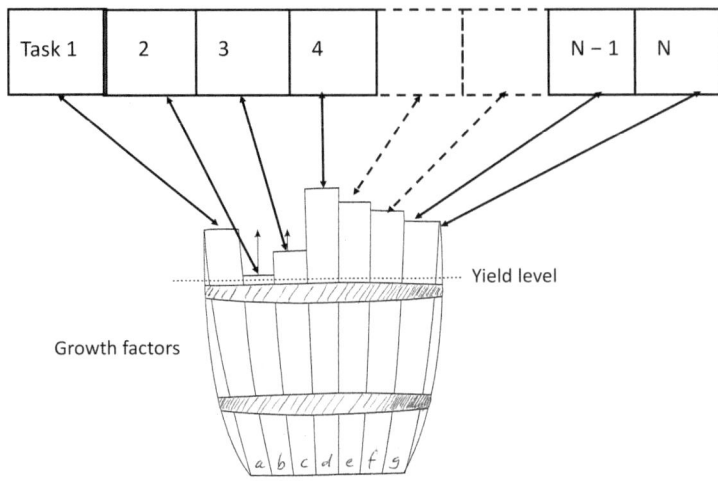

Figure 2.3 The *beta-gamma* translation (author's own elaboration).

the capacity of the roots to absorb the available nitrogen – is a third one. Then comes the availability of water, the absence of weeds that could asphyxiate the germinating plant, the absence of pests and diseases, etc., Central, then, in Von Liebig's approach is that the yield level of the plant (i.e., the result of the bio-physical process of growth) critically depends on the growth factors – and, more specifically, on the most limiting one. The totality of growth factors can be represented as a barrel and each growth factor as a stave. Together the staves constitute the barrel, and the amount of water it can contain (i.e., the yield) evidently depends on the shortest stave.

The shortest stave determines the yield to be obtained. It's as simple as that. Nonetheless, at the time this insight was developed, it represented a somewhat revolutionary thought. It definitely implied that (1) yield levels are not given but are potentially variable because they depend on a range of growth factors that are changeable, (2) the increase of yields assumes the identification of the growth factor that fails to meet the mark, and (3) the identity of the 'shortest stave' cannot be assessed beforehand, nor can it be remedied with a generic solution (of the 'blueprint' type). Instead, one needs to figure out, time and again, which is the limiting factor, which will depend on the local situation, just as the solution will necessarily be local.

The growth factors identified by Von Liebig did not only exist in his laboratory. They are universally present in the fields as well. In the latter they exist as the object, and result, of the farm process of labour. The availability of nitrogen in the subsoil is not fixed in eternity, nor is it a static characteristic of the soil (determined by the laws of physics, chemistry and biology). It is the object (and outcome) of a crucial task within the farm labour process; that is, *fertilization* (or 'the reproduction of soil fertility'), just as transportability depends on *ploughing* (or, more generally: the preparation of the land or seedbed), the capacity of the plant to absorb nitrogen depends on *selection*, the availability of water depends on *irrigation* and *drainage*, the absence of weeds depends on *weeding*, etc. In short: in and through the farm labour process, the different growth factors are regulated, changed and mutually coordinated (see Figure 2.3). Where the Von Liebig analysis operates with nouns, the labour process analysis operates with verbs. It centres on *praxis*; that is, on the goal-oriented and knowledgeable activities that together regulate the relevant growth factors.

Figure 2.3 unravels the specificities of co-production. It shows how nature is affected, regulated and thus transformed (towards more productive constellations) by, and through, farm labour. It also shows how nature informs the actors involved in co-production. It shows the length of the different staves and can also give, over the course of time, feedback about the effects of interventions and adaptations. Thus, the knowledge and experience of the actors involved is – possibly – enriched. The already discussed cycle of observation, interpretation, comparison and subsequent translation into new interventions, corrections and adaptations evidently is central here. Without such cycles, farming would fall flat or be, at best, a stationary endeavour, working to never-changing rules.

Understanding agriculture as co-production or, more specifically, farming as the goal-oriented and knowledgeable regulation of growth factors also opens the door for urgently needed changes in agricultural sciences. Agricultural sciences are now divided into so-called *beta* disciplines (that understand nature as a biophysical entity) and *gamma* disciplines that study human activities, social relations, institutional patterns, etc. In short: on the one hand, nature is studied, and on the other, man. Thus, co-production as the interaction of man and nature appears to be torn between two segments that compose, at best, an unhappy marriage. The conceptualization of farming as the *social regulation of growth factors* (as synthesized in Figure 2.3), however, opens the door for a fruitful cooperation between the two. It allows, indeed, for a true *social agronomy*.[4]

The further unfolding of co-production allows *new* growth factors to be identified which might change the *interplay* of all of the growth factors together.[5] In this respect, the interplay between (agricultural) sciences and farmers' knowledge can be highly helpful and productive. This also indicates, at least theoretically, that there is no 'law of diminishing returns' which would 'exhaust' the interplay shown in Figure 2.3, meaning that at some point the search into further increases of yields would become futile.

For those involved, the capacity to design, adapt and further improve the farm labour process and the results of so doing (as well as the process) are almost always a source of joy, satisfaction and pride and confirm their autonomy. Mastery over the different labour tasks and, especially, the capacity to coordinate the different tasks in such a way that progress is wrought are important outcomes of previous dedication, involvement and insights. In this sense, independence is far from an abstract notion – it is solidly anchored in the mastery of the farm labour process. This same independence and the possibility of mastery it brings compose a sharp line of demarcation with, for example, factory workers, especially since industrial management took over the component of mental labour that once was also part of the daily routine of wage workers (see Harry Braverman, 1974).

The mastery over the labour process (and especially the unity of mental and mental labour) also generates specific types of knowledge (which I will discuss later as *art de la localité*), skills and craftmanship.

The 'function of production'

An important theoretical cornerstone, if not doctrine, of modern agricultural sciences concerns the so-called function of production. Such a function specifies the relation(s) between input(s) and output(s) in agricultural production. Graphically it is usually represented as an S-shaped curve that shows increased returns at low input levels whilst at high input levels increases bring decreasing returns. Mathematically this is formulated as $Y = f(C, L, I)$, in which Y is realized production (or output), C is capital, L is labour, and I is inputs. Building on this basic formula, more sophisticated functions can be elaborated.

The function of production is thought to be grounded in bio-physical realities (which in turn are mediated by technologies). Thus, the function 'governs' agricultural production. Going 'beyond' this function is thought to be materially impossible (the function of production is viewed as a 'technical ceiling'), while producing at levels 'below' the function is inefficient. The efficient production possibilities are located *on the function*. The projection of reigning market relations (prices for input and output) on the function specifies the optimum. Thus, markets and technology determine agricultural production.

As opposed to this deterministic view, the notion of *regularity* can be used. In the space defined by inputs and outputs there are regularities: patterns of frequently occurring interdependencies between inputs and outputs. But these regularities are the *outcome* of farm labour processes structured in a particular way, and if these farm labour processes change, new regularities will emerge. Mostly, these regularities 'show up' as a *cloud* that contains considerable diversity (see, e.g., upper left quadrant of Figure 10.3). This diversity, in turn, reflects the different positions that are taken and which depend on strategies, particular situations, etc. The 'upper side' of the cloud (the 'frontier function' in the language of economics) shows the technical possibilities already available (already 'put into practice') and which probably highlight the margin for farms located elsewhere in the cloud to move ahead. This margin for endogenous development (discussed later in this chapter) is sometimes considerable.

Externalization

In certain epochs and places, the range of tasks and subtasks that together constitute the farm labour process has been widened by adding new tasks to the work that is to be done. In others, the range is actively reduced by externalizing particular tasks; for example, delegating them to outside agencies. This is illustrated in Figure 2.4 (which builds on Figure 2.2).

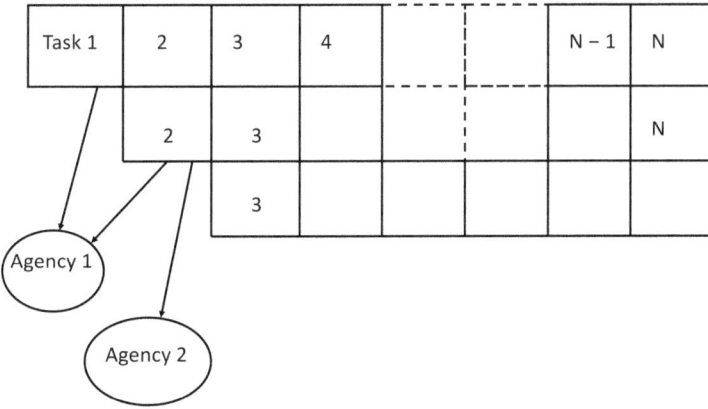

Figure 2.4 The process of externalization (author's own elaboration).

Let task 1 again be the fertilization of the fields (this task includes a wide range of subtasks). Fertilization can be an integral part of the on-farm labour process (meaning that all of the work is done by members of the farming household, all of the relevant decisions are taken by those involved in this task and most, if not all, of the means used are produced within the farm itself). However, the task can also be delegated to one or more external agencies. Instead of using compost or manure from one's own farm, chemical fertilizers (or manure produced on other farms) may be acquired and used. Instead of relying on one's own knowledge of the soil and the associated need for nutrients, a specialized company can be contracted to perform a soil analysis and translate the outcome into advice for fertilization (composition, quantities, timing, sequence, etc.). Finally, a custom worker can be contracted to do the job. If task 2 is the production of feed and fodder, a similar pattern of externalization can occur: the production itself (including conservation, etc.) can be delegated to other farms (or specialized companies) and the knowledge needed to assess the quality of roughage and calculate the supplementary doses of concentrated (or industrial) feed required to provide optimal alimentation for the cattle can be contracted out as well.

The degree of externalization can be low or very high. In the example used so far – a dairy farm – nearly all tasks can be delegated: only milking remains (although even milking can be robotized: then only the supervision of milking remains). Another well-known example of extreme externalization is large-scale soy producing farm enterprises that operate in the Amazon area of Brazil and the *Llanos* of Colombia. The only instrument used by the 'farmer' is his mobile telephone. It is used for contracting a loan and ordering seeds, fertilizers, herbicides and, if needed, pesticides. Then a contract worker is contacted (by phone evidently) to do the field work (from soil preparation and seeding to harvest), after which different companies are contacted to sell the harvest (or, in case of a failed harvest, government agencies are called to arrange a pardon for the debt engaged with the bank and contract worker).

Externalization implies a rupture in a once organic unity of different and carefully coordinated tasks and subtasks – all located within the farm. It also implies that the externalized tasks (which are shifted from the farm to outside agencies) need to be recombined again. This occurs through a double set of relations: commodity relations and techno-administrative prescriptions.

Externalization (whether through buying, leasing, renting, etc.) commodifies the items that were an integral part of the farm labour process as a whole (whether artefacts, blocks of knowledge, and/or particular activities). What were formerly non-commodities now enter the farm labour process as commodities. They represent an exchange value that needs to be earned back again through the operation of the farm. This may seem to be a subtle, if not negligible, difference in the eyes of outsiders, but in the practice of farming it makes a huge difference whether one uses self-produced feed and fodder or feed and fodder acquired on the market. As farmers sometimes say: 'Having enough feed and fodder produced in your own farm makes cattle feeding into a pleasure, you can give them what they need. Then you can care far better for them'. (This quote is from a research programme carried out among dairy farmers in northern Italy.) The same research also showed that dependency on the *market* for feed and fodder meant quantitively lower feed rations per cow: $r = -0.42$ ($P = 0.001$).[6] In Ireland, the local cultural repertoire was very explicit about having sufficient amounts of good quality hay. It was like an insurance policy for those who lived off the land (see also Text Box 3.3).

Together with new commodity relations, externalization also – and simultaneously – brings about new techno-administrative relations. The farmer knows the feed and fodder produced in his own farm like nobody else (the more so because he or she will mostly have many years of experience in doing so). Buying feed from elsewhere (the specific place of origin is often unknown) suddenly raises a new series of questions: how is it composed, what is its particular history, is there any danger that it carries unknown additives, how is it to be combined with other ingredients, how will the animals react to it, does it represent sufficient continuity with previous (as well as future) cycles? The same applies for other items such as seeds, fertilizers, heifers, etc. Their acquisition needs to be accompanied with information (techno-administrative prescriptions) that specifies how the different 'things' are to be used. In short, they come with a manual or with a consultant (linked to the supplying agencies) that helps the farmer to integrate the newly acquired elements into the farm labour process (which often implies adapting and changing the farm labour process).

Thus, with increased externalization, a widening range of outside agencies begin to function (according to the Italian rural sociologist Bruno Benvenuti) as a techno-administrative task environment (usually abbreviated as TATE) which increasingly prescribes the farm labour process (Benvenuti, 1982; Benvenuti et al., 1983). This comprises, in the words of the French rural sociologist Placide Rambaud (1983), a 'nomenklatura': the assembly of specialists, accountants, bank employees, state officials, etc., that specify how the work on the farm is to be done.

Externalization and the subsequent division of labour between farms, agricultural industries, banks and providers of services are the main vehicle for the penetration of capital relations into the agricultural sector. This implies (1) a slow, gradual but nonetheless persistent re-constitution of the farm labour process and (2) a re-distribution of the wealth produced. Agribusiness, banks and providers of services increasingly come to function as a social relation of production. This is especially the case where externalization becomes a seemingly irreversible process and/or in situations where agribusiness, banks and service providers are able to operate as obligatory passage points providing means and services without which no agriculture seems to be possible.

Empirically, the process of externalization has different drivers, rhythms and results. As empirical analysis shows, it is a highly unequal and differentiated process. Alongside highly market-dependent farms (which have externalized many tasks) one also finds relatively autonomous farms (showing low degrees of externalization). This allows for meaningful comparative research (which often is of considerable socio-political relevance). In given empirical situations it sometimes is also possible to encounter and describe processes of re-integration that exist alongside processes of (ongoing) externalization. The former occurs when farmers try to re-integrate particular tasks in their farms that were once externalized. On-farm processing of agricultural produce into food products is just one (albeit important) example. I will come back to such re-integration (which often implies enlarging the autonomy of farming with the farmers once again taking the reins in their own hands) in Chapter 7.

Externalization is not only driven by state, capital and science. Farmers' search for system integration (mentioned in Chapter 1) can also have a strong impact, and a powerful momentum can be created when internal and external drivers interact.

Endogenous development

The analysis of the farm labour process is also useful insofar as it reveals several possibilities for *endogenous development*; that is, development generated in and through the farm labour process itself. Endogenous development is, as the phrase implies, generated *from within*, whilst the contrasting notion, *exogenous* development, refers to development that is generated by external drivers and also critically dependent on them. This is the case with, for example, 'Green Revolution schemes' that centre on the introduction of new technologies, associated inputs, capital (in the form of credit) and externally developed knowledge introduced by extension officers.

A first mechanism of, and for, endogenous development resides in the organic unity of production and reproduction as entailed, especially, in peasant agriculture. Through the reproduction of objects of labour, small improvements might be realized (e.g., better yielding and/or more resistant seeds; cows that outperform the previous generations, etc.). Mostly this requires dedication, care, perseverance and a 'sharp eye' to notice what initially might seem an irrelevant deviation. Breeding and selection are far from easy. But there might be luck as well: co-production sometimes comes with surprises. Be this as it may, the accumulation of small improvements may, over the decades, bring considerable progress.

An important advantage of this kind of *novelty* production[7] is that the improved objects of labour and/or instruments seamlessly fit in the specific context within which they have been generated, whilst the application of innovations (that stem from outside sources) nearly always requires modifications in the organization of the farm (see Chapter 3). The latter might bring prohibitive costs or other difficulties – these will, in the end, hinder or even block a successful adoption of innovations.

An additional but highly interesting issue is how this first mechanism interacts with the internal differentiation of the peasantry. Having a good cow that promises to bring very good offspring is often seen, by poor peasants, as too high a risk: losing that particular animal would be too great a loss. Consequently, they will sell her. Maintaining the promising element was and is the privilege of the better-off (or rich) peasants. They can afford such a risk. In most sub-cultures, breeding and selection are consequently defined as being the *duty* of rich farmers. This duty includes that, once proven to be successful, the novelty (the promising sire, productive new seeds, etc.) is to be *shared* with all other member of the local society.

However, the local cultural repertoire might also hinder, if not paralyze, the operation of this first mechanism for endogenous development. This can apply in the case of the 'image of the limited good'.[8] If some members of a peasant community or village are better off than others, this is understood as the cause of others being worse off. The apparent advantages of the better yielding plants, for instance, have been 'robbed' from the others. In short: progress is beneficial for some but detrimental for most (which is also reflected by theories on the 'technological treadmill', in which early adopters get the benefits, whilst the negative effects that emerge are paid for by the 'laggards').[9]

A second mechanism of endogenous growth and development resides in the fine-tuning of the labour process in the farm (see Figure 2.2). The growth factors that don't come up to the mark are identified and corrected through ongoing cycles of observation, interpretation, analysis and comparison. The resulting process of small but cumulative corrections can contribute considerably to a steady, but ongoing, growth, especially if the experiences obtained are shared in enlarged networks of farming people.

Most farming communities know their experienced farmers, whom they consult with in order to identify and overcome an underperforming growth factor in their own farms. Formal extension services often use these local insiders as highly effective sources of information. Classical agronomists, and especially the 'walking professors', constitute another important modality for the socialization of this second mechanism for endogenous development. In his book *Social Agronomy*, Chayanov (1924) promoted such an approach, which is currently being re-developed by the agroecological movement. *Campesino-a-campesino* (from peasant to peasant) is an important key word in this movement (Rosset et al., 2011; Val et al., 2019). Organizing 'lotteries' and 'price contests' were other modalities that have been used in the recent past (see Text Box 2.1). *Inter alia* the lottery and contest show that fostering progress was far from alien to peasant societies, defined by many as 'stuck in the mud'.

Text Box 2.1　The contest and the lottery: Socializing the cycle of observation, comparison, interpretation, intervention, re-organization

---

In the second half of the 19th century, farmers' organizations in the province of Groningen, in the northern Netherlands, started to collect questions from farmers all over the province and bundled them together in a small booklet that was then sent to all of the organizations' branches in the different villages all over the province in order to be discussed. In this way, a question from a farmer in the southern part of the province ('I have tried all different possible quantities of manure on my fields in order to get better barley yields, but time and again in vain. What can I do?') could be answered by a farmer in the north who knew from experience that fields recently conquered from the sea, with many shells in the subsoil, would react positively to higher doses of manure and thus bring better barley yields. If this farmer were illiterate, the local schoolmaster would note down his answer. All of the answers obtained in this way were brought together again in another little booklet and again distributed across the province. Thus, the initial farmer would receive several suggestions and possibly start to explore them on his own farm. Positive results were reported back and each year the provider of the 'most stimulating response' was awarded a prize. This widening of farmers' horizons through the sharing of experiences was called 'the prize contest'.

The 'lottery' was another important mechanism for widening the horizon and sharing obtained experiences in a far wider group. When new machinery became available that aroused curiosity, all of the farmers of a local branch would put together small contributions (say 10 cents each) to buy the new equipment. Then there was a lottery and the farmer who won obtained the equipment, with the duty to try the new thing out. This had a double advantage. It eliminated the economic risk of buying an expensive instrument as well as the social risk of being the 'idiot of the village' for having bought something that was possibly useless. At the same time, the new equipment was tested, and possibly adjusted, and its advantages and disadvantages were shown to all of the villagers.

Social mechanisms like these strongly spurred endogenous development in Groningen in the 19th century and the first half of the 20th century.

'L'art de la localité' *(peasant knowledge)*

The farm labour process is both informed by and generates knowledge. This knowledge is closely tied to the place(s) where it is generated and used. It mostly is extremely rich and full of details (Conklin, 1954, 1957). It has specific features, a specific scope and reach, specific shortcomings and specific dynamics. It also has its particular grammar. That means that it is constructed, and operates, in a particular way. The knowledge generated in the farm labour process is highly relevant to the job: it is knowledge that specifies how the job of farming is to be done. It is what the French beautifully call *le savoir-faire paysan*. It is 'knowing how to do things' or, probably even better: 'knowing how to farm'. It is knowledge that stems from practice and that is used to inform practice. It is partly knowledge without words (Darré, 1985): by mutual cooperation it passes – without words – from one actor to another. Insofar as it uses words, these often are ambiguous if not downright enigmatic (at least to outsiders), barely standardized and sometimes metaphorical. It is also very much knowledge derived from and passed to others through *learning by doing*. Thus, farmers' knowledge is grounded, in a manifold way, in experience; it stems from being engaged in practicing farming in a knowledgeable and curiosity-driven way. For farmers themselves, the language used to express this experiential knowledge is very much to the point. Farmers' knowledge is accumulated experience (and it is as variegated as practice itself). It is also a guideline (on how to perform), just as it is a kind of roadmap: it often specifies where to go and how to arrive there. It reflects history (as much as it is built on it), but it is, at the same time, future oriented. However, in farmers' knowledge systems there is usually not one single future but a range of possible futures. This implies that choices are always possible and that they have to be made. This, in turn, requires criteria for judging – there are many such criteria in most farmers' knowledge systems, but they are usually a bit ambiguous. There are clear values and norms, but there are also counterpoints.

A comparison between farmers' knowledge and scientific knowledge might help to clarify at least some of the decisive features of the former. Such a comparison is also helpful because both blocks of knowledge often engage in uneasy forms of interaction in today's agriculture. Table 2.1 offers an overview of the main differences.

Farmers know their fields by means of the crops (and herbs) that grow well (or not, or not at all) on their fields, just as they come to know the quality of their fodder crops by observing how their animals eat and digest them, with digestion being understood through observation of the structure and composition of the dung and the appearance of the animals' skin (and other factors). That is to say, farmers' knowledge is *relational*. Each object and each activity is understood through its effects on other objects, behaviours, and activities – just as the object itself is understood as being partly explained by preceding activities and interventions. This is in stark contrast with scientific knowledge (as embodied in modern agricultural sciences). Scientific knowledge focusses on the *intrinsic* aspects of each object (and on the 'optimal' way of organizing each activity), and by doing

Table 2.1 A schematic comparison of farmers' knowledge and scientific knowledge

Feature	Farmers' knowledge	Scientific knowledge
Focus	The *relational* properties of the elements involved	The *intrinsic* properties of elements involved
Grammar	*Subjunctive*	*Imperative*
Reach	In principle limited to the *local*	Claims *universal* validity
Object	*Living* nature	*Controllable* matter
Subject-object relation	*Organic unity*	*Separation*
Balance	*Reciprocity*	*Instrumental utility*

so it takes these objects (and activities) in isolation[10]: avoiding 'disturbances'. This approach can render useful insights and help to improve farming. Its weak point, though, is that it is, firstly, very difficult to translate the obtained insights into farming practices as compound wholes. Knowledge of the intrinsic properties of a particular segment is obtained at the expense of insights into, and knowledge about, the relations with other segments (specific, or optimal, relations are, at best, assumed – an issue to which I will return in Chapter 3 when discussing potatoes and in Chapter 10 when discussing taste). Secondly, new techniques (based on the knowledge thus obtained) are sometimes hardly applicable in the practice of farming (simply because the assumed relations are not available).

Theoretically speaking, the combination of farmers' and scientific knowledge would make for a very strong combination. In practice, though, the intertwinement or even the interaction of the two turns out to be very problematic.

There is, within agricultural sciences, a kind of 'subterranean' movement of scholars who put co-production centre stage. It acknowledges the existence of 'geno-forms' whose intrinsic properties might very well be explored, whilst simultaneously recognizing that, in practice, we only meet, and deal with, 'pheno-forms', which are understood as the outcome of co-production and thus show a range of relational properties.[11] In and through co-production, these geno-forms are moulded and re-moulded into particular pheno-forms. Agroecology is a current, and very strong, expression of this new, subterranean, approach.

Another important difference between the different knowledge systems resides in their grammar. In most knowledge repertoires used in peasant agriculture, the *subjunctive* is the main form (Kessel, 1980). This focusses on the possibility of achieving certain desirable results that *might* be possible,[12] although there is no certainty of doing so. One has to wait and see. It is based on experimentation and hope. There is little certainty. It is the language of *craft* that reflects the interaction with living nature, which is, by definition, unpredictable. Science, by contrast, uses the imperative. Things are as they are, and results are as they ought to be. It is the language of *control*. It often takes the *nomological* form: 'if certain conditions are met, then intervention A will necessarily produce outcome B'.

A third difference concerns the reach of knowledge. Farmers' knowledge stems from, and concerns, the local interaction of different growth factors as shown through the relational impacts they produce. It is *l'art de la localité*. The carriers of this knowledge are very much aware that their knowledge is especially, if not only, valid in the place where it has been elaborated – that is, one's own farm or farming community.

> Every year that I live in the country, and every year that I know more of what the people who work the land of the United Kingdom are doing, I realise more fully the profound agricultural truth underlying the remark of a skilled Dutch farmer to an English landowner: if you should come to Holland to farm, you would imitate me, but if I were to go to farm in England I would imitate you. (Robertson Scott, 1912:ix)

Scientific knowledge, on the other hand, claims universal application (which reflects the notion that intrinsic properties represent universal, objective and value-free truths). Hence, scientific knowledge is supposed to be applicable everywhere. It is also thought to be incremental, implying that the newest knowledge (and the newest technology derived from it) is always the best. This view is sometimes countered with the thesis that scientific knowledge is as 'local' as farmers' knowledge. It reflects (and is, in a way, bound to) another locality: the laboratory, the experimental plot and the digitized models used for producing simulations,[13] none of which may reflect the realities under which actual farmers practice agriculture.

Subject-object relations (and the nature of language)

The organic unity of mental and manual labour renders a knowledge repertoire that reflects the local conditions from which it emerged (although, as noted above, the horizon of the local can be increased considerably). In, and through, the labour process, the land, the fields, the animals and the farm as a whole are moulded in a specific way, just as it renders specific knowledge. The thus-produced knowledge intimately corresponds with the work done and the way the farm as a whole (or the local farm style as shared by farmers in a particular area) is moulded. This tight unity between practice and knowledge (between object and subject) comes with a kind of self-affirmation that often is completely mis-understood. When asked why they do things in a particular way, farmers often respond that they 'have always operated this way' and possibly add that, anyway, this is 'the best way to perform'. *Detached* from their particular farm and the par-ticular organization of the labour process it contains, such statements seemingly express nothing but tradition (and a corresponding unwillingness to change). However, when *related* to their farm and farm labour process, such statements are both true and insightful. Farmers deal with their resources in a manner that fits the way these resources are shaped – just as the resources indicate, through many feedback mechanisms, the best manner to work them. So, it is not tradition that is unveiled: these expressions highlight the fine-tuning real-ized through the labour process and, consequently, the close coherence between practice and knowledge.

The strong unity between manual and mental labour also explains why words and expressions that at first sight seem terribly ambiguous and vague allow, nonetheless, for a very precise communication. In the Andean moun-tains, for instance, local peasants use up to 40 contrasting combinations of words to describe the quality of their fields: *high* or *low*; *hot* or *cold*; *black* or *red*; *deep* or *shallow*; *grateful* or *tired*; *hard* or *soft*; *willing to give* or *stubborn*; etc. Together these oppositional pairs compose a vocabulary that in the eyes of outsiders is extremely imprecise, overlapping and probably a source for chronic misunderstandings. It is not very clear, for instance, which dimension the notion of *height* refers to. It is even less clear when high stops being high and being low starts to be the case. Neither is quantifiable, nor is it clear how very high relates to high. Is it twice as much, or more, or less? This raises lots of questions – and this is the case nearly everywhere. The way Dutch cattle breeders talk about their cows is as confusing as the way their Peruvian counterparts speak about potatoes or their plots. Nonetheless, the former are able to communicate with each other as much as Peruvian potato growers do. In Peru, every year farmers exchange some fields for others and, in doing so, they carefully communicate the properties of these fields to other farmers using the vocabulary indicated above. For those who share *both* specific practices and the corresponding block of knowledge, these terms are highly meaningful and precise (Darré, 1993). The description of a particular field as high, but black, deep and grateful is, for them, very precise. For them it is as much as a biography of the field: it informs in a synthetic way about how it has been worked and the problems it might bring. In this vein, a field being 'high' does not inform about the mathematical distance from sea level but about the susceptibility of the field to frost during the night (which is a feature that can be corrected through, e.g., the application of considerable amounts of manure).[14] Thus, the description becomes a meaningful and precise script on what to do with the field, which varieties are best grown in it and how its fertility is to be maintained or increased. In exactly the same way, Dutch breeders understand each other when they talk about a 'noble cow' or about a colleague who is able to 'look underneath the skin of the animal' (thus assessing its potential). Different groups who share the same practices share the same language, and for them the language is crystal clear. The precision and power of the language used are the outcome of co-production and the associated unity of manual and mental labour.

The reciprocal relation between peasants and their objects of labour (between those who know and that which is known) is an important dimension of farmers' knowledge. It means, in the end, that living nature is actively involved in the construction of knowledge. Farmers' knowledge repertoires are co-constructed by both farmers and living nature. The reciprocal relation underlying this co-construction is part of a wider reciprocity that defines the relations between 'man and land' (see Text Box 2.2). Reciprocity introduces equity in non-commoditized forms of exchange. This exchange may involve commensurable as well as incommensurable products and services ('you help me one day, I will help you another day': 'I lend you my tractor' and in return 'you allow me to mow one of your fields and use the obtained hay for my animals'). There might be a considerable time span between giving and receiving, just as the return of the gift may take a very different form. Reciprocity is governed by a 'moral economy' (Scott, 1976); that is, the values, norms and rules shared in local economies.[15] For further reading on reciprocity, the outstanding work of Eric Sabourin (2009, 2011, 2012) is recommended.

Text Box 2.2   On reciprocity

When Italian peasants discuss the way they relate to their fields, their cows and their crops, they probably will use the word *cura*. This expression refers to 'care', just as the verb (*curare*) refers to giving care. It is, essentially, about reciprocity. Within the framework of reciprocity, the animals and/or the crops will only render a nice production (*una bella produzzione*) after giving care to them. Giving care is far from just an instrumental activity. It supposes, within the discourse of Italian peasants, the presence of *passione* (passion), *impegno* (the willingness to dedicate yourself completely to what you are doing), and it requires *conoscenza*: you have to be knowledgeable. Finally, there is the requirement of *autosufficienza*: that the resources used in the process of production are owned by the farming family itself. Tight dependency relations with markets at the input side of the farm are to be avoided, precisely because they would induce 'the logic of the market' into the heart of the farm. The 'logic of the market' would threaten, if not exclude, working with *cura*.

The concept of *cura* defines, and simultaneously reflects, a reciprocal relation between the farmer and his objects of labour. It is about giving and getting back. It is, as it were, about gifts that flow in two ways. The farmer raises and takes care of the calf, gives her shelter and the possibility to develop into a good milking cow, and then he will feed her, probably with a diet that is carefully adapted to the individual needs of the animal. In return, the cow will give the farmer new, hopefully promising calves and a rich flow of milk that might continue for many years. It is, as Victor Toledo (1992) would put it, a non-commodity exchange between farmer and living nature (see Figure 1.1).

This type of relation underlies many farming systems around the world. The reciprocity of it is strengthened, according to van Kessel, an anthropologist who worked for many decades in the Andean region, through 'metaphoric connotations' that assume, in the end, a kind of personification: land, crops, lakes, wells, but also the light, the rains, the frost and other meteorological phenomena are perceived and understood as living beings that provide all kinds of signs. In this context, it becomes nearly self-evident to say, for example, that 'this piece of land is grateful' (for all the care it received) and that consequently 'she [land is notably feminine] is generous' (i.e., willing to give back). This does not imply that Andean peasants are dreamers – on the contrary: 'the norms for their technical operations in the field are dedication, comprehension and affection [*compromise, comprensión, cariño*]' (Kessel, 1990:92). These latter concepts strongly coincide with the Italian notions (*impegno, conoscenza, passione*) discussed above – just as the Frisian saying '*as jo lân hâlde wolle dan moat it sines ha*' (if you want to stay on the land, you have to give it what it needs) echoes the give-and-take relation found among both Italian and Andean farmers (Ploeg, 2003:94). The Chinese have a similar saying: 'if man works hard, the land will not be lazy' (Arkush, 1984:461).

Since ancient times, farming has been embedded in a wide range of reciprocal relations. These not only concern the interactions between, and mutual transformation of, man and living nature. They go far further. They equally embrace:

1   The interrelations within the farming family, between father, mother and children (and probably grandparents, uncles, nieces, etc.).
2   The interrelations between generations. Indeed, 'fathers work for their sons', as the beautiful title of a book by Sarah Berry (1985) says.

3   The interrelations between neighbours. These are governed through well-known mechanisms such as *ayni* (labour exchange, or the exchange of labour for oxen, etc.) and *compañia* (one person contributing the land and/or inputs and the other contributing labour, after which the harvest is divided in two equal parts) or, as it goes in Italy: helping during the olive harvest and receiving, later on, some bottles of very good olive oil.

4   The interrelations at the level of peasant communities as a whole, which assume on the one hand that one behaves 'as a good member of the community', whilst they offer, on the other hand, many services and secure many rights (such as access to land and water, protection when injustices occur, a platform for conflict resolution, etc.; Boelens, 2008).

5   And, finally, the interrelations with markets (or more specifically: with and within marketplaces) can be embedded in a reciprocal framework. This is reflected in, for instance, the Italian saying '*il mercato tira*'. This expression refers to a market that offers good prices, to a market that literally 'draws': it draws you ahead – consequently, you 'reply' by increasing your production (reciprocity relations *within* marketplaces are discussed by Dessein [2002], who discussed Western African rural markets as representing flows and moves – the opposite of stagnation and regression).

Experiments

The many small differences between fields, varieties (and even plants), animals, types of dung; the many small differences that come with slightly different ways of working (a different performance of the many tasks and subtasks); and the many small differences that come with different cycles, seasons, years and atmospheric conditions are all carefully observed and interpreted as possibly entailing indications for improving the farm labour process. But one never blindly follows such indications. Instead, they are to be tried out and to be repeated under different conditions and in different years. Thus the typical farmers' experiment arises. Each farmer embarks on such experiments, which are often hidden in the apparent confusion that some farms seem to offer at first sight (just as they contribute, in the eyes of the outsider at least, to the seeming confusion).

The archetypical form of farmers' experiments is illustrated in Figure 2.5. It shows a field worked in a particular way that is synthesized here as A. B represents a slightly modified way of working (for instance, ploughing deeper, or earlier, etc.). X represents the standard modality for fertilization (or the variety always used) and Y is another deviation: a slightly (or radically) modified way of fertilization (or planting another variety).

Thus, a three-level experiment is constructed. First, there is the standard approach; secondly, the impact of shifting from A to B and from X to Y can be checked; and finally, there is the interaction of B and Y. BY informs the farmer about the joint effect of B and Y (as compared to AX). BY might be superior to (or inferior, or the same as) either BX and AY. If, however, BY is superior ('is looking interesting', as farmers would say), it might be the starting point for further steps

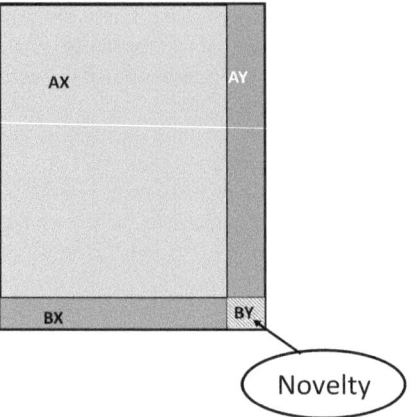

Figure 2.5 A hidden experiment in a field (author's own elaboration).

(repeated in subsequent years to check for the additional effects of yearly changes in climatic conditions, repeated in other corners of the same field, repeated in other fields, etc.). The same will also apply for AY and BX. Thus, BY might turn up as a novelty: a new artefact, a new way of working, a new idea put into practice, a new sequence, a new timing, etc. –something *new* that *promises to work well* (Roep and Wiskerke, 2004). As yet the farmer does not understand why, but it works. According to the *effects* it renders (higher yield, for instance, or a stronger crop) it is promising: in the *relational whole*, it brings *positive effects*.

Novelties and niches

Novelties (often rooted in on-farm experiments as outlined above but which also emerge accidently) represent another mechanism for endogenous growth. Novelties differ from innovations (as discussed above). The latter are mostly designed and tried out in scientific institutions (agricultural universities or organizations for applied research) and produced by agro-industries, and then they enter, *ready-made*, into the farm. Their performance is well understood (and explained in clear scientific terms) and the way they are to be handled and used is synthesized in a clear manual. Novelties, on the other hand, emerge out of the practice of farming itself; they need to be unfolded further, and their performance is (as yet) not well understood (let alone in terms of their 'intrinsic properties').

Novelty production is neatly interwoven with farming. Because farming is based on co-production, unexpected outcomes emerge every now and then. Co-production also allows for making small modifications in the farm labour process (such as the farm experiments, outlined above). But for novelty production to have a chance to render positive contributions (of whatever kind), some specific conditions need to be met. The deviations need to be recognized and to be treated with care. Brush et al. (1981) give a meticulous sketch of the little *jardincito* (the little garden) that is part of most small potato-producing farms in the Andean mountains. This garden typically contains 31 different varieties planted in 13 rows each having 10 holes for seedlings. There are, in the presented case, 57 holes that each contain two varieties and 73 having one variety. Together this makes for an impressive diversity. The *jardincito* is, as current language would have it, the *seed bank* of the peasant farm. These little gardens are nearly always managed by the farm women (Howard, 2003). The way the *jardincito* is organized allows for cross-fertilization and for new varieties to emerge. The problem is that it is already very difficult to distinguish 30 or so different varieties from each other (the mix mostly changes from year to year), but distinguishing a *new* variety is *extremely* difficult. Small, if not very small, differences in flowering, in colours, leaves, roots, taste, overall appearance, and reactions to different soil conditions, temperatures, etc., need to be recognized and added together to compose a new image: the image of a new variety that can help to route the new variety to the next stages.

But novelty production needs more than 'having an eye', as the capacity to perceive and distinguish the exceptional or the deviation is often called in the countryside. There also needs to be a *niche*: a protected space where novelties can emerge and germinate, where they are cared for, allowing them to unfold further and show their potentials. Following the example given above, new varieties can possibly emerge (and be recognized) in this niche (shaped by and as the *jardincito*), but in large fields for production this evidently is impossible. In the latter there will be monocropping (just one variety because otherwise the harvest cannot be sold), and the fields are far too large to recognize any deviation (especially when it is just one plant) whatsoever. Ironically, with increased efficiency (as symbolized in, and by, the extended fields dedicated to one variety only), the production of new, novel elements that might renew and dynamize production as a whole goes down. The production of renewal then shifts to external agencies. It is, indeed, externalized, and the new has to be brought in as innovation and, later on, to be acquired as commodity.[16]

Niches, such as the one described above, are often somewhat hidden. This is sometimes even on purpose. Many dairy farmers refer to their own breeding activities as their 'hobby'. It is often expressed as a kind of excuse: 'breeding is just my hobby'. In the language of the countryside, this means that questions that are usually asked (is this activity profitable, is it organized in an efficient way, are the experts consulted, etc.), do *not* apply. Thus, the farmers create a protected space against the asphyxiating norms that normally apply. At the same time, this language reflects the hegemony of the techno-administrative task environment (TATE) within which breeding stations play an important role. These breeding stations like to claim that breeding has to be done by 'professionals' and that it is too important to be 'left to farmers'.

With the shift from farm-born novelties to scientifically designed innovations, the nature of the objects concerned structurally changes, and this has far-reaching consequences, which I will discuss in the next chapter.

The farm labour process is, as I have argued throughout this chapter, an important engine for development and growth. We should be aware, though, that this only applies when the steps forward help to improve the farmer's own situation. The same search might also become blocked or even be turned into its opposite. When there is generalized despair (due to, e.g., unequal power relations that imply that the benefits of whatever improvement will be grasped by others or highly unequal market relations that imply a siphoning of benefits away from those who produced them), the construction of steps forward may fall flat (if not be actively brought to an end) and farming families may even start to 'eat their own farm' (consuming the resources instead of using them for production). Stagnation may also occur when communication between farmers (and the associated exchange of knowledge) is blocked. In the dominant discourse, such situations are mostly represented as due to traditional farmers, underdeveloped farming methods, non-economic behaviour, etc. Wherever this occurs, the sociology of farming has to carefully probe into the social relations of production that reduce or eliminate the room farmers need and thus generate stagnation and/or regression. Making such relations visible might be a crucial step in changing them.

Notes

1. The problem evidently is that the costs incurred are missing. I will come back to this issue in Chapter 4 (see Table 4.3) when discussing agroecology. In Chapter 9, agroecology is discussed as part of political struggle.
2. This equation is derived from Hayami and Ruttan (1985), who applied it in a world-wide comparison of agricultural systems.
3. Dry matter refers to a measurement method that indicates the total production per grassland field (whether it is harvested as hay, grass or silage). For example, in the Netherlands a good grassland can render between 10 and 12 metric tons of DM (dry matter) per hectare, but extremes of up to 20 DM/ha are also possible. The DM production partly depends on the way grazing and mowing are organized (when, how many times, etc.).
4. This echoes the title of a well-known book by Chayanov (1924). The expression was also widely used in Western Europe in the 1930s and the early 1950s (see, e.g., Vries, 1947; Timmer, 1949).
5. For instance, when use of chemical fertilizers is stopped, the capacity of the soil to autonomously deliver nitrogen increases again.
6. See Bolhuis and Ploeg (1985), especially the section on 'meagre and well-fed cows' on pp. 120–123. See also Ploeg (1990:73-76) and Benvenuti et al. (1988).
7. A *novelty* is a new artefact, a new object of labour or a new insight that promises to work better than previous ones. Mostly it is not clear why it functions, but experience shows positive and enduring effects. The notion of novelty represents the opposite pole to the concept of innovation. An innovation is 'ready-made'. Its operation can be explained in scientific terms.
8. This concept was developed by the American anthropologist Foster (1965). In today's language, one would probably refer to it as a 'zero-sum game'.
9. 'The greatest profits go to the first to adopt ... the innovator gains financial advantage through his innovations' (Rogers and Shoemaker, 1971:187).
10. This is reflected in the disciplinary organization and segmentation of agricultural sciences. Soil science studies the soil (it goes even further: soil physics studies the physical aspects of soils, soil chemistry the chemical aspects, soil biology the biological ones), animal science studies the animals, crop science studies the crops, etc.

11. For an excellent application to soil sciences, see Sonneveld (2004); for dung and manure, see Reijs (2007); for crop cultivation, see Adey (2007). In a way, these studies are a return to, and simultaneously a further elaboration of, classical agronomy; see, for example, Visser (2010).
12. This reflects the possibility of unfolding a geno-form into different and mutually contrasting pheno-forms.
13. It is partially true that scientific knowledge is, in a way, just another local knowledge system. However, if it is partly true, this implies that the claimed universal applicability is exaggerated. It can also be argued that scientific knowledge (even if it has very 'local' origins and applicability) *has been made true* by subordinating the rest of the world to the insights elaborated in and propagated by science.
14. This is also done in cold areas in Europe, such as Tras-os-Montes (Portela, 1994).
15. An important aspect of reciprocity is that it allows for a combination of resources that would otherwise remain separated and, therefore, most probably, remain unused. Thus, acts of reciprocity converge in inducing development. Reciprocity also helps to reset local balances and equilibria.
16. Tellingly, new, endogenously produced varieties circulate in the Peruvian mountains as gifts. They cannot be sold or bought – the seedlings are treated as non-commodities.

Bibliography

Adey, S. (2007), *A Journey without Maps: Towards Sustainable Agriculture in South Africa*, Ph.D. thesis, Wageningen University, Wageningen, the Netherlands.

Arkush, D. (1984), 'If man works hard the land will not be lazy: Entrepreneurial values in North Chinese peasant proverbs', *Modern China* **10** (4), pp. 461–479.

Benvenuti, B. (1982), 'De technologisch administratieve taakomgeving (TATE) van landbouwbedrijven', *Marquetalia* **5**, pp. 111–136.

Benvenuti, B., S. Antonello, C. de Roest, E. Sauda, and J. D. van der Ploeg. (1988), *Produttore Agricolo e Potere; Modernizzazione delle Relazeioni Sociali ed Economiche e Fattori Determinanti dell' Imprenditorialita Agricola*, CNR/IPRA, Roma.

Benvenuti, B., E. Bussi, and M. Satta. (1983), *L'imprenditorialitá Agricola: A la Ricerca di un Fantasma*, AIPA, Bologna, Italy.

Berry, S. (1985), *Fathers Work for Their Sons: Accumulation, Mobility and Class Formation in an Extended Yoruba Community*, University of California Press, Berkeley.

Boelens, R. (2008), *The Rules of the Game and the Game of the Rules: Normalization and Resistance in Andean Water Control*, Ph.D. thesis, Wageningen University, Wageningen, the Netherlands.

Bolhuis, E. E., and J. D. van der Ploeg. (1985), *Boerenarbeid en Stijlen van Landbouwbeoefening*, Ph.D. thesis, Leiden University, LIDESCO, Leiden, the Netherlands.

Braverman, H. (1974), *Labor and Monopoly Capital: The Degradation of Work in the 20th Century*, Monthly Review Press, New York.

Brush, S. B., J. C. Heath, and Z. Huaman. (1981), 'Dynamics of Andean potato agriculture', *Economic Botany* **35** (1), pp. 70–88.

Chayanov, A. (1924), *Die Sozial Agronomie, ihre Grundgedanken und ihre Arbeitsmetoden*, Verlagsbuchhandlung Paul Parey, Berlin.

Conklin, H. C. (1954), 'An ethnoecological approach to shifting agriculture', *Transactions of the New York Academy of Sciences* **17**, pp. 133–142. https://doi.org/10.1111/j.2164-0947.1954.tb00402.x

Conklin, H. C. (1957), *Hanunoo Agriculture. A Report on an Integral System of Shifting Cultivation in the Philippines*, Vol. 2, FAO, Rome.

Darré, J. P. (1985), *La Parole et la Technique: L'Univers de Pensée du Ternois*, Editions L'Harmattan, Paris.

Darré, J. P. (1993), 'Production des connaissances dans les groupes locaux d'agriculteurs', in *L'innovation en Milieu Rural: Synthèse des groupes de travail de la Table-Ronde du LEA et textes des contributions au séminaire du LEA, session 1991–1992*, edited by J.-P. Chaveau, M. C. Cormier Salem, and E. Mollard, 167–174, ORSTOM, Montpellier, France.

Dessein, J. (2002), *Het Stremmen en Stromen van de Markt*, Ph.D. thesis, Universiteit van Leuven, Belgium.

Foster, G. M. (1965), 'Peasant society and the image of the limited good', *American Anthropologist* **67**, pp. 293–315.

Hayami, Y., and V. Ruttan. (1985), *Agricultural Development: An International Perspective*, John Hopkins University Press, Baltimore, MD.

Howard, P. (2003), 'Women and the plant world: An exploration', in *Women and Plants: Gender Relations in Biodiversity Management and Conservation*, edited by P. Howard, 15–35, Zed Books, London.

Kessel, J. van. (1990), 'Produktieritueel en technisch betoog bij de Andesvolkeren', *Derde Wereld* **1990** (1–2), pp. 77–97.

Liebig, J. von. (1840), *Die organische Chemie in ihrer Anwendung auf Agricultur und Physiologie* [Organic Chemistry in Its Application to Agriculture and Physiology]. Republished as Schwedt, G. (2021), *Justus von Liebig, Die organische Chemie in ihrer Anwendung auf Agricultur und Physiologie* (with an introduction by George Schwedt), Springer, Hamburg, Germany.

Ploeg, J. D. van der. (1990), *Labour, Markets and Agricultural Production*, Westview Press, Boulder, CO.

Ploeg, J. D. van der. (2003), *The Virtual Farmer: Past, Present and Future of the Dutch Peasantry*, Royal van Gorcum, Assen, the Netherlands.

Portela, E. (1994), 'Manuring in Barroso: A crucial farming practice', in *Born from within, Practice and Perspectives of Endogenous Rural Development*, edited by J. D. van der Ploeg and A. Long, 59–70, Van Gorcum, Assen, the Netherlands.

Rambaud, P. (1983), 'Organisation du travail agraire et identités alternatives', *Cahiers Internationaux de Sociologie*, **75**, pp. 305–320.

Reijs, J. (2007), *Improving Slurry by Diet Adjustments, a Novelty to Reduce N Losses from Grassland-Based Dairy Farms*, Ph.D. thesis, Wageningen University, Wageningen, the Netherlands.

Robertson Scott, J. W. (1912), *A Free Farmer in a Free State, a Study of Rural Life and Industry and Agricultural Politics in an Agricultural Country*, Heinemann, London.

Roep, D., and J. S. C. Wiskerke. (2004), 'Reflecting on novelty production and niche management in agriculture', in *Seeds of Transition: Essays on Novelty Production, Niches and Regimes in Agriculture*, edited by J. S. C. Wiskerke and J. D. van der Ploeg, 341–356, Royal van Gorcum, Assen, the Netherlands.

Rogers, E. M., and F. F. Shoemaker. (1971), *Communication of Innovations; a Cross-cultural Approach*, The Free Press, New York.

Rosset, P. M., B. M. Sosa, A. M. Roque Jaime, and D. R. Avila Lozano. (2011), 'The campesino-to-campesino agroecology movement of ANAP in Cuba: Social process methodology in the construction of sustainable peasant agriculture and food sovereignty', *Journal of Peasant Studies* **38** (1), pp. 161–191.

Sabourin, E. (2009), *Camponeses Do Brasil: Entre A Troca Mercantil e a Reciprocidade*, Editoria Garamond Universitario, Rio de Janeiro.

Sabourin, E. (2011), 'Teoria da reciprocidade e sócio-antropologiado desenvolvimento', *Sociologias* **13** (27), pp. 24–51.

Sabourin, E. (2012), *Organisation et Sociétés Paysannes: Une Lecture par la Réciprocité*, Editions Quae, Versailles, France.

Scott, J. C. (1976), *The Moral Economy of the Peasant*, Yale University Press, New Haven, CT.

Sonneveld, M. P. W. (2004), *Impressions of Interactions: Land as a Dynamic Result of Co-production between Man and Nature*, Ph.D. thesis, Wageningen University, Wageningen, the Netherlands.

Timmer, W. J. (1949), *Totale Landbouwwetenschap*, Wolters, Groningen/Djakarta.

Toledo, V. (1990), The ecological rationality of peasant production, in *Agroecology and Small Farm Development*, edited by M. Altieri and S. Hecht, 53–60, CRC Press, Ann Arbor, MI.

Val, V., P. M. Rosset, C. Zamora Lomelí, O. Felipe Giraldo, and D. Rocheleau. (2019), 'Agroecology and La Via Campesina I. The symbolic and material construction of agroecology through the dispositive of "peasant-to-peasant" processes', *Agroecology and Sustainable Food Systems* **43** (7–8), pp. 872–894. https://doi.org/10.1080/21683565.2019.1600099

Visser, J. (2010), *Down to Earth, a Historical-Sociological Analysis of the Rise and Fall of 'Industrial' Agriculture and the Prospects for the Re-rooting of Agriculture from the Factory to the Local Farmer and Ecology*, Ph.D. thesis, Wageningen University, Wageningen, the Netherlands.

Vries, E. de. (1947), *Problemen van de Javaanse Landbouw* (inaugural address), Landbouwhogeschool, Wageningen, the Netherlands.

3 Markets and technology: A space for manoeuvre

Overview: Main concepts discussed in Chapter 3

Space for manoeuvre
Commodity
Non-commodity
Commoditization
Degree of commoditization
Use value
Exchange value
Homo economicus
'Make or buy'
Transaction costs
Governing costs
Technical efficiency
Entrepreneurship
Craftmanship
The social life of things
Patrimony
Mediating commodity relations
Technology as organic unity of artefact and knowledge
The code of a technology
Multifunctional technical artefacts
Skill-oriented technologies
Mechanical technologies

Markets and technology are of utmost importance in todays' agriculture. Yet, they do *not determine* how farming is organized and developed. The impact of markets and technologies is not direct, nor the same for all farms. Instead, it is the *relations* between markets and the supply of technology on the one hand, and the farm on the other, that are decisive. The farm can *relate in different ways* to the markets, and different market relations bring different degrees of autonomy and dependency. The same applies to the supply and use of technologies. New technologies

DOI: 10.4324/9781003313274-3

can become, and operate as, an important ordering principle within the farm, but it is also possible for the farmer to make selective use of the available technologies, guided by his or her own strategic orientation.

Together markets and technologies provide a room for manoeuvre[1] – a space that allows for different positions. Each position brings its own transaction costs, governing costs and rationale, just as it is associated with a specific ordering of the farm labour process and, consequently, with a consistent whole of internal relations within the farm. In short, the positioning of the farm vis-à-vis markets and technology and the structuration of the processes of labour and production (i.e., the style of farming) are very much interwoven. Each assumes, brings and results in the other.

Relations with markets

Farms are linked to markets through two sets of relations. One set considers what are known as 'downstream' relations. These are the relations with the different output markets. Through this set of relations, the marketable surplus of the farm is sold. On the other side of the farm there are the 'upstream' relations. These include the relations with markets for factors of production (land, capital, labour) and for non-factor inputs (feed and fodder, fertilizers, energy, etc.).

Nearly all farms are linked with at least one output market. Only farms that exclusively focus on domestic production (i.e., production that is exclusively meant for consumption in the farming household) have no such relationship. But as soon as there is, next to production for the household, a marketable surplus (i.e., a surplus brought to and sold on the market), the farm is tied to a market. Throughout history, nearly all farms had their *culture d'or*[2]: the households needed money for paying taxes, for buying goods they could not produce themselves, etc. The oft-mentioned 'subsistence farming' is merely a fiction created in modern agricultural sciences.[3] It suggests a remnant of a 'natural economy' that was definitely disconnected from economic progress. In this framework, subsistence farming is considered to be awaiting civilization and development – it is something that should be eliminated or profoundly transformed into something different and superior.

In the discourse about markets, there is an important difference between markets as a system and marketplaces. The latter are concrete places where concrete transactions of concrete goods and services take place between concrete sellers and buyers. These marketplaces are governed in concrete ways and by concrete institutions. There are many such marketplaces, and new ones are being created all the time. In the case of farming, the construction of new markets is an important feature, to which I will return in Chapter 8.

On the input side of the farm (the upstream side), many relations can be distinguished. The impact these relations have, singly and together, can vary considerably. Theoretically speaking, all, or nearly all, resources needed for farming might be produced and reproduced within the farm (or in the wider rural community, after which they enter the farm through socially regulated exchange). However,

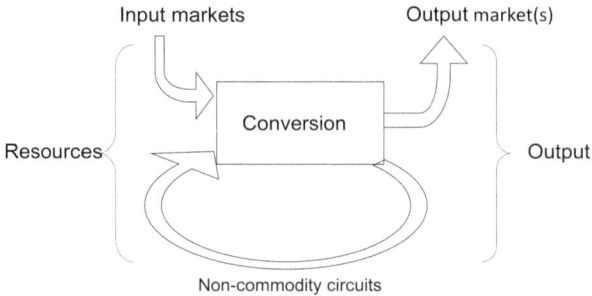

Figure 3.1 Farming and resource flows (author's own elaboration).

the production and reproduction of these resources can also be externalized (see the previous chapter), and this may occur to different degrees. Technically, it is easy to calculate, for single farms as well as for an agricultural sector as a whole, the *degree of commoditization*. This measures the proportion of all needed resources that the farm purchases as commodities: these are the resources acquired on the respective markets. The other resources are produced and/or reproduced within the farm. This is illustrated in Figure 3.1. The degree of commoditization can be calculated for each distinguishable resource (for cows, feed and fodder, machines services, labour, land, short term, medium- and long-term capital, etc.), but a synthesized degree is also possible (by adding the monetary values together). This is further detailed in Methods Boxes #6 and #7 at the end of this chapter.

Commoditization evidently influences the organization of agricultural production and its development. The bought-in resources are no longer present as *use values* to be operated according to the insights of the direct producers. They also become *exchange values* whose use is tied to following the logic of the markets (especially when the degree of commoditization is elevated). A high degree of commoditization also implies that market risks are introduced into the farm and felt in the labour process itself: they need to be dealt with seriously. The precise impact of commoditization depends, of course, very much on the conditions reigning in, and between, the different markets. The relation, for instance, between prices paid for the outputs and those of the bought inputs can be favourable (Italian farmers call this *un mercato che tira* [a market that draws you forward]), but the margin can also be very meagre. Markets may be volatile or stable.

Figure 3.1 summarizes the basic structure of farming. Farming is a process of conversion: resources are converted into an 'output' – a range of useful products. This conversion is grounded on a double flow of resources. Some resources are mobilized in the respective markets, and others are produced and reproduced within the farm itself (or within the wider farming community and obtained through socially regulated exchange). That is, some resources are *bought*, and others are *made* (in the farm itself or the locality). Once the resources are converted into products, this output is again divided into two flows. One part is sold in the market, and another part is re-used in the farm itself.

The shift from self-produced (and reproduced) resources to resources obtained in the markets was discussed in the previous chapter as the process of externalization.

If the flow of non-commoditized resources dominates the flow of resources obtained in the markets, we talk about *peasant* (or peasant-like) agriculture. If, on the other hand, the flow of resources that enter the farm labour process as commodities dominates the flow of non-commoditized resources (that is, if the degree of commoditization is high), we talk about *entrepreneurial* agriculture.[4] It is noted here that the wording (peasant-like and entrepreneurial) does *not* refer to the personal attributes of the people involved but to the structure of the farm and the relations between farm and markets.

Externalization is not a unidirectional process. It occurs with different rhythms, sometimes slowly, at other moments abruptly. It may proceed in one way but also be redressed, and it can take many different forms. It is also a differential process. While some farms pursue rapid and massive externalization, other farms may develop the other way around: re-integrating previously externalized tasks and sub-tasks (or adding completely new tasks to the farm). More generally speaking, in agriculture there is no well-delineated and straightforward process that goes from an assumed natural economy towards a fully commoditized economy. Nor is there an unstoppable movement that goes towards an ever further evolving division of labour. Consequently, there are empirically different degrees of externalization in nearly all farming areas in the Global North as much as in the Global South, and these differences have important consequences.

I have presented and discussed Figure 3.1 many times in lectures given in the countryside. It was always well recognized and often translated to the kinds of metaphor that abound in the countryside. 'It is like the human body and the blood circulating in it', a farmer's woman once said. 'The flow of one's own blood' (she pointed to the flow of self-produced and reproduced resources), 'needs to be the main one; it is the one that carries you forward'. And then she added, very tellingly, 'that upper flow at the left is like a blood transfusion: it is sometimes needed but you should never depend on it'. Male farmers, at least in the Netherlands, mostly use another metaphor: the dam or causeway. Most Dutch farms are divided by a ditch that separates the farmyard from the main road. Their comment is brief but also very clear: 'There should not be too much cargo passing the dam'.

A neglected field of interest

Intriguingly, agricultural sciences (of the hegemonic type) have hardly paid attention to differences in externalization. Just like the farmer himself (or herself), the relations between farming and markets figure, within theory, as a *black box*. The real farmer is replaced by an assumed *Homo economicus*. This means that there is no need to study the farmer, for it is already known how he or she will think, decide and operate: as *Homo economicus*. If he or she is not doing so, he or she is operating wrongly, because 'good farmers' ought to perform as specified in the model of economic rationality. It is as simple as that. The same applies for the relations between the farm and the markets. An important axiom of neo-classical economics postulates that farms should be operated *as if* all of the resources (all factors of production and non-factor inputs) were acquired in the market. It does not matter whether a particular resource (say hay) is produced in the farm itself or acquired on the market. It is to be dealt with as if it were bought. Thus, the reigning market price for hay is projected onto the hay produced in the farm. Once the same price is attributed to the two types of hay, they are equal. This applies not only to hay but to all resources. In this way, the relations between farms and markets are made into a black box. They are unknown, and because there is no need to know, they remain unknown.[5]

There is one very important exception: neo-institutional economics. Developed by Coase (1988, 1992), Williamson (1981, 1985) and others, this approach puts one question centre stage: *make* or *buy?* This particular question was raised in order to understand big industrial conglomerates that sometimes *outsource* many productive activities and so have to *buy* the needed artefacts whilst, at other times, they try to integrate as many productive activities as possible within their own company (i.e., to *make* everything themselves). To probe into this issue, two crucial concepts were developed: *transaction costs* and *governing costs*. In each and every exchange there is a price paid for the exchanged good or service. This price is the market price. However, on top of this market price there are the transaction costs. These are additional costs related to, for example, the exploration of the market (where to obtain the best quality/price ratio), risks (the product functioning badly), etc. In other words, transaction costs specify the costs *of using the market*. If, however, the market is not used – that is when the product is *made* within one's own industry – there will be *governing costs*. By making everything needed within one and the same enterprise, there is a far higher need for coordination, management becomes more complex and additional training is needed. Governing the more complex and extended enterprise thus brings increased governing costs.

Neo-institutional economics argues that the relations between transaction costs and governing costs are decisive in the strategic choice between making or buying. This insight is used to explain the relations between farms and markets (where the question of making or buying is as strategic as in industry). These relations compose a theoretically relevant and empirically researchable field of interest. As far as I know, Vito Saccomandi, an unorthodox and gifted agricultural economist from Italy, was the first to apply the guiding principles of neo-institutional economics to agriculture (see Saccomandi, 1995, 1998). It was the starting point for a fruitful cooperation of this type of economic analysis and the sociology of farming.[6]

Changing schemes of reproduction

Although there is no unilinear process of externalization, we can very well distinguish different *degrees* of externalization (and, consequently, different degrees of commoditization). A high degree of externalization implies a change in the

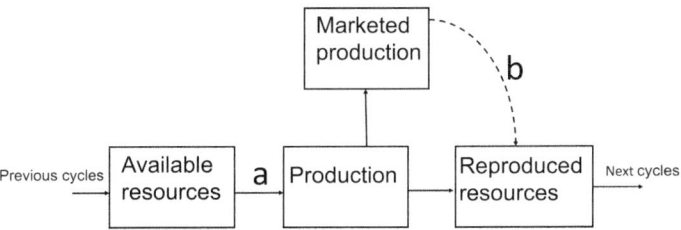

Figure 3.2. Historically guaranteed, relatively autonomous reproduction (author's own elaboration).

structure of reproduction compared to a low degree of externalization. This is summarized in Figures 3.2 and 3.3.

Figure 3.2 presents the reproduction scheme for situations characterized by a low degree of externalization. Most resources needed in the productive cycle to come are available within the farm itself or locally. They have been produced and/or reproduced in the previous cycles. Thus, these resources enter the farm labour process as non-commodities (relation a in Figure 3.2). They were not acquired in the market but have been produced in the farm itself.

With the available resources, the farm realizes a production that is divided in two parts. One part will be re-used in the farm (as the resources for the next cycle). Another part will be sold in the market. The set of reproduced resources might be further increased if part of the money obtained by selling the marketable surplus is used to acquire additional resources (relation b in Figure 3.2). Then these additional resources enter into the next cycle of production as 'already paid for'. They become an integral part of a historically guaranteed reproduction. Every cycle of production builds on the previous ones, just as it composes the basis for the next cycles. Such a scheme means that production is a relatively autonomous process. In this model, production interacts with the markets but is not dependent on them.

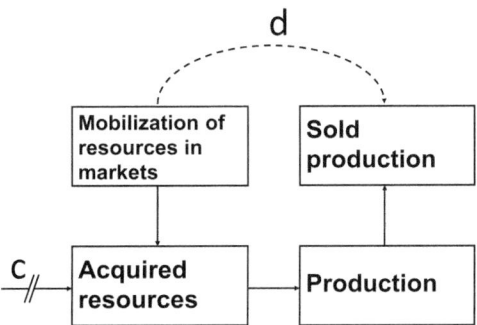

Figure 3.3 Future and market-dependent reproduction (author's own elaboration).

In this scheme of relatively autonomous and historically guaranteed reproduction, the markets represent an *outlet*. The markets for outputs are crucially needed, but as outlets: as the place where (part of) the produce is to be sold. The markets (for outputs and inputs) do *not* act as *ordering principle*. Instead, the relation with the output market is strategically governed by the farming household itself – and in this respect, cultural repertoire, collective memory and institutions play important roles (Text Box 3.1).

When farming is grounded on this specific scheme of reproduction, the relation between the resources used and the sum of sold production and reproduced resources (i.e., *total* production) becomes strategic. The stronger this relation, the more the farm will advance (i.e., produce an adequate income as well as good

---

Text Box 3.1   Hay or butter (and how much butter)?

Across the centuries there have been large and important markets for hay trading (this is still reflected in street names like 'Hay Market'). Armies needed hay for their many horses, as did stud farms and, in later times, large animal breeding companies ('feed lots') that depend on delivery of silage and hay from elsewhere.

Producing hay for the market is a devilish issue, especially in dairy farms. Selling too much hay brought 'immediate and easy money', but it prevented its conversion into milk and, subsequently, butter. With the production of butter, a far better valorization of the available resources was possible. Thus, the laws of butter-producing provinces such as Friesland (in the northern part of the Netherlands) strictly prohibited exports of hay (and, for that matter, of manure as well). Hay was to be used within one's own farm (or at least in the province) to generate more wealth (which would also render more tax payments). Later this rule became internalized by the dairy farmers of the area. It became part of the moral economy: 'a good farmer never sells his hay' (if there was too much hay, his farm was not well balanced). Hay was definitely considered to be a non-commodity. It equally applied that 'a good farmer does not buy hay'. The boundary between making and buying was clearly drawn here.

On the other hand, the focus on butter also involved extra economic considerations. Feeding all of the good hay just to the dairy cows (as was done in the early 19th century) implied that calves and heifers would only get the low-quality remnants. This again would bring relatively quick money, but it would simultaneously be detrimental for the quality of the offspring; that is, the animals that in the future would replace the milking cows. Small calves and bony heifers would threaten the quality of the herd (and future butter production). This meant that production would start to threaten reproduction – another devilish problem. This triggered another important guideline entailed in local repertoire: 'a good farmer is not a merchant' (he or she should focus on building a beautiful farm and not sell the needed building blocks). So even the production of the commodity called butter was embedded in precise non-commodity considerations.

prospects for the further development of the farm). Analytically this relation is referred to as *technical efficiency*: the quantitative ratio between total production and the resources used to produce this production (Timmer, 1970).[7] The technical efficiency depends on the *quality* of the resources used and the *craftmanship* of the actors involved in the farm labour process.

The relation between the part of total production sold on the market and the part to be re-used in the next cycle is, within certain limits, flexible. This contributes to the (relative) autonomy of the farm.

Figure 3.3 represents the opposite image. It describes the reproductive scheme that arises with high levels of externalization. In the ideal-typical situation of an extremely high degree of externalization, nearly all of the resources needed in the cycle of production are mobilized on the relevant markets. There is no substantial flow of resources delivered by the previous cycles (relation c). Instead, there is a rupture (which may be due to a wide range of possible causes, both internal and external) which results in the needed resources being mobilized through the markets. As a result, the resources enter the farm labour process as commodities, and each one represents a cost that needs to be repaid (otherwise, the farm will not be able to continue to operate).

The acquired resources are 'set in motion' and converted into production to be sold. The marketed production remunerates the costs made during, and with, the previous acquisition of the resources. Within this scheme, relation d is decisive. The obtained benefits need to compensate for the financial obligations the farm has taken on. As a result, it is essential to organize production (and farm development) in such a way that it aligns as much as possible with reigning market relations. The capacity to do so is referred to as *entrepreneurship*.

With the shift from the situation outlined in Figure 3.2 to that of Figure 3.3, the market changes from being an outlet towards operating as the ordering principle (Methods Box #6, located at the end of this chapter, shows how such a shift can be measured). In Figure 3.3, the farm is literally driven by the input and output markets and thus it has to follow *the logic of the market*. Beyond that, the farm will also face higher transaction costs (see Text Box 3.2). Consequently, the farmer will most likely try to structure the process of production to minimize the risks that come with the high integration in (and dependency on) the markets. How this translates in practice will depend on the concrete conditions that reign within and between the markets.

Reproduction as shown in Figure 3.3 is not only market dependent but is, in a way, *dependent on the future* as well. The costs that come with the mobilization of resources are incurred at the beginning of the cycle. The benefits, though, will only show up at the end of the cycle; that is, in the (near) future. However, at the beginning of the cycle, the conditions that will be present at the end of it are (mostly) not known with certainty. This introduces extra risks and requires extra margins. Formulated differently, the transaction costs can easily escalate, sometimes considerably. If the initial expenses have been financed with credit, this pinches even more.

Text Box 3.2   More about hay (on transaction costs)

Even if there is a global market for hay (with, e.g., farmers in the Basque Country in northern Spain producing Lucerne hay and large Chinese and Dutch dairy farms importing it), buying hay is not easy. It comes with considerable transaction costs. The two (the difficulty of buying good hay and the high level of the transaction costs) are, in this case, intimately related. The hay that is on the market is, in principle, anonymous. It might come from anywhere and is produced under circumstances that are not known. This brings considerable risks. The hay might have been harvested in fruit yards (or vineyards, for that matter) after treatment of the trees (or vines) with pesticides. Hence, the hay might carry (invisible) quantities of poisonous remnants. This can, potentially, bring diseases (or even death) to the cattle fed with this hay or cause significant drops in milk production. It can also severely contaminate the milk. Whether this will occur is not known (at least not beforehand), but it could happen. It is a risk. This risk can be minimized, but doing so requires time to search for the origin of the hay, how it has been grown, etc. Alternatively, one can buy certified hay, but certification comes with a price as well. Finally, an assurance can be negotiated, but once again there are costs associated with this. In short: there is not only the price that is paid during the transaction (the price of the hay as such) but the same transaction brings additional costs: transaction costs (i.e., the costs that come with the transaction or, as some economists would say, the cost of using the market). Such transaction costs are related to the (calculated) risk, the costs of gathering information and/or the costs of assurances and certification – and they can be considerable.

To *make or buy* is a permanent dilemma. If one opts to buy one has to reckon with the (partly unknown) transaction costs. Buying can bring comfort (and help to avoid drudgery) but, in turn, can be quite dangerous as well. That is precisely the reason why local knowledge repertoires (the commonly shared *l'art de la localité*) are rich in proverbial sayings and messages that indicate that what is locally made and known is often preferable to the unknown (interestingly, the sharpest vetoes are expressed when it comes to borrowing money, which is mostly represented as dealing with the devil). Such messages are often misunderstood as expressions of a traditional (if not downright stupid) form of provincialism. But the mistrust in things coming from faraway and/or unknown places might also be seen as accumulated experiences condensed in the message that the danger of high transaction costs should be avoided.

Take, for instance, the pruning of fruit and olive trees and vines. If this is not done in a knowledgeable way, it can incur considerable transaction costs: damaged trees and vines, reduced harvests and sorrow in the heart. Therefore, farmers nearly always prefer to contract workers from their own locality: they know the job and they themselves are also known in the locality. Damaged trees or vines would put shame on them and make it difficult for them ever to be contracted again in the village or its wider surroundings.

Long before Coase, Williamson and others started to develop their theories on transaction costs, Italian peasants would agree that *moglie e buoi, paesi tuoi*. That is, you are better off with a wife or a pair of oxen from your own village. They know how to skilfully work according to the local rules, just as the locals know how they have been raised.

A self-controlled resource base

There is yet another important difference between the ideal-typical situations outlined in Figures 3.2 and 3.3. The 'available resources' (mentioned in Figure 3.2) constitute as much as the resource base of the farm. It combines the resources produced and/or reproduced in the previous cycles of production. It is a resource base that allows the farming household to engage, in a relatively autonomous way, in agricultural production. It is a self-controlled resource base: the seeds selected from last year; the calves, heifers and cows; the barn full of hay; the family members knowing how to do the job; the networks they build with neighbours (who may help in case of need), traders and clients; the fertile fields; the well-maintained irrigation system; etc.[8] On one hand, this is all terribly self-evident (even to the degree that science hardly pays any attention to it) but, on the other hand, it is highly exceptional, for it is a resource base that has been actively constructed over the decades – probably even by different generations – and it is highly appreciated by farmers themselves. It is what they themselves have constructed; it is their insurance (see Text Box 3.3), and their promise for the future. However, it is equally true that this resource base might get slowly (or even quickly) eroded, thus giving rise to the situation of dependency portrayed in Figure 3.3.

This difference (having a self-controlled resource base or not) relates to identity. Having a self-owned and self-controlled resource base (a *patrimony*) translates into confidence: it allows facing the future in a self-confident, albeit modest, way: after all, agricultural production is co-production, and one never knows what nature will bring. By contrast, being strongly dependent on resources mobilized in and through the markets turns farmers into people whose fate is basically pending on others. Their fate critically depends on what the markets will bring. 'They are', as the Dutch saying goes, 'deliberating but in the end the market is determining'.[9] Whether or not the sold production covers the costs of the acquired resources is decisive, and this depends on the market situation at harvest time. Once debts are created, the question of whether they can be paid for with and through the production of *next* year or not becomes paramount.

Text Box 3.3   Facing the hostile environment

In animal husbandry, having sufficient feed and fodder (of the needed quality) is an important expression of a self-controlled resource base. It is at the same time insurance. It helps to carry on through rough times and to face a hostile environment. As was said in Ireland: 'a good abundance of quality hay was like an insurance policy for those who lived off the land. Come rain or hail, sleet or snow, if the haggard was full the farmer could say: "Awful weather, but shure we never died a winter yet"' (John Arnold, 'Tay, dinner and toil: reflecting on the heyday of haymaking', echo.live.ie/nostalgia/arid-40106573.html).

Having one's own self-controlled resource base does not exclude uncertainties. These include many variables, such as climate, health issues, etc., and also the market. But even if a complete harvest is lost (due to misharvest and/or very low prices), *one's own resource base is still there*, and it allows one to take up production in the next cycle and thus for a (step-wise) recuperation of the losses suffered. In the case of a highly market-dependent reproduction (Figure 3.3), low prices and/or a misharvest mean that only a *deficit* remains. Whether or not farming can be 're-started' critically depends on others.

Iron: The seeming exception

Seeds, fertilizers (in the form of manure, compost, intercropping, green fertilizers, whatever), animals, feed, fodder, buildings, irrigation systems, horses or oxen (for traction), labour force, savings to pay for daily expenses: these can all be produced and reproduced within the farm (or in the wider farming community). The ostensible exception seems to be iron, or all of those artefacts that contain iron, from hoes to sophisticated tractors and equipment – these all need to be acquired outside of the farm. Such artefacts cannot possibly be produced within the farm (with a few exceptions). Consequently, relations with (input) markets become necessary. At this point, the *conditions* to which I already referred become relevant – just as *dealing* becomes important.

First, there is the choice between contracting a custom worker (mostly a specialized enterprise that supplies machine services, but it could also be other farmers) to do all of the work that requires 'iron' versus the possibility of having one's own machine park. Using the custom worker (whose services will need to be paid for according to standard market prices) represents externalization and commoditization. On the other hand, having one's machinery will involve capital outlay and spending time (and having the knowledge) to maintain it properly.

When it comes to one's own machinery there is a basic difference between a tractor that has been paid for by savings from previous cycles of production and a tractor acquired on credit (or a leased tractor). In the former case, the tractor is already paid for. At the moment of acquisition it was a commodity, but once it is in the farm it can be used in the farm labour process as a non-commodity. Things have a social life, as Appadurai (1986) argued.[10] Consequently, they have a biography. During their 'life', their condition changes. The tractor is bought (during the transaction it is a commodity) but, thanks to the use of savings, it can be considered, from then onwards, as part of the patrimony of the farm – it can be used according to the insights and goals of the farming family. When obtained with credit, the situation is different. Then the tractor comes with financial costs: every year the interest rates and redemptions have to be paid. The tractor enters the farm labour process with exigencies that relate to the upstream markets. And these exigencies need to be met.

Thirdly, it is nearly always possible to buy secondhand machinery. This will greatly affect the acquisition price and mitigate the consequences that might arise in the organization of the farm labour process.

Fourth, maintenance and repair can be well organized in the farm itself. If this is done well, the technical life span of the tractor can be prolonged. This implies that replacement costs can be distributed over more years and thus be lowered. The differences thus created can be considerable.

Fifth, it is possible to share specific equipment with other farmers (as French farmers do within their machine cooperatives, called CUMA: *coopérative d'utilisation de matériel agricole*: [cooperative for the use of agricultural equipment]).[11] This can sharply reduce the costs of acquiring this equipment, so the impact it has *as commodity* is actively diminished. On the other hand, when jointly using machinery, there need to be clear agreements that ensure that all members have timely access to the machinery (and which also specify rules for maintenance and other responsibilities). Often membership in these cooperatives is a catalyst for further forms of farmer cooperation.

Buying a brand-new, expensive tractor *translates* into a series of exigencies *into* the farm (especially if it is bought on credit). The yearly depreciation (plus interest payments plus costs for maintenance and repair) might very well imply a need to enlarge the farm or change the cropping scheme in order to earn the extra income to cover the costs associated with the new tractor. Buying a secondhand tractor, or using the cooperative solution, will strongly mediate such exigencies (or requirements). On the other hand, these solutions imply higher governing costs. At the end of the day, the 'weight' of these costs will depend on how easily the farmer relates to colleagues, his ability to maintain and repair, the pleasure experienced with being involved in an extended variety of tasks, etc.

Conditions do matter (the relative price of machinery, the interest rate, whether or not there is a good supply of secondhand machinery, etc.), just as the *dealing* (making the acquisition dependent on one's own savings, actively engaging in maintenance and repair, becoming member of a CUMA, etc.) is important.

In short: 'iron' is not the proverbial exception to the rule. It can be a vehicle for commodity relations to enter into and restructure the farm labour process, but such effects can also be mediated and mitigated to a considerable degree. Here, the agency of farmers and the conditions reigning in the different markets are decisive. Theoretically speaking, farmers can make mechanization into an integral part of a historically guaranteed and relatively autonomous reproduction. However, careful empirical research is needed to assess the extent to which this occurs in reality.

From markets to technology

Farmers can adopt *different* positions when dealing with markets. These can range from relative autonomy to high dependency. The same is true of farmers' relations with the available technologies.[12]

Technical artefacts (or instruments) only function as technology when combined with the specific knowledge about how to operate these artefacts. One needs to know where and when to apply these technical artefacts, how to manage and control them, what they should be used for, who should operate them, how they should be integrated in a wider sequence, how to evaluate the results, what pre-requisites need to be fulfilled, etc. This applies as much to a seemingly simple instrument such as a *radu* (see Figure 3.4) as for complex machinery such as a laser-guided scraper (a rice field can be levelled with either).

The answers to these questions (*where, how, for what, by whom, when, what is needed, etc.*) form a kind of *code: the code of a technology*. All technologies come with a code, and those operating the technical artefact need to thoroughly know this code. Without knowledge of the code, the technical artefact is useless, the more so because these codes are often *materially built into* the artefact: they can only be used for particular objectives, in a particular way, etc. They cannot be operated in an alternative way or used to reach other goals.

It is always important to be aware of who (or what) built a specific code into a particular technical artefact. It is equally important to analyze which (and whose) interests and prospects are reflected in this code. The code of a technology might be an important carrier of specific social relations of production. The in-built code of particular technical artefacts order (at least partly) the farm labour pro-cess and also specify (at least partly) how the produced wealth is distributed.

Technical artefacts, together with their code, are a concentration and materi-alization of a wider set of social relations. As a result, such artefacts are not neu-tral. They are like 'a language'[13] that specifies what is to be done and how it is to be done. Some of these artefacts allow for considerable control at a distance. By issuing particular seeds (with an in-built code, a clear manual and an extended set of legal prescriptions), some companies are able to govern agricultural production in large parts of the world. Nonetheless, we have to be very clear that there never is just one technology or just one way to produce.

Another feature worth mentioning here concerns the potential use(s) that can be made of a specific technical artefact. Some instruments are multifunctional: they might be used for many different activities or in ways that were not originally intended. Others are highly specialized and only allow for one type of operation. In the latter case, the feature of functioning as language evidently is far stronger. The example *par excellence* of a multifunctional artefact is the tractor. It can be used to transport all kinds of loads, it can drive a pump or a generator and, with specific equipment attached, it can be used for ploughing, harrowing, sowing, cul-tivating, harvesting, etc. Multifunctional machinery (such as the three-wheeled diesel tractor in China and the Api in Italy) has brought radical and far-reaching changes in many cases. Peasants were able to adapt this machinery into the many and complex operations that, together, compose their farm labour processes. They could also use the same machinery in several non-farming operations. This facilitated a transition to pluri-activity (or multiple job holding).

In her study on Asian rice economies, Francesca Bray (1986) developed a pair of concepts that are very helpful for further distinguishing different technologies. These are *skill-oriented technologies* and *mechanical technologies*. Skill-oriented technologies are relatively simple artefacts but need considerable and detailed knowledge to operate them (see Text Box 3.4). Mechanical technologies represent

Text Box 3.4   The *radu*, an example of a skill-oriented technology

The *radu* (a Criollo word derived from the Portuguese *arado*, meaning plough) is an instrument used in the tropical rice polders (*bolanhas*) in Guinea-Bissau. It is also called a *kayendo* (in the Balanta language). The *radu* is used to construct huge dikes, for irrigation work and for the preparation of the soil before sowing rice. It consists of a spoon-like wooden blade (15 by 50 or 60 centimetres), covered with a horseshoe-shaped iron protection and connected to a long rod (up to 2 metres long). At first sight it looks like an impossible instrument: it is pushed in the mud and then, seemingly against all physical logic, lifted and turned so that the elevated mud is put in place. The secret of the *radu*, though, is its in-built elasticity. It is cut out of special trees (*pao negro*) in such a way that there is flexibility in the blade: when pushed, the handle and blade veer back. Hence, instead of lifting, one has to slightly push down on the rod. This activates the in-built elasticity and the *radu* with the mud on it comes up by itself. One has to know *how* to use the *radu*; if not, it is an impossible task.

But the required knowledge does not end here. Making ridges for rice cultivation requires a detailed knowledge of the fields, the herbs that grow there, the slope of the terrain and the variety of rice that will be grown. One also has to know how to maintain soil fertility (the straw and roots of the previous season end up in the heart of the newly made ridge) and how to organize a working group. Knowing local culture (in order to assess the right moment for starting the field work) is another important block of knowledge. By working well with the *radu*, young men show their strength and can establish a good reputation. This is reiterated in wrestling games. When the horseshoe-shaped iron protection is worn, it is left in little 'temples' to show the Gods (*iraos*) that one has been working well.

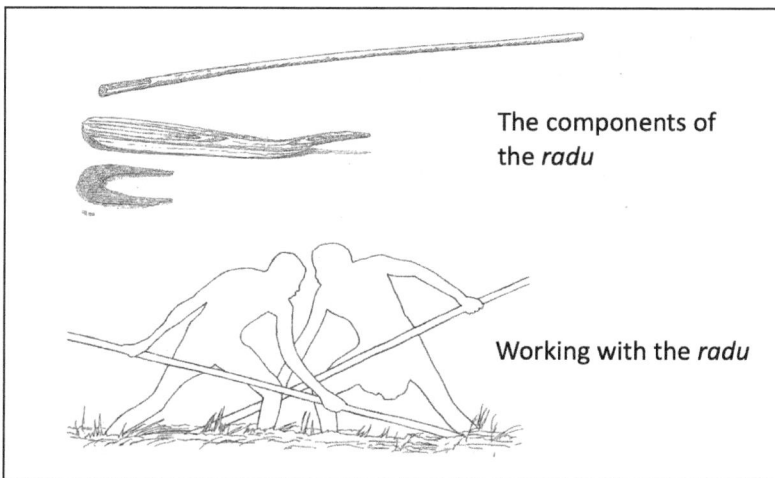

The components of the *radu*

Working with the *radu*

Figure 3.4 Radu.

the opposite: highly sophisticated artefacts which require relatively little knowledge to operate. Examples *par excellence* of the latter include automated feeding machines, robotized milking equipment and the laser-guided scraper for rice fields, mentioned before. They reduce the farm labour process to a limited set of mechanical operations that follow a script developed elsewhere. Skill-oriented technologies, on the other hand, rely on the experience, skills and decisions of the direct producers.

Another important conceptual difference is the one between *locally adapted* technologies and *universally applicable* technologies.[14] Analytically, this distinction differs from the one developed by Francesca Bray, although in practice the two often overlap or even strongly coincide, for skills are, almost by definition, locally bounded.

Locally adapted technologies are typically developed within farms with the aim of fitting as well as possible within the farm context. Universally applicable technologies are supposed to perform well, even under very different conditions. They are mostly designed on the drawing boards of agro-industries or agricultural universities and tested in experimental fields. They are well adopted to the typical conditions of those experimental fields or research stations, but meeting other conditions (those of the fields 'out there') can be more troublesome. Nonetheless, the universally applicable technologies might contain the potential for making considerable steps forward (in yield levels, for instance), whilst the locally adapted technologies mostly only allow for small steps forward.[15] Text Box 3.5 gives an example by comparing potato seedlings selected in Andean farms and scientifically designed potato seedlings.

Universally applicable technologies assume a range of conditions that need to be met in order to operate in a satisfactory and profitable way (these normally are the conditions of the experimental field). This implies that the farm, the layout of the fields, the architecture of farm buildings, the accessibility of the farm, the farm labour process itself, etc., need to be structured in a way that meets these conditions. The 'miracle seeds', for instance, that were and are at the core of Green Revolution technologies, critically assume specific (and relatively high) levels of fertilization, the application of herbicides and pesticides, well-controlled irrigation, certain levels of mechanization, etc. (which is why these seeds mostly come as part of a comprehensive 'package'). Refrigerated milk containers that allow for the storage of 2½ to 3 days of the obtained milk assume the presence of a high-power energy supply, the possibility for large tankers to arrive at the farm, etc.

Thus, at the farm level, the application of new, universally applicable technologies almost inevitably brings the need to make a series of adjustments within the farm in order to be able to 'plug in' the new artefacts. The farm needs to be 're-built' (at least partly) according to the scientific design (see, e.g., Roep, 2000). This brings additional transaction costs. It is not only the new artefact that is to be paid for but also the necessary adjustments (this type of transaction cost is sometimes known as a *transformational cost*). In the case of locally adapted technologies, such costs are far lower or even absent.

Text Box 3.5   Potato seedlings

The typical peasant farm in the Andean mountains has access to a wide range of different potato varieties. This is the available geno-typical material (partly entailed in the 'seed bank' discussed in Chapter 2). The farm may have 10, maybe even up to 20, small different fields, all representing specific pheno-typical conditions (known by the farmers in terms of 'hot' or 'cold', 'red' or 'black', etc.). The selection of seedlings occurs in a seemingly simple way. The women, who know and manage the available stock of genotypes, select those varieties that fit best with a particular field. In practice this implies ongoing processes of trial and error – also because the fields are continuously being improved (if possible).

If the chosen variety performs well, the best tubers are taking from the strongest and best yielding plants and re-used in the next cycle. In this way, the varieties selected are those that best meet the different field (or pheno-typical) conditions. Those varieties that do *not* meet the characteristics of the fields are removed from the stock. As a consequence, the varieties used carry the code of the particular eco-systems, fields and, most probably, the preferences of the farming household. This is synthesized in Figure 3.5.

With the introduction of new, scientifically designed potato varieties (of the 'Green Revolution' type), these dynamics are radically altered.

First, an ideal-plant type is defined. The new variety needs to show its 'superiority' compared to existing ('traditional') varieties: that it is better able, for example, to convert the available solar energy into higher yields. With such goals in mind, the new genotype is materially constructed (by means of classical breeding or through genetic modification). However, the success of this proposed conversion, in actual farmers' fields, depends on a number of preconditions. It is only possible *if* sufficient nutrients are available in the subsoil, *if* they are transportable to the roots, *if* enough water is available, and so on. Mostly the required growth factors (requirements 1 to N in Figure 3.6) are specified and tested in experimental plots (or, as is increasingly the case, specified with simulation models). Now, the obvious point is that in order to be successful in the fields, these requirements need to be present in

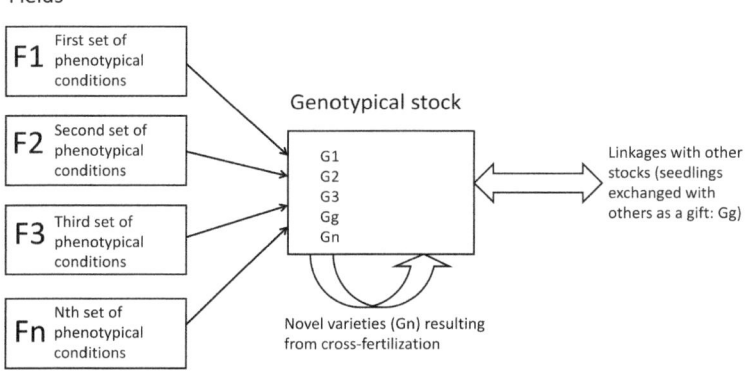

Figure 3.5 The dynamics of peasant-organized potato breeding and selection (Ploeg, 1990).

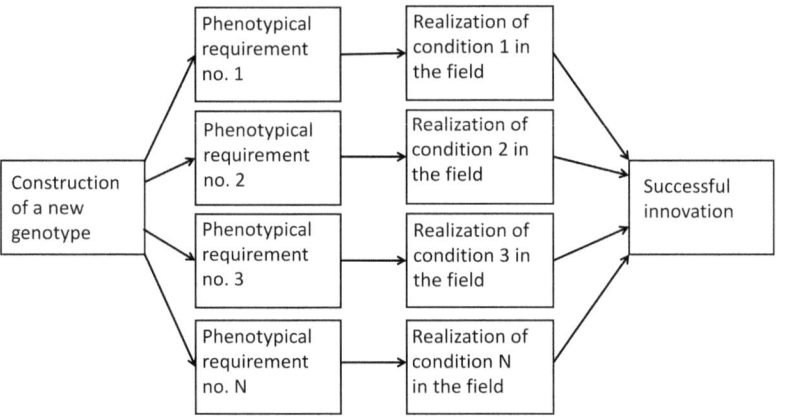

Figure 3.6 The structure of scientific plant breeding (Ploeg, 1990).

real farmers' fields. Thus, the farm labour process becomes externally prescribed. To meet requirement 1, fertilization needs to be organized in a specific way (quantity, composition, timing, etc.). The same applies to requirement 2, which can only be met if a tractor is used, etc.

The scheme entailed in Figure 3.6 clearly shows the intrinsic difficulty of successfully adopting new innovations. Just putting new seedlings in the soil is useless. In order to be successful, the in-built code (i.e., the specified growth factors) needs to be materially created in the fields. The fields need to be 're-built' to meet the requirements of the new potato variety (and, for a variety of reasons, this might be too difficult, too costly, time-consuming, etc.). As compared to 'traditional' breeding and selection, (see Figure 3.5) the new, scientifically designed genotype introduces a new sequential logic that goes *from the new variety to the fields* (and implies a need to 'rebuild' them). This represents a strong contrast with the sequential logic entailed in peasant production which goes *from fields to varieties* (that meet the conditions of these fields). This does not exclude improving the fields, but this occurs in a step-by-step way (and thus also shows potentials that had, so far, laid hidden in the existing genetic stock). By contrast, the 'scientific' method excludes such a step-by-step approach. All of the specified requirements need to be met simultaneously and immediately. If not, the 'improved' potato will underperform and quickly degenerate.

For further reading about this case, see Ploeg (1993) and Scott (1998). The underlying problem is discussed, in general terms, in Latour (1983, 1993) and in Law (1994). Latour (1983) carries a telling title: 'Give Me a Laboratory and I Will Raise the World'.

A specific problem that comes with many agricultural development programs in the Global South is that the high-yielding varieties and the associated requirements assume, and bring about, a range of new dependencies. This directly relates to the expensive nature of these new varieties and, especially, of the many inputs, machine services and additional labour needs (at peak moments) that come with them. The new varieties are designed without properly taking local socio-economic and agronomic conditions into account (I will return to this issue in Chapter 10). The high costs associated with such Green Revolution type of agricultural development programmes bring about abrupt increases in market dependency (i.e., abrupt changes from the situation shown in Figure 3.2 to that in Figure 3.3) which impacts strongly and often adversely on the organization and development of the farms participating in these programmes (Hebinck, 1990). Methods Box #7 (at the end of this chapter) gives an example and also shows how empirical research might capture the effect of these newly emerging dependencies on markets.

In between the locally adapted and universally applicable technologies there is the possibility to deconstruct available technologies (of whatever type) into separate elements, select some particular elements out of the wider array and then to recombine them with elements already available within the farm into a new assemblage that functions well. Figure 3.7 gives an illustration. It shows how in one Italian farm the small *motofalciatrice* (motorized mowing machine used for mowing daily portions of fresh roughage for the animals) is combined with a powerful tractor for transporting heavy bales of straw and other bulky and heavy items. From a financial point of view, having this double type of mechanization makes little sense. In terms of *care*, though, it allows for giving the best possible rations to the animals.

Figure 3.7 A remarkable combination (photograph by the author).

How markets and technologies provide spaces for manoeuvre

Together, markets and technologies provide a space for manoeuvre that allows farmers to adopt different positions (see Figure 3.8), in which the horizontal dimension shows the mobilization of resources (remember Figure 3.1), while the vertical dimension centres on the conversion of resources into outputs (which might be summarized in terms of technology).

The relations between farms and markets (the horizontal dimension of Figure 3.8) might be structured in a way that secures a considerable degree of autonomy (that is, reproduction is historically guaranteed) or, on the opposite side, a high degree of market dependency. With the different positions along this dimension, both the magnitude of, and the balance between, transaction costs and governing costs will vary considerably.

Along the vertical dimension, which shows the technologies used, there is, again, a range of possibilities. As discussed before, these range from skill-oriented (and locally adapted) to mechanical (and universally applicable) technologies with different possibilities for de-construction and re-combination. Once again, the transaction costs and governing costs will vary considerably along this axis.

Within this space, different positions – that is, different styles of farming – are possible. Figure 3.8 indicates two positions: peasant-like agriculture and entrepreneurial agriculture, which are linked to markets and the available technologies in different and mutually contrasting ways. Because the current supply of technologies centres on 'packages' (sets of interrelated innovations), new technologies are usually expensive and also impose new credit relations. The (already mentioned) case of high-yielding varieties is, in this respect, exemplary – but the same applies in the Global North. Therefore, the main trend shown in Figure 3.8 goes from the bottom-left to the upper-right position. From the bottom-left position, other trends are possible as well: to the bottom right, for instance. This could occur when peasant agriculture becomes increasingly impoverished and is driven into dependency relations.

Figure 3.8 intersects (in a diagonal way) with Figure 1.4 presented and discussed in Chapter 1.

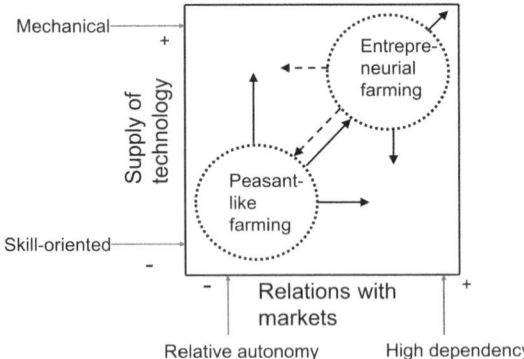

Figure 3.8 The space for manoeuvre within markets and technologies (author's own elaboration).

The wider repercussions of technological change

Do the changes summarized in Figure 3.8 matter? I think so. In the second half of the 1990s, the Experimental Dairy Farm Station (Proefstation voor de Rundveehouderij) near Lelystad in the Netherlands started a highly interesting multi-year comparison of two farms that were operating alongside each other. One was referred to as the Low-Cost Farm and the other as the High-Tech Farm. The first one typically represented the peasant-like farm indicated in Figure 3.8 and the other the entrepreneurial farm. It was assumed that these represented two of the main farming styles in Dutch dairy farming (Kamp and de Haan, 2004; Evers et al., 2007).

Both farms were designed in such a way that they could and should render the same income (identical to the salary of a skilled urban worker). All of the work, on both farms, was done by one person working normal working hours. The difference resided in the technologies used. Following the concepts discussed above, the Low-Cost Farm applied skill-oriented technologies, whilst mechanical technologies were central to the operation of the High-Tech Farm. In practice, this meant automated feeding, the use of milking robots, automated cleaning of stall floors, etc. There was a corresponding difference in the breeds used: highly productive Holstein Friesians in the High-Tech Farm (needing high levels of industrial concentrates, etc.) and Montbéliardes in the Low-Cost Farm, a breed that is more robust and dual purpose: providing lower milk yields but good quality meat. The architecture of the buildings also differed: the Low-Cost Farm was simpler and built with cheaper materials.

All of these different elements were combined in two coherent and well-functioning patterns. However, in order to reach the indicated objectives (comparable income and working time), the High-Tech Farm needed a quota twice as high as the Low-Cost Farm: 800,000 kg of milk per year versus 400,000 kg per year. Table 3.1 reports some of the relevant economic data obtained after some years of operation.

Table 3.1 Some economic data from the low-cost and high-tech farms (author's own elaboration)

	Low-cost	High-tech
Labor units	1.0	1.0
Working hours/year	2,500	2,490
Hectares of land	32	35
Cows	53	81
Production/cow	7547	9673
Production per farm (kg)	400,000	783,515
Concentrates/100 kg milk (€)	3.8	7.5
Labour/100 kg milk (€)	13.0	6.7
Machine costs/100 kg (€)	5.4	7.1
Total production costs/100 kg (€)	34.5	34.7
Income/hour (€)	19.20	16.36

The differences in technological profiles clearly comes to the fore in the higher level of machine costs (+31%), concentrate use (+97%), higher milk yields (+28%) and the lower labour input (−47%) in the High-Tech Farm. Nonetheless, both farms produced the same income, while the income per hour worked was even slightly higher in the Low-Cost Farm.

The *total* costs of production (calculated labour cost included) per 100 kg of milk were identical. In the High-Tech Farm, the monetary costs (related to the acquisition of external resources) made up a far higher portion of the total costs than in the Low-Cost Farm. As a consequence, *labour income* per 100 kg of milk (see Figure 3.9) in the Low-Cost Farm was twice as high as in the neighbouring High-Tech Farm. The latter compensated the low VA/GVP ratio by means of the magnitude of total production (800,000 kg versus 400,000 kg).

At the time this comparison was made, the total national milk quota of the Netherlands equalled 10 billion kg of milk. If this total production was produced by low-cost farms, there would have been space for 25,000 dairy farms (and an employment level of roughly 25,000 full-time equivalents). If, however, all of the production was realized by high-tech farms, 12,500 farms would have sufficed, and employment would have been halved. Total sector income (remember that income per person is the same in both types) would also have gone down by 50%.

In this case, technological change is neither augmenting the incomes earned nor does it increase total production. It just implies a shift of 50% (!) of the total sector income from the agricultural sector to the industries providing the technical means implied by this technological change and a loss of 50% of farm-based jobs. In such a scenario, agriculture is fuelling accumulation in other sectors with obvious negative consequences for the rural economy.

As shown by this analysis, it is hard to sustain the position that technological change is, *in itself*, beneficial. There need to be additional concrete arguments to make a positive case for technological change – such as a scarcity of labour supply in other economic sectors or a need for an increase in production that cannot be

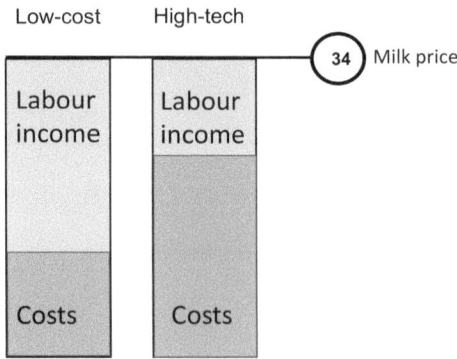

Figure 3.9 Labour income and monetary costs for different technological profiles (author's own elaboration).

realized otherwise, etc.[16] Few such arguments have validity today, whether in the Global North or Global South.

There is one other aspect that should be mentioned here. At the time of this multi-year experiment (which ran throughout the 1990s and early 2000s), the milk price was fairly stable due to price protection and import-export regulations. Later, the milk market became much more volatile. In 2008–2009, for instance, the price paid for milk fell from 34 euros per 100 kg to less than 26 euros. This produced unexpected consequences. When projecting the lower price level on the bars in Figure 3.9, the High-Tech Farm would render a *negative cash flow*, whilst the Low-Cost Farm would face a strongly reduced labour income but could, nonetheless, continue its operation.[17] By that time, however, the multi-year comparison had already been finished, because it was argued that it was no longer rendering any relevant insights. In retrospect, it is clear that shortsightedness governs, at least sometimes, the considerations of the political elites who decide about agricultural policies and research.

Methods Box #6   Mapping market dependency in dairy farming, Emilia Romagna, Italy

Assessing the degree of commoditization for the different resources used in farming is simple. For the different resources one needs to ask which part is mobilized in the relevant market and which part is produced and reproduced in the farm itself. This allows us to calculate the degrees of commoditization as (resources acquired in the market)/(total amount of resources), in which the total amount of resources is (resources produced and/or reproduced in the farm itself plus resources acquired in the market). For dairy farming in Emilia Romagna, this rendered the following results (M = average, (s) = standard deviation; Ploeg, 1990).

Market dependency for	M	(s)
Labour	9.1%	22.8%
Contract work	30.7%	28.5%
Credit (short term)	4.6%	16.3%
Credit (medium term)	11.1%	50.5%
Credit (long term)	2.4%	3.4%
Land	28.7%	37.8%
Fodder and feed	43.8%	18.2%
Cattle	7.2%	9.0%
Overall dependency	26.0%	25.0%

As the table shows, there were high degrees of commoditization for feed and fodder, machine services (contract work) and land (short-term leasing). The overall dependency amounts to 26%. However, the standard deviations show that there was considerable variation around the average levels (especially for medium-term credit and land). Thus, dairy farming here partly tended towards a relatively autonomous and historically guaranteed scheme of reproduction and partly towards the opposite: a market and future dependent reproduction (see Figures 3.2 and 3.3).

Methods Box #7   Mapping market dependency in potato production

---

The resources needed in, and for, potato production in the Andes can be obtained in different ways. Following the logic of Figure 3.1, land can be property of the farming household (and be passed down from one generation to the next), but it can equally be obtained through exchange (one particular plot for another), be purchased or rented and/or be passed to the household by the peasant community. If the labour available within the farming household is insufficient, mechanisms for socially regulated exchange, such as *ayni*, *companía*, and/or *faenas* can be applied – otherwise, labour can be mobilized in the labour market. Capital (i.e., here the money needed to buy all of the needed items that the farm cannot produce) can be available in the form of savings (from previous cycles) or obtained through other economic activities (e.g., weaving) or wage labour elsewhere. It might also stem from loans from friends and family or be part of a *companía* deal. Capital can also be mobilized on the capital market (through banks). Traction power can be delivered by one's own pair of oxen (*yunta*), access to a communal tractor or through *ayni* (exchanging the use of oxen for one day's labour). In addition, machine services (tractor plus implements plus operator) can be hired from specialized enterprises. Fertilization can be realized with manure produced on the farm itself or obtained through socially regulated exchange. Communal schemes for field rotation also play an important role. But these strategies can also be partially or completely replaced by purchasing fertilizers in the markets. The same applies for potato seedlings: they can stem from production and selection in the farm (see Text Box 3.5), but they can also be purchased (partly or entirely). These factors all imply a specific degree of commoditization, which can be quantified and (if needed) also added together in order to assess the overall market-dependency.

The following path diagram (constructed with multiple regression techniques) shows the impact of market dependency (specifically the markets for labour and capital) on the structure of the farm labour process and, subsequently, on the yields obtained.

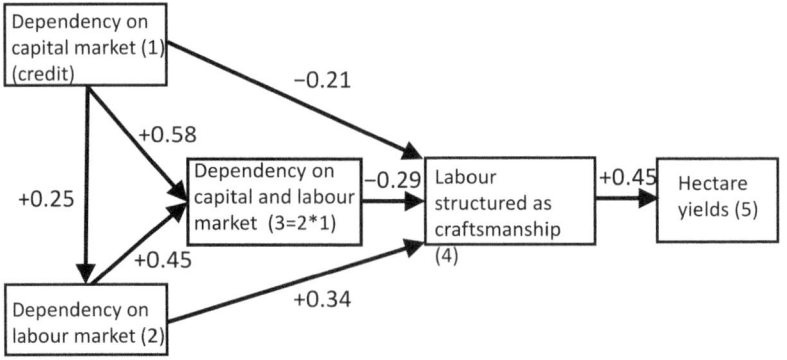

An increased dependency on the capital market (realized through the credit mechanism) translates here in a restructuration of the farm labour process and thus exerts a downward pressure on yields. Comparative research has shown that this is a recurrent pattern, in both the Global South and Global North. The hidden implication of this is that while many Green Revolution programs offer the promise of yield increases, they simultaneously induce downward pressures on yields (as a result of greatly increased market-dependency). For further reading, see Ploeg (1990, chapter 3) and Hebinck (1990).

Notes

1. Norman Long (2001) gives an excellent and very accessible discussion of this concept.
2. This concept, coined by the French agrarian historian Maurice Bloch (1939), refers to the crop that was produced in order to be sold so that the required money (*or*; i.e., gold) could be obtained. Bloch argued that there never has been an *economie-nature* – a kind of natural economy that operated without exchange or money. The history of agriculture is the history of the *economie d'argent*.
3. *Subsistence farming* is a pejorative expression meant to denounce types of farming that do not fit well in the Western agricultural paradigm. It is thought to express irrational and non-economic behaviour that is at odds with national economic development. The term neglects the possibly considerable magnitude of the marketable surplus; it equally neglects the risky context in which many farming households operate, which leads them to opt to prioritize the security of their own supply of food. Typically, 'technical assistance' and 'development cooperation' have nearly always tried to directly reduce the part of the farm used for domestic consumption and increase the amount of land and labour dedicated to the 'cash crop'. Experience, though, has shown that it is far better to first improve the productivity of land and labour in the segment for home consumption. When it comes to the assumed contradiction between partly producing for home consumption and the overall development of agriculture, the Chinese experience is fascinating. In nearly all of the 200 million peasant units of production in China, the production for home consumption (directly or indirectly through barter) is strategic. Nonetheless (or due to this), national agricultural production and productivity have, over the last 7 decades, grown more than in any other country in the Global South.
4. With such a classification, I follow, in a way, the scientific and agro-political discourse of the 1950s, 1960s and 1970s (the epoch of agricultural modernization, as it was called at that time). It was assumed that together farm enlargement, spurred mechanization, systematic use of credit and farm management guided by modern farm accountancy methods would make for a definitive shift from peasant farming to an entrepreneurial style of farming. These assumptions are amply documented in, for example, Hofstee (1953) and Mendras (1967). I disagree with this modernization discourse in that 1) I maintain that such modernization has been a selective, differential (and partly reversible) process. This implies 2) that considerable parts of the agricultural sector are composed ('still' or 'anew') by peasant-like farms and that 3) these are not per se inferior to entrepreneurial ones. My position also implies that the difference between peasant and entrepreneurial farming is not dichotomic (not a black-or-white issue) but a matter of degree. Beyond that, the differences are multi-dimensional.
5. Practitioners (such as farm accountants) and agrarian economists close to empirical reality knew, of course, that the specific relations between farms and markets matter very much. This knowledge, however, has never been built into the theoretical models of agricultural economics.

6. Exemplary are Ventura (2001) and Milone (2009); see also Ploeg and Saccomandi (1995) and Marsden (2009, 2012), who built on the concept of 'economies of scope' (as opposed to economies of scale) that is derived from neo-institutional analysis.
7. See Yotopoulos (1974) for a theoretical discussion and application. Salter (1966) showed how this technical efficiency is constantly improved by those involved in the labour process getting more experience and better grips on the process of production. This is called 'disembodied' technical change.
8. We have to take into account that building this resource base represents, for those who construct it, a process of emancipation. It helps them to escape from the yoke of often asphyxiating relations of dependency and exploitation. This is also reflected in the comparative analysis of farm accountancy data. Over the ages, farmers improved the value added as part of gross value of production (VA/GVP) – also referred to as 'the clean part' –through the construction of an autonomous resource base.
9. In Dutch: 'de boer wikt, de markt beschikt'.
10. If your mother buys a nice thing for your birthday, it is, at the moment of acquisition, a commodity. When she hands it over to you as a gift, it stops being a commodity. And most probably you will not sell it: moral economy excludes selling a gift you got from your mother. As simple as this example might be, it perfectly illustrates what is meant by the social life (or biography) of things.
11. There are 13,000 CUMAs in France, and one in two farmers (225,000 farmers) are involved in this kind of cooperative.
12. Throughout this text I use technologies and techniques as synonyms. Some authors differentiate between the concepts of technique (those artefacts that compose an organic unity with farmers' knowledge) and technology, which assumes the presence and embodiment of scientific knowledge.
13. An expression from Benvenuti (1982).
14. Typical examples are the cubicle stall for dairy cattle, Green Revolution packages, the Holstein Friesian breed, large and partly robotized combines, etc. They look (and, as a matter of fact, are) the same all over the world.
15. Over longer periods of time, though, the addition of many small steps forward might be considerable (even if the individual steps themselves are small). The cumulative effects of slow yearly growth over a longer period might even be equal to (or greater than) a sudden increase associated with new, scientifically designed technologies. See, for example, Bennett (1981).
16. In their comments on these two farms and the results obtained, some farmers argued that 'the High-Tech Farm is better, because it represents the farm of the future'. Others (and especially foreign farmers visiting the experimental station) opted for the Low-Cost one: 'why produce twice as much when it does not render you any extra income?' Comments like these show the strength of the normative frames operating in the agricultural sector.
17. Empirical analysis shows that this did occur in practice. It required massive interventions from the banks involved to bail out the large-scale, high-tech farms (see Dirksen et al., 2013).

Bibliography

Appadurai, A. (1986), *The Social Life of Things: Commodities in Cultural Perspective*, Cambridge University Press, Cambridge, UK.

Bennett, J. (1981), *Of Time and the Enterprise, North American Family Farm Management in a Context of Resource Marginality*, University of Minnesota Press, Minneapolis.

Benvenuti, B. (1982), 'De technologisch administratieve taakomgeving (TATE) van landbouwbedrijven', *Marquetalia* **5**, pp. 111–136.

Bloch, M. (1939), 'Economie-nature ou économie-argent: un pseudo dilemme', *Annales d'Histoire Sociale* **1** (1), pp. 7–16.

Bray, F. (1986), *The Rice Economies: Technology and Development in Asian Societies*, Blackwell, Oxford, UK.

Coase, R. H. (1988), *The Firm, the Market, and the Law*, Chicago University Press, Chicago.

Coase, R. H. (1992), 'The institutional structure of production', *The American Economic Review*, **82** (4), pp. 713–719.

Dirksen, H., M. Klever, R. van Broekhuizen, J. D. van der Ploeg, and H. Oostindie. (2013), *Bouwen aan een Betere Balans: Een Analyse van Bedrijfsstijlen in De melkveehouderij*, WUR/DMS, Wageningen, the Netherlands.

Evers, A. G., M. H. A. de Haan, K. Blanken, J. G. A. Hemmer, C. Hollander, G. Holshof, and W. Ouweltjes. (2007), *Results Low-cost Farm 2006*, Report No. 53, ASG/WUR, Lelystad, the Netherlands.

Hebinck, P. (1990), *The Agrarian Structure in Kenya: State, Farmers and Commodity Relations*, Verlag Breitenbach Publishers, Saarbrucken/Fort Lauderdale, FL.

Hofstee, E. W. (1953), *Sociologische Aspecten van de Landbouwvoorlichting*, Bulletin 1 van de Afdeling Sociologie, Landbouwhogeschool, Wageningen, the Netherlands.

Kamp, A. van der, and M. de Haan. (2004), High-tech farm and low-cost farm in the Netherlands: What is the solution?, paper presented at Djurhälso & Utfodringskonferens 2004.

Latour, B. (1983), 'Give me a laboratory and I will raise the world', in *Science Observed, Perspectives on the Social Study of Science*, edited by C. D. Knorr-Cetina and M. Mulkay, 141–169, SAGE, London.

Latour, B. (1993), *We Have Never Been Modern*, Harvard University Press, Cambridge, MA.

Law, J. (1994), *Organizing Modernity*, Blackwell, Oxford/Cambridge.

Long, N. (2001), *Development Sociology: Actor Perspectives*, Routledge, London.

Marsden, T. K. (2009), 'Mobilities, vulnerabilities and sustainabilities: Exploring pathways from denial to sustainable rural development', *Sociologia Ruralis* **49** (2), pp. 113–131.

Marsden, T. K. (2012), 'Towards a real sustainable agri-food security and food policy: Beyond the ecological fallacies?', *The Political Quarterly* **83** (1), pp. 139–145.

Mendras, H. (1967), *La Fin des Paysans – Innovations et Changement dans l'Agriculture Française*, Futuribles/SEDEIS, Paris.

Milone, P. (2009), *Agriculture in Transition: A Neo-institutional Analysis*, Royal van Gorcum, Assen, the Netherlands.

Ploeg, J. D. van der. (1990), *Labour, Markets and Agricultural Production*, Westview Press, Boulder, CO.

Ploeg, J. D. van der. (1993), 'Potatoes and knowledge', in *An Anthropological Critique of Development: The Growth of Ignorance*, edited by M. Hobart, 209–227, Routledge, London and New York.

Ploeg, J. D. van der, and V. Saccomandi. (1995), 'On the impact of endogenous development in agriculture', in *Beyond Modernization, the Impact of Endogenous Rural Development*, edited by J. D. van der Ploeg and G. van Dijk, 10–27, Van Gorcum, Assen, the Netherlands.

Roep, D. (2000), *Vernieuwend Werken, Sporen van Vermogen en Onvermogen (een Socio-Materiele Studie over Vernieuwing in de Landbouw Uitgewerkt voor de Westelijke Veenweidegebieden)*, Studies van Landbouw en Platteland 28, Circle for Rural European Studies, Wageningen University, Wageningen, The Netherlands.

Saccomandi, V. (1995), 'Neo-institutionalism and the agrarian economy', in *Beyond Modernization: The impact of endogenous rural development*, edited by J. D. van der Ploeg and G. van Dijk, 1–9, Van Gorcum, Assen, the Netherlands.

Saccomandi, V. (1998), *Agricultural Market Economics: A Neo-institutional Analysis of Exchange, Circulation and Distribution of Agricultural Products*, Royal van Gorcum, Assen, the Netherlands.

Salter, W. E. G. (1966), *Productivity and Technical Change*, Cambridge University Press, New York.

Scott, J. C. (1998), *Seeing Like a State. How Certain Schemes to Improve the Human Condition Have Failed*, Yale University Press, New Haven and London.

Timmer, C. P. (1970), 'On measuring technical efficiency', *Food Research Institute Studies in Agricultural Economics, Trade and Development* **9** (2), pp. 99–171.

Ventura, F. (2001), *Organizzarsi per Sopravvivere, un Analisi Neo-institutionale dello Sviluppo Endogeno nell'Agricoltura Umbra*, Ph.D. thesis, Wageningen University, Wageningen, the Netherlands.

Williamson, O. E. (1981), 'The economics of organisation: Transaction costs approach', *American Journal of Sociology* **87**, pp. 548–577.

Williamson, O. E. (1985), *The Economic Institutions of Capitalism*, The Free Press, New York.

Yotopoulos, P. A. (1974), 'Rationality, efficiency and organizational behaviours through the production function darkly', *Food Research Institute Studies* **13** (3), pp. 89–103.

4 Styles of farming

The farm labour process not only results in the making of a range of useful (if not essential) products; it also produces different styles of farming: cohesive patterns that strategically tie together the many elements that make up the farm and which guide the many decisions that need to be taken. A style of farming gives coherence to the farm, its internal organization, the way it interlinks to the external context as well as to the way it develops. A farming style is a specific way of regulating the many factors that contribute to growth (of the plants, livestock and the farm itself). It is also an organized flow of activities through time. Each farming style relates, and ties together, past, present and future according to specific modalities. A style of farming is like a narrative that tells how farming should be done. It is based on, and part of, a cultural repertoire just as it reflects, to a degree, collective memory (or the denial of it). It is also a routine and a set of

DOI: 10.4324/9781003313274-4

rules that summarize how to avoid 'devilish problems' (see Text Box 3.1) and keep transaction costs under control. A farming style specifies a series of balances and the best possible equilibria. In this way, the different tasks and domains are tied together in a coherent whole (see Figure 1.5 in Chapter 1).

A style of farming is, in short, a strategy that informs and guides the organization and development of the farm. Farming styles research has produced a rich documentation on such strategies, just as it reflects the impressive inter-regional diversities that (still) exist in our increasingly homogenized world (see, for example, Wiskerke, 1997; Roep, 2000; Commandeur, 2003; Dominguez Garcia, 2007; Paredes, 2010; Langthaler, 2012). While there obviously are spatial and temporal differences in farming styles (and their preponderance), more interesting, from an analytical point of view, is that contrasting styles can also be found within each single time-space location. Comparing these styles, their outcomes and their underlying *raisons d'etre* can be a fruitful exercise.

Specifications of how the farm work is to be done and how the farm is to be developed carry counter-images about what should be avoided. Sometimes the validity of other peoples' ways of operating is acknowledged, but very often it causes perplexity and rejection: 'they [the others] don't know how to farm well'.

Everywhere in the world the local agricultural sector is always considerably diverse. This diversity is multi-dimensional and covers many different aspects of farming and is never just random or chaotic. It stems from human efforts to bring nature, agriculture and one's own livelihood more or less in line. Thus, different patterns of *coherence* are produced: co-production is structured in specific and well-balanced ways. We cannot know a priori what types and what degrees of coherence will be created: this can only be assessed through careful empirical research. Farming styles research tries to identify strategically constructed coherence. This is a main difference from other research traditions.[1]

Farming styles analysis is the systematic inquiry into patterns of coherence at the level of agricultural production and development that proceeds from the assumption that human agency, together with nature, co-produces these patterns of coherence.

Heterogeneity within, and of, agricultural production, is also highly relevant for society as a whole (see Chapter 6). Within rural areas, this heterogeneity is strongly linked with other socially relevant issues, such as gender, sustainability, levels of overall production, etc. Farmers often embrace the outcomes of farming styles research – experiencing these outcomes as a confirmation of the logic, coherence and justification of their own practices (their own 'way of doing', as Lola Dominguez García [2007] argued).

When preparing a farm styles research, it is important to identify, right from the beginning, what will be understood and delineated as *relevant* diversity. This will determine the variables that need to be taken into account. There are different methods available that allow for the identification and exploration of farming styles. Methods Box #8 gives an overview.

Methods Box #8   Exploring and identifying farming styles

There is a wide range of methods that can be used for the identification and exploration of farming styles but no single best method. The choice of an appropriate method should be guided by the (secondary) material already available, the objectives of the research project and the time that can be spent. In general terms, a combination of qualitative and quantitative methods is preferable.

Semi-structured interviews are very appropriate for exploring the narrative that accompanies and guides farming practices. This narrative is, as it were, a synthesis of how farming ought to be done and the farmer's own experiences, views and prospects. The reconstruction of the labour process (see Methods Box #5) can be very helpful in this respect. Another easy entrance is a reconstruction of the farm's history: how has it developed over time, and what has guided the different key-decisions along this trajectory? The normative framework ('how farming should be done') can also be discussed by using contrasting examples as shown below (these examples need to reflect the major dimensions of the observed diversity) and then asking whether the respondent *recognizes* these different situations, *why* some farmers tend towards a particular approach and others to another, who are the ones who earn the best *incomes*, how the respondent would *label* such farmers, which example would be close to one's *own practice*, etc.

Farmer A	**Farmer B**
20 milking cows	30 milking cows
5,000 kg milk/cow	4,000 kg milk/cow
Total production: 100,000 kg	Total production: 120,000 kg

These examples are, on purpose, very simple and give the respondents the space to drive the conversation wherever they want it to go. The examples allow discussion of issues that may be avoided if directly talking about their own farm. The example given above was used in research undertaken in 1979/1980 in Emilia Romagna in Italy. The data used in the two examples are now very much outdated, but at that time they inspired long and passionate discussions.

The extensive answers given helped, at that time, to construct different *calculi*. A *calculus* is as much as the 'backbone' of a farming strategy: it is the grammar used by farmers to structure the organization and development of their farms. It guides their choices and even helps them to calculate costs, benefits and risks. In this particular research programme (see Ploeg [1990] for an extended description), such calculi were reconstructed with the help of 'folk concepts' (expressions used by the involved actors to reflect on, and to order, their own practices; see also Methods Box #1). One of these *calculi* (i.e., farmers opting for ongoing intensification) was synthesized as follows:

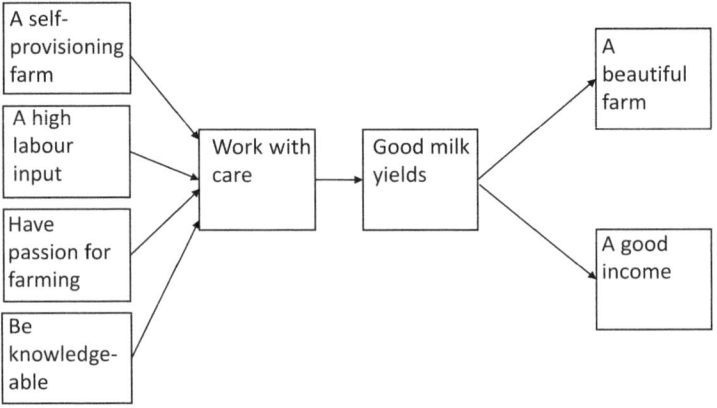

Figure 4.1 A calculus (Ploeg, 1990).

This calculus shows that working with care and obtaining good yields is at the centre of this farming style. This style provides a good income and helps to construct, in the long run, a 'beautiful farm'. The requirements for working in this way are also specified: having a self-provisioning farm (i.e., a farm grounded on a historically guaranteed, relatively autonomous reproduction), with a high labour input, and work that is done in a dedicated and knowledgeable way.

It is remarkable that by operating in this way the farm reproduces itself over longer time spans. The 'beautiful farm' is the one that, in the eyes of these farmers, will in the future have a self-produced resource base, whilst the good income allows for a high labour input and working in a passionate way. That is, the outcomes of this kind of operation create the conditions for doing so in the future as well.

In *surveys*, different approaches are needed. One possibility is to formulate short 'portraits' that describe different farmers (the building blocks for these portraits should be obtained through extensive interviewing). I take the following example from the PhD thesis of Cees Leeuwis (1993:200–201):

PORTRAIT 1: I like a nice-looking cow. Milk yield is not without importance; turnover and accretion [i.e. the production of meat], however, are also very important indicators to me. They are the indicators that I attune my breeding to. By not trying to increase my milk yields too much, I can keep more cows, and thereby increase turnover and accretion.

PORTRAIT 2: I try to farm in as economical a way as possible. I reduce costs as much as possible, and I minimize indebtedness. In this manner, I manage to get a good income and maintain prospects for the future.

PORTRAIT 3: I try to take very good care of everything I do. The art of running a farm is in fine-tuning. In developing one's farming enterprise, one has to be careful not to shoot beyond the possibilities. One needs to find a practical balance.

PORTRAIT 4: I very much enjoy breeding, and to me the sweet things in life are to take care of the animals and see the milk flow. This is why I have to pay much attention to the production of roughage and the fine-tuning of fodder and feed rations. In order to allow for this way of working, I needn't have too many cattle, since that would be at the expense of individual care and attention.

PORTRAIT 5: I most enjoy working with machinery, both on the land and in the barnyard, doing maintenance and repairs. The most important thing for me is to do the work on the land and in the stable as efficiently as possible. I do not aim at the highest production per cow; that is not much of a problem, for the masses will make up for it.

PORTRAIT 6: In order to have a good income, one needs to first invest firmly and spend a lot of money. It means that one has to work hard and really push it. That is why they sometimes call me a fanatic; but one has to be like that if one wishes to survive.

These portraits described what farmers refer to as *double goalers* (first portrait), *thrifty or economical farmers* (2), *practical farmers* (3), *cow men* (4), *machine men* (5) and *fanatical farmers* (6). The respondents were then asked to indicate whether they fully identified, partially identified or did not identify at all with each single portrait. With discriminant analysis, all of the answers (including the non-identifications) were taken into account. This gave a cognitive map of different (and sometimes overlapping) farming styles that in turn helped to explain many other meaningful differences.

The identification of different styles can also be grounded on *farm accountancy* or *census* data (provided that data on single farms are available). Meaningful variables are put together in a single database to which principal component analysis, factor analysis or cluster analysis can then be applied. The first two types of analysis reveal different patterns of coherence, and the third type gives clusters of farms that very much resemble each other (within each cluster) or strongly differ from each other (between clusters). An example (of principal component analysis applied to a data set of 113 Friesian dairy farms) is given in Table 4.1.

Table 4.1 Farming styles as seen from principal component analysis

Variable	Style 1	Style 2	Style 3	Style 4	Style 5
Total production/ha	**0.87**	−0.33	−0.03	0.11	0.00
Additional feeding costs/cow	**0.81**	0.08	−0.12	−0.10	−0.36
Total costs/ha	**0.79**	−0.37	−0.29	−0.17	−0.19
Animals/ha	**0.73**	−0.46	0.03	−0.24	0.04
Applied nitrogen/ha	**0.71**	−0.31	0.16	−0.12	0.35
Costs of mechanization/ha	**0.70**	−0.34	−0.12	−0.06	0.01
Milk yield/milking cow	**0.70**	0.33	0.06	0.54	−0.17
Costs of fertilizers/ha	**0.68**	−0.36	0.10	−0.23	0.28
Industrial concentrates/milking cow	**0.65**	0.08	0.08	0.19	−0.36
Number of calves/10 milking cows	0.38	**0.74**	0.06	−0.29	0.07
Meat and offspring/milking cow	0.44	**0.70**	0.05	−0.22	0.26
Heifers/10 milking cows	0.47	**0.63**	0.06	−0.35	0.04
Costs for artificial insemination/cow	0.47	**0.50**	−0.06	−0.10	0.17

(Continued)

Table 4.1 Farming styles as seen from principal component analysis *(Continued)*

Variable	Style 1	Style 2	Style 3	Style 4	Style 5
Economic size units	0.10	−0.14	**0.68**	**−0.40**	0.19
Number of milking cows/farm	0.15	−0.18	**0.63**	0.22	0.34
Fat content of the milk	0.10	−0.28	**−0.59**	0.20	**0.28**
Protein content of the milk	−0.07	0.06	**−0.58**	0.07	**0.49**
Veterinary costs/milking cow	−0.07	0.06	−0.48	**0.34**	**−0.13**
Labour income/milking cow	0.44	0.46	−0.13	**0.65**	0.18

These patterns of coherence were subsequently 'verbalized' and then discussed with larger groups of farmers asking them (1) whether they recognized these patterns; (2) whether there are farms in their neighbourhood that could be regarded as typical examples; (3) whether they could explain the underlying logic or strategy; (4) whether they could characterize the people on these farms (this would result in obtaining proper names for each style); (5) which pattern would be the most profitable, sustainable, etc.; and (6) which pattern best resembles their own farm (for further details, see Ploeg [2003:119–125] and Paredes [2010:161–223]).

One can, of course, also use field data on levels of intensity, scale and other important variables (as shown in Methods Box #6 in Chapter 3).

If further analysis is needed, it is very helpful to have the factor scores calculated (this is very easy with programmes for statistical analysis such as SPSS and SAS). The factor score gives the 'position' of each single farm on each factor rendered by factor analysis. It indicates how much a farm is structured according to a specific style. It also clearly shows those farms that are combining different styles.

Miriam Paredes (2010) undertook extensive research into farming styles in Carchi in the highlands of Ecuador. Her research centred on pesticide use and how to reduce it, because it is the cause of many casualties, and institutional programmes aiming to reduce pesticide use had, so far, chronically failed to achieve this (as demonstrated by Steven Sherwood, 2009). Instead of isolating pesticide use from the farm labour process as a whole, Paredes developed a meticulous farming styles analysis and then related pesticide use (amounts, types, timing, quantities, protection, etc.) to these different styles and underlying strategies. She identified four styles. The *tradicionales* (traditional farmers) applied an ancient potato growing technique, the *wachu rozado*. These farmers saw themselves as *tradicionales* 'out of a strong sense of pride in maintaining their "old" way of producing potatoes' (Paredes, 2010:84). Their way of working was characterized by a high labour input, good yields and meticulous attention to the growing plants. The *seguros* (secure ones) tried to avoid monetary risks, whilst the *arriesgados* (risk takers) relied on high levels of mechanization and fertilizer use. They were often wealthy farmers 'who had accumulated their own capital in occupations other than farming (e.g., as traders). [...] Most of the inputs were market-acquired [...]

and the large investments were financed by means of bank loans. [...] These farmers considered potato production as a "lottery" in which "lucky farmers" could make a fortune' (Paredes, 2010:85). And, finally, there were the *experimentadores* (experimenters) who only had small fields but farmed them intensively. They 'lacked capital but considered family labour as "the capital of the poor". They put more care into tasks such as soil preparation, seed selection, planting and ridging-up [...] and reincorporated weeds and leftover crops into the soil after each production cycle in order to boost their yields by maintaining soil quality and minimizing erosion' (Paredes, 2010:87). There is a lot more on the different styles in Paredes' book, but most important here is that she showed that there are several highly interesting relations between these styles and 'pesticide use in the field and farmers' perceptions of poisonings' (Paredes, 2010:225). When it comes to the use of carbofuran (a carbamate), she found 'that *tradicionales* apply the lowest quantities of active ingredients and that *experimentadores* apply the highest quantities [...] even higher than recommended [...] and this increases the workers' exposure to high concentrations of carbamates' (Paredes, 2010:255). More precisely, *tradicionales* not only use the lowest quantities of pesticides 'but [also] apply them with a high degree of fine-tuning' (based on frequent observations in the fields; Paredes, 2010:256). In synthesis:

> certain styles of farming in Carchi are more sustainable than others in terms of pesticide use and natural resource management. They also contribute more significantly to the food security of the peasant families. This contribution involves increased yields, better quality potatoes and improved benefits. The advantages of these styles only become apparent when looking at [...] heterogeneity. They are not evident when peasant farming is seen as a uniform whole. (Paredes, 2010:257)

In short, when taking into account the different styles of farming, it is often possible to design, or adopt, different ways forward that will probably function far better than a generic approach.

In farming styles research carried out in another potato-producing area, the peasant community of Chacán (in the Antepampa near Cuzco in the South of Peru), it was found that the community members here (loosely) classified themselves (and others) in term of *ricos*, *pobres* and *medios*; that is, rich and poor farmers and an in-between group. The interesting point is that the notion of 'richness' referred to having a self-controlled resource base: enough land at one's disposal, enough working capital, enough seed potatoes and one or more pairs of oxen. Scarcity in this group often resides in the labour that the farm household is able to supply. When it is not enough to do all the work properly, additional labour needs to be contracted (or obtained through socially regulated exchange). The poor, on the other hand, do not have sufficient land, nor do they have the necessary means of traction. However, they do have a surplus of labour, so they contract this to others (sometimes also in nearby cities) or mobilize the lacking means (land, traction, capital) through local exchange mechanisms, such as *ayni* (labour for labour or labour for a pair of oxen) and *compañía* (one person bringing in the land, the other the labour, after which the harvest is jointly shared). The in-between category, the *medios*, occupies a more or less intermediary position: overall there is an equilibrium, but due to the nature of the labour process, this can be temporarily out of balance, with them suffering shortages at some moments and having a surplus (of labour, land, traction, etc.) at others. An important finding of this study (see Methods Box #7 in Chapter 3) was that existing formal credit mechanisms turned out to be totally inadequate for farmers to address disequilibria. Dependency on credit induced relations that blocked the search for high and sustainable yields. According to the farmers who had entered into credit relations, they 'turn the world crazy' (*el mundo se ha vuelto loco*).[2] Striving for high yields is a challenge and having them is a relief. Nonetheless, working with credit pushes farmers in an opposite direction: to enlarging the area cultivated with potatoes and accepting lower yields.[3]

Going beyond merely production

Styles of farming can embrace all domains of farming (see Figure 1.5 for a synthesis) and extend, as argued, into specific fields of interest as sustainability, gender relations, sire selection, quality of grain and/or cheese making (just to mention a few).

Catharine Laurent et al. (1998) developed an interesting classification in France concerning the linkages between the domain of the family and the domain of production. The 'farm' can mean[4] different things within and for the farming family. It might represent, first and foremost, a *patrimony* that was inherited from previous generations and will be passed on to those of the future. The members of the household work in the farm and the farm provides them with an income. It is also a *domus*.[5] But it is equally possible that a farm is owned by a family but that the work is done by employees with the profits going back to the owners. Or that the farm is mainly used to qualify for subsidies, to have homegrown food and/or for leisure. In this vein, Laurent et al. (1998) distinguished 11 different institutional patterns. These are (together with their relative presence in French agriculture):

- employee-run companies (1%);
- capitalist agriculture (3%);
- agriculture as a structural profession (20%);
- agriculture based on a traditional farming logic (21%);
- rural enterprises (8%);
- non-integrated multi-activity (7%);
- subsistence farming for retired farmers (13%);
- qualifying for social welfare coverage/pension payments (9%);
- agricultural production for home consumption and barter (2%);
- luxury agriculture (4%); and
- small scale recreational agriculture (12%).

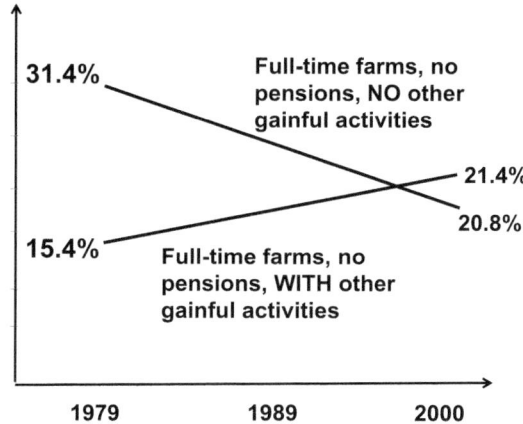

Figure 4.2 Land-based activity systems in France and their development over time (author's own elaboration).

Together with these patterns, the objectives, relevant contextual elements (policy included) and the concrete design of agricultural activities *sensu strictu* differ considerably. In an associated study, Laurent and Remy (1998) also showed that these patterns are historically variable. Two broad categories are defined and compared: full-time farms having other gainful activities next to, but probably connected to, agricultural production and a second group lacking such additional activities. Figure 4.2 shows how both types have developed over time. In Chapter 7 I will go into the probable reasons that underlie these differential trends.

Other comparable research has been done in China where 'multiple job holding' is a widespread feature: most farming households in China are engaged, in one way or another, in other economic activities (frequently in faraway places and for considerable time). However, they also continue farming, and this raises the question (just as in France): *what does the farm mean for the farming family?* Or, to be more specific: how much of the family's income is earned in the farm (self-provisioned food included) and how much is earned elsewhere? Applying these questions in village studies provides images such as the one shown in Figure 4.3.[6]

Figure 4.3 represents the balance, in one Chinese village, between the income from farming and income obtained elsewhere, yet it also reflects the considerable diversity that exists in most Chinese villages. One segment (the majority) is strongly anchored in 'other economic activities' (having a non-agricultural rural enterprise, migrant labour, etc.) but also continues with farming activities. The maize produced on these farms is mainly meant for self-consumption (although on average 18% of the harvest is sold); the same applies to other food products. Production for self-consumption is considered to be strategic, and the farm is 'kept in the family' for a variety of reasons. Another, contrasting (and in terms of numbers, far smaller) segment depends mainly on incomes from agriculture.

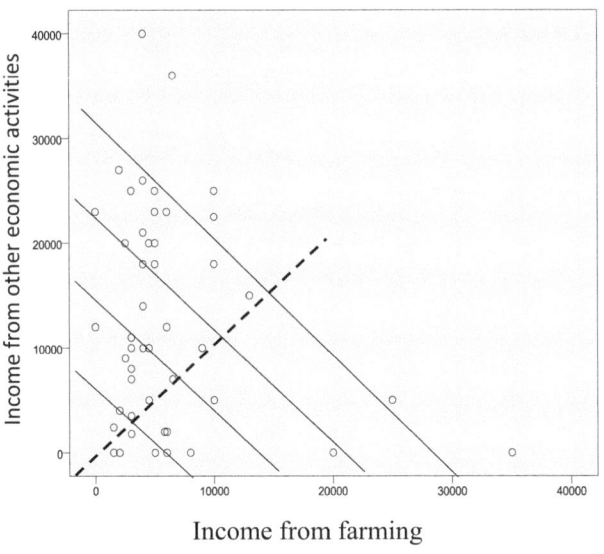

Figure 4.3 Diversity in income sources in a Chinese peasant village (Ploeg and Ye, 2010).

Maybe surprisingly, these incomes are not obtained only by means of the standard cash crop (maize) but also through novel agricultural activities (such as processing glass noodles, having herds of goats, producing tree seedlings, developing agro-tourism, specialized fruit production, etc.). In a way they represent the Chinese 'model' for rural development (as I will discuss in Chapter 7). Yet another segment is located in the lower left-hand corner of Figure 4.3. It basically contains the poor: elderly, handicapped and/or left-behind people. In contrast, there is the segment that obtains more or less half of their total income from farming and the other half from non-agricultural activities. They represent the classical peasants who have developed their farms according to their well-established logic of farming. Typically, this segment has the highest sales of maize (53% of the total maize production; the rest is for feeding the animals, barter, seeds for the next cycle and home consumption). The different concrete conditions delineated here translate into the way farming is organized and developed, in the objectives of farming and in the prospects of those involved.

Contrasting models for generating incomes

Different styles of farming not only represent specific, and mutually contrasting, productive constellations – they also contain a range of specific models for generating adequate incomes. Highlighting such possibilities ('many different roads lead to Rome') has been one of the main objectives of farming styles analysis in the Netherlands. Thus, farming styles analysis emerged as an implicit but clear

critique of modernization policies that focused on technology-driven intensification and scale enlargement as the *only* viable farm development trajectory.
There are, theoretically speaking, several other possibilities, including achieving increases in labour productivity through further mechanization, seeking cost
reductions, improved yields obtained through fine-tuning, the development of
new economic activities and their integration into the farm and/or part-time fam-
ing (or pluri-activity). Each possibility has specific consequences, and each has its
own balance of benefits and costs. Empirical research can inform us about all of
this and the net results of each strategy.

Each style has its own specific criteria that allow the farmer to assess his or
her relative success or failure (and which help to compare one's own farm with
others). In the style of farming economically (characterized by low external input
use and debt level), the level of external inputs used is the telling beacon – the
lower the better. In dairy farming, this basically comes down to bought-in feed
and fodder. In the high input/high output style of farming (in the example summarized in Table 4.2, this style consists of farms with a high cattle density[7]), the
'margin per hectare' (or labour income per hectare) is the self-evident criterion for
optimization. For breeders who, next to the production of milk, also focus on the
production of heifers to be sold (often for export), the sales of animals and growth
of the herd ('turnover and accretion') are what matters. For cow men (who aim for
high milk yields) it is the margin per 100 kg of milk (or labour income per 100 kg
of milk) and for large farmers it is the magnitude of the farm (expressed here in
economic size units).

Table 4.2 shows that (at least in this case) the process of production in these
farms is moulded in such it way that it renders the best outcomes according to
the style-specific criteria for optimization. Each style has its own specific way of
performing, and this is reflected in specific criteria for success and in the empirical
scores on these criteria. In practice, the different trajectories ('the different roads
towards Rome') that are crafted over time all promise to bring adequate results.
Here it is not just the case that the walker creates the road (*'caminando se hace el*

Table 4.2 Criteria for optimization and empirical results (western clay grounds of Friesland,
1990, n = 113; all data in guilders) (Ploeg et al., 1992)

	High-input/ high-output farmers	Breeders	Large farmers	Cow men	Economical farmers
Labour income per hectare	9,580	6,906	7,939	8,719	7,550
Sales of animals and growth of herd per 100 kg milk	12.66	14.15	11.19	9.54	13.31
Economic size (in SFU)	184	185	233	207	209
Labour income per dairy cow	4,809	4,899	4,561	5,237	4,624
Total additional costs for feed and fodder	1,363	1,135	1,053	966	845

SFU (standard farm units) is a measure for the economic size of farms.

camino', as is said in Spanish) but also that the road makes the walker. The walker knows how to walk (using specific criteria for assessing his direction, speed, etc.) and the chosen road allows him to do so. This is what congruence (or coherence) is supposed to be: practice and strategy fit very well together; the one moulds the other and vice versa.

Whatever the reasons, the existence of several 'walking trails' has gone unnoticed, not by those who construct and follow them but by the main agronomic (and related) institutions, such as applied agrarian sciences, agricultural policies and the hegemonic farmers' unions. One of the important benefits of the sociology of farming is that it is able to develop the concepts and methods that help to highlight the merits of different styles of farming that have been, to date, obscured and ignored. The sociology of farming is probably better equipped than other disciplines to do so because (1) it puts actors' perspectives and experiences centre stage in the analysis of agricultural practices (novel practices included) and (2) it understands development as a differential and man-made phenomenon (it is not simply 'dictated' by markets and technology, nor just interpretable in terms of 'right' or 'wrong'). I will give three examples, one of which is a little bit dated; the other two are recent.

When the first examples of farmers developing so-called side activities on their farms started to attract attention (in the late 1980s and early 1990s), this novel and unexpected trend[8] was not understood (let alone through the concepts used by the actors involved themselves). It was, instead, quickly categorized as a 'last and desperate step' of supposedly marginal farmers ('those who could not earn a living from milking alone'), and it was thought to be especially pronounced in marginal areas. It was assumed that these farmers (and their additional activities) would disappear anyway, which meant that it was not worth studying these newly emerging styles. Thus, this novel trend was virtually ignored by agricultural sciences and institutes as a topic worthy of applied research. As a consequence, these new realities fell below the radar of agricultural policies. The sociology of farming was the first (and initially the only) approach that started to pay attention to this new phenomenon[9] and, especially, to explain that this new trend was not only about adding something 'additional' to an otherwise unchanged farm (let alone to a supposedly 'disappearing' farm) but about *transformation*: about creating a new style of farming that was able to face the newly emerging conditions of the 1990s (I will come back to this phenomenon in Chapter 7 where I discuss rural development processes).

From more or less the same moment (the late 1980s onwards), another process started to germinate, first in the Global South but in several 'hidden'[10] places in the Global North as well: it aimed to counteract the growing dependency on external resources (chemical inputs included) and to simultaneously improve the quality of resources internal to the farm (visualized in Figure 3.1 as a shift from the upper-left flow of commoditized resources towards the underlying flow of resources produced and reproduced in the farm). In the Global South this process became increasingly known as the struggle for a new agroecological way of farming[11]; in the Global North, it remained, until recently, hidden and unnoticed (although

Table 4.3 Comparison of grassland-based and conventional farms in Bretagne, France (Ploeg et al., 2019)

	Conventional farms	Grassland-based farms	Difference
GVP/worker	118,281 €	86,837 €	−27%
VA/GVP	33%	51%	+54%
VA/worker	38,884 €	44,179 €	+14%
Family income/family worker	15,797 €	27,271 €	+73%

it became clear, in hindsight, that here as well, many new 'proto-agroecological' farming systems had been actively developed in the meantime).[12]

The resulting impasse was recently broken, at least partly, with the publication of a lengthy article in the *Journal of Rural Studies* written by 29 European scholars (Ploeg et al., 2019), many of them rural sociologists. It brought together a range of scattered research findings (often elaborated by farmers' networks) and proposed a theoretical model[13] that put the concept of value added as part of total gross value of production (VA/GVP) centre stage. This concept is central in the strategy used by agroecological (and proto-agroecological) farmers to organize, manage and develop their farms and strongly reflects the classical peasant notion of 'the clean part' (that is, 'what remains for the farmer him- or herself').[14] Thus, theory was adapted (by introducing an important concept used by the involved actors themselves) and the merits of novel practices, which otherwise would have remained in the dark, were shown. This is illustrated in Table 4.3, which compares two samples of dairy farms in Bretagne in France.

The data show that when the magnitude (or economic size) of the farms is taken into account, the conventional farms outperform the others: the grassland-based (or agroecological)[15] farms are 27% smaller than the conventional ones. Their hidden strength, though, resides in the far higher VA/GVP ratio that they achieve. This makes for an income that, at the end of the day, is significantly higher than that of the conventional farms. Similar relations were found for 10 other European countries. This helped considerably in demonstrating, and explaining, 'the economic potential of agroecology' (the actual title of the article).

A last example regards the new tendency for young people to do the seemingly impossible: start a new farm. This is very difficult for young people with an agrarian background (even if they inherit the land from their parents), but entry for non-agrarian youngsters seems to be totally blocked, by several barriers. Nonetheless, the numbers of young entrants are growing. Young scholar-activists such as Neus Monnlor from Spain and Kevin Morel from France engaged with the movements of these youngsters and started to outline the prospects that these new peasants were beginning to construct. Kevin Morel et al. (2017, 2018) combined field data with an intelligent model (that reflects the new realities created) and convincingly demonstrated that after a three-year period of building the new farm and developing the knowledge (of the *art de la localité* type) needed, even small farms of 1,000 square metres (!) can generate an average monthly income of 1,500 euros. In her turn, Neus Monllor demonstrated that the new entrants bring refreshing new views into the rural world and introduce promising ruptures in the vested routines, although not without encountering some resistance (see Monllor and Fuller, 2016).

Moulding nature, constructing new socio-material constellations

Farming styles are both a strategic repertoire and a specific set of practices. Each informs and confirms the other. In each style of farming, nature (the fields, the soil, the animals, the plant varieties) and the other material resources (the buildings, technologies, connecting infrastructure) are co-shaped by the farmers operating according to a specific strategy. Thus, nature and other material resources are (just as the social resources) developed in a style-specific way: they become differentiated. Within each style the resources are moulded and brought together in ways that are style specific. This also explains the continuity of styles: they cannot be changed overnight for they are materialized in the specificity of available resources (and the networks within which these farms are engaged). Thus, styles of farming are anchored, as it were, in the materiality of things.

Using large data sets from cattle breeding associations, Ab Groen et al. (1993) investigated the breeding values of cows and the sire selection for a large number of farms that were known in terms of having a specific style. They showed, in the first place, that farmers shape their cattle (through selection and the making of a specific 'phenotypical environment') in such a way that the cattle fit as much as possible with the farmer's overall strategy and objectives. Take, for example, *double goalers* (farmers who bet on both milk and meat production: see Methods Box #8) and *cow men* (farmers who aim to realize, in a craftlike way, the highest possible milk yields). This difference leads to different preferences in the breeds of cattle: *milk-specific* breeds (especially Holstein Friesian that deliver very high milk yields) and *dual-purpose* breeds (rendering both good milk yields and a good amount of high quality beef), such as MRY cattle.[16] Now, the interesting finding was that whatever the specific breed, *cow men* mould them into milk-specific animals (even if they are of the MRY breed), just as *double goalers* are able to remould Holstein Friesian cattle into animals that also render good meat production. Another intriguing finding is that cows on the farms of *machine men* (see once again Methods Box #8) show the best breeding values for legs: their legs are stronger and longer than those of cows developed in other styles. This is related to the nature of farming mechanically (which is what *machine men* do). There is no summer feeding inside the stable. Cows have to do their own job and graze in the fields. This favours the selection of those animals best able to do so. Those with weak legs are replaced (whilst in other styles, farmers continue caring for them because they have, for instance, good milk yields). In the same vein, several other statistically significant and theoretically meaningful effects of style on breeding values were found. This is also reflected in sire selection: farmers following different styles tend to opt for different sires (although the interrelations are not always easy to understand), and this contributes both to a further differentiation and to a more solid anchoring of farming styles in the materiality of 'things' (living nature included). I will come back to this in Chapter 10 when discussing the socio-materiality of agriculture.

Gender relations and farming styles

The organization of the farm labour process also translates into, and is affected by, gender relations. As men and women mostly work together in the farm (evidently the case in family farms but, of course, not limited to these), both share in the results it produces (albeit sometimes in highly unequal parts), and both are involved in the decision-making processes (at least theoretically) just as they both feel the impacts and consequences of the decisions taken. As such, gender relations are a central element of the farm labour process and thus also affect the style of farming.

In her Ph.D. research, Sabine de Rooij (1992) made a comparative analysis of gender relations in differently structured farms. First, she distinguished farms in which the produced milk is processed into cheese (i.e., the mixed farms) from farms that deliver the produced milk to the dairy industry for further processing (specialized farms). On-farm cheese making mostly was, and is, the task of the woman, just as milking typically is the man's job (at least in northwest Europe). The women involved in cheese making regard this as their own field of responsibility: governed by the women themselves and a source of pride and identity. It offers them a 'silent equality' vis-à-vis the men. This contrasts with the situation in highly specialized farms which just deliver the milk to the dairy industry. Women may be very much needed in these farms (for all kinds of tasks and also as a 'helping hand' for a range of other tasks), but they do not have a clearly delineated responsibility of their own. They perceive their work as being of 'secondary relevance': it is often invisible; does not come with its own, clear agenda; and is often overlooked by others. Secondly, de Rooij distinguished between small- and large-scale farms (whereby scale explicitly refers to the relation between the work to be done and the available labour force). Taken together, she argued, these two dimensions not only describe some of the main differences in Dutch agriculture but also touch on the main developmental trend of recent decades: from small-scale mixed farms to large-scale specialized farms. Thus, a comparison was made that possibly reflects, at least partly, the recent historical changes.

When combined, scale enlargement and specialization bring a reduction of labour time spent in the farm by women: from 35% to 20%. Total time (per week) spent (in both the farm and the household) goes down less: in absolute terms is goes from 64.5 hours spend in the small-scale and mixed farms to 57.0 hours in the large-scale specialized farms[17] (for documenting time spent on different tasks, see Methods Box #9). But above all, it is the nature of tasks done that changes the most. This goes from having women having their own, clearly delineated tasks that they design, perform and evaluate themselves (and thus build up considerable experience and augment their own qualifications and skills; see de Rooij, 1992:92), to a range of sub-tasks that are, just as household work, mainly

Methods Box #9   Time-writing

Registration of the different tasks and sub-tasks and of the time spent on each of them is an important research method in rural gender studies. Farm women often greatly underestimate the time they contribute to the farm. Getting a clear and empirically correct estimation can be, along with other instruments, an important tool for empowerment. 'Time-writing' is a method for doing so. It starts with developing a clear list of all of the possible tasks farm women might be engaged in. This is especially important for the many tasks done on behalf of the farm but not physically within the farm itself but, instead, in the house. This may include dealing with all of the telephone calls that relate to running the farm; receiving the technicians, veterinarian, bookkeeper and all of the other external agents who relate to the farm; having coffee and bread ready when the men have a break; doing shopping for the farm; keeping the farm accountancy up to date; washing the clothes used on the farm; etc. On a schedule (one for each day), these activities are noted by the farm women themselves, together with the time spent. This is repeated for several days/weeks and then discussed and, if possible, extrapolated to the year as a whole. One important finding has been that farm women have, just as the men, extremely long working days and a working week that goes well beyond the 40 hours/week. Nonetheless, women's work is typically perceived as 'not really a job'. 'You are busy all day, but at the end of the day you have the feeling that you haven't done anything' (partly because there are no visible results).

What follows is an excerpt of one 'time-writing sheet' from de Rooij's research (de Rooij, 1992).

6:00		8:40	Connect mixing machine
6:10		8:50	Drain whey from cheese
6:20		9:00	Have breakfast
6:30	Get up	9:10	Continue draining
6:40	Put cheese milk in the containers; add	9:20	Clean table and kitchen
6:50	rennet; de-cream; circulate; pump to	9:30	Prepare rennet
7:00	heating container	9:40	Put cheese forms in large container
7:10	Heat cheese-milk, prepare furnace	9:50	Clean cheese-making room
7:20	Wake up children	10:00	Start cleaning the house
7:30	Prepare breakfast for the children and	10:10	Continue doing so
7:40	their lunch for school	10:20	Turn and clean cheeses in the
7:50	Bring children to school	10:30	storage cellar
8:00	Continue with cheese making	10:40	Drink coffee
8:10	Clean pump	10:50	Check incoming mail
8:20	Spin-dry the laundry	11:00	Start preparing lunch
8:30	Put laundry on line for further drying		Etc.

invisible and limited to 'supporting the man' (de Rooij, 1992:80). This later position within the overall farm labour process is associated with a reduced say in the decision-making process. In the small-scale cheese making farms, most decisions are jointly taken by husband and wife together. In large-scale and specialized farms, it is more often than not just the man who takes the major decisions. This strongly suggests that the modernization of agriculture comes with a marginalization of farm women. They lose their fields of responsibility within the farm; the quality, 'importance' and visibility of their jobs are reduced; and they are moved to the margin of the decision-making process.

This is not to say that gender inequalities were (or are) absent in the cheese making farms. On those farms, taking care of the household and children is delegated almost entirely to women (thus bringing a double workload). Sabine de Rooij delved into this by means of a novel method: comparing the studied family farms with others run by men only (two brothers, for instance). Whilst the division of labour within the typical family farms could be defined as *sex-segregated* (this occurs in cheese making farms) or as *sex-sequential* (in specialized farms where the sub-tasks done by women 'follow' the men's activities)[18] – in farms 'where there is no woman', the division of labour is typically *rotational*. There is no hierarchy of tasks and all tasks (including household labour) are equally shared by the partners. Another solution is *equivalence*: one man specializes in specific tasks, the other in others, but both sets (and consequently both partners) are defined and seen as equal to each other. The tentative conclusion is clear: the typical *family* farm comes with specific gender relations and gendered identities. The nature, degree and dynamics of these relations are, however, historically variable. Gender relations may also be affected (and changed) by farm women struggling for equality.

Finally, it should not be forgotten that the emancipatory drives of farm women are not only manifest in overt struggles and/or novel practices. They are often present as a silent and mostly invisible force that, regardless, should not be ignored. Asked whether they agreed with their men when the latter proposed a further expansion of the farm by, say, 100 units, many women indicated that in order to avoid major disruptive effects in both the family and the farm, they opted instead for, say, 50 units – and mostly succeeded in getting actual expansion limited to socially acceptable proportions (de Rooij et al., 1995). Rural women represent the axis that brings family and farm together: thus, they defend the family when the farm becomes a threat (see Figure 1.5). Probably the most extreme form of this has occurred in rural areas with a strong patriarchal tradition. Here farm women told their daughters: 'marry whomever you like as long as it is not a peasant'. This has resulted in a massive depopulation of large rural areas, particularly in Spain.

Women reclaiming lost domains

As part of a wider process of rural development (which I will discuss in detail in Chapter 7), farm women increasingly started, from the 1990s onwards, to construct, once again, their own field of responsibility in the farms that, at least nominally, also belong to them. This took many forms, including integrating care services (for young children, elderly and/or handicapped people) in the farm; organizing agro-tourist activities; re-integrating processing activities such as cheese making, butter production, making jams and chutneys; the direct marketing of food products; etc. This new trend evidently was a response to the growing crisis that resulted from new neo-liberal policies developed and implemented from the beginning of the 1990s onwards. It also stemmed from the growing awareness of farm women of their particular position and the desire to become more visible as farm women (contributing in clearly visible ways to the family income). The interaction of these two factors has proved to be a strong driver for a range of changes in the agricultural sectors of Western and Southern Europe (see, e.g., Figure 4.2 where 'other gainful activities' includes the newly constructed fields of responsibility mentioned above).

Interestingly, the development of new on-farm activities by farm women has also affected the organization and development of the 'male part' of the farm enterprise. If having, for example, a mini-campsite alongside a (fruit producing) farm, it is not attractive to repeatedly warn young children (staying with their parents on the camping site) that they should not touch and eat the fruits because they have been sprayed with insecticides. Thus, a change towards organic production became more than probable.

There are, of course, also problems linked with the development of new on-farm activities. Their very success can be a reason to stop again. This is, for example, the case when the new activities bring prolonged working days, heavy workloads and the need to be permanently available.

The reclamation of previously lost domains (and claiming new ones) is a key element of farm women's search for, and construction of, (new) identities. Farm women's identity used to be, at least in the Global North, strongly related to entering the farm through marriage. Being a farm woman was equated to, if not derived from, being married to a (male) farmer, and farming was first and foremost portrayed as a male occupation (see Shortall [2006] for an overview). However, the many-sided transformation of agriculture and the empowerment of (rural) women have brought about a wide series of changes in gender roles, positions and identities in which farm women have come to the fore as important actors in on-farm diversification (Bock, 2004; Brandth and Haugen, 2010; Seuneke and Bock, 2015). They take part now in rural organizations (Teather, 1996), participate in joint farm ventures (Cush et al., 2018), have off-farm work (Kelly and Shortall, 2002; Shortall, 2006) or run farms on their own (Whitley and Brasier, 2020). In the Global South, agrarian change is having similar effects, although there are dissimilarities due to, for example, male out-migration, relatively low educational levels, women's restricted mobility and other factors (Sachs and Garner 2017; Sachs, 2019; Adesesugba et al., 2020). Nonetheless, traditional gender roles and gender ideologies prove to be very persistent. In the Global North it is still women who do the majority of domestic work and childcare, whereas 'being a farmer' is still associated with men (see, e.g., Brandth, 2002; Kelly and Shortall, 2002; Shortall, 2006). Adesesugba et al. (2020) argued that this is also true for farm women in the Global South (and probably more so). Women remain responsible for domestic work and childcare, whilst they are simultaneously engaged in farm and/or other work that generates extra income (often in the informal sector; see in this respect Sachs and Garner, 2017).

Since the mid-1990s, research on farm women has focused strongly on issues of identity and perceptions of femininity and masculinity. In addition, the concept of intersectionality has been introduced into rural gender studies (see, e.g., Leder and Sachs, 2019; Sachs, 2019). These studies make clear that, despite changes in the work of farm women and men, identity shift is a slowly evolving process that takes time and in which social interactions and social relations play a crucial role (see, e.g., Kelly and Shortall, 2002; Bock, 2006; Shortall, 2006, 2017).

Gender relations and the domains of farming

Figure 1.5 (see Chapter 1) specified different domains of farming. Farming is not just about the production of specific products (milk, potatoes, etc.). Farming also implies that those who are producing are actively engaged as well in other domains, such as the organization of specific relations with markets, the organization of household and family and, of course, the reproduction of the farm and the resources within it. Farming includes, and assumes, the mutual coordination and corresponding fine-tuning of these domains, thereby shaping them into a specific constellation: a style of farming.

The dominant pattern of gender relations allocates men and women different spaces within the four domains. Men and women each have specific responsibilities delineating the specific fields that belong to the men or the women. Each field of responsibility includes tasks that need to be aligned (i.e., designed, done, evaluated, corrected and attuned) to the other fields. This distribution of responsibilities can be influenced by legislation, tradition, religion, emancipatory movements, etc.

In some parts of the world, the responsibility of farm women is reduced to the domain of household and family (and even here this is often conditioned by the men deciding on the portion of the overall farm income spent in supporting the household and the portion that is to be reinvested in the farm). This represents a remarkable contrast with situations described above in which farm women have well-outlined responsibilities in the domains of production, reproduction and, sometimes, marketing as well.[19] It also differs from the current situation (with farm women [re]claiming [often novel and] clearly outlined responsibilities beyond the household). Alongside this pattern there are seeming exceptions. One is farms that are mainly run by women. This situation is encountered nearly everywhere in the world: in Latin America, in China, throughout Europe (but especially in the Mediterranean area) and in Africa. The men are away (in the past because they had to fight wars, nowadays because they work in faraway places as labour migrants) and the farm women are running the farm in its entirety.[20] They operate in all domains. Nonetheless, they mostly do so according to the rules defined by the men (which are also the rules transmitted by tradition). Meng Xiandang (2014) undertook a meticulous study of farm women in China, showing that they are active in all four domains of farming (with just a few 'heavy' tasks, such as the preparation of the land before sowing being done by the men when they return to their village for the spring festival). All of the other tasks are done by the women but mostly according to the (internalized) instructions of the husband.

Meng's study on Chinese women running the farm equally showed that the tenacity of gender relations also stems from the cultural repertoire of the countryside. Specific tasks, such as the preparation of the seed beds, are defined as 'heavy tasks' (if not as 'important and heavy'). This implies that it is mostly the men who perform such tasks – which in turn defines the men as being the 'strong' ones because they are able to do the 'heavy' work. This is reaffirmed by the accepted views on income flows and expenses (see Figure 1.6 for the analytical background). There are two flows of income in Chinese peasant households: one generated by the farm (and mostly associated with the woman) and the other stemming from involvement in 'other economic activities' (and mostly coming from the man). The ratio between the two flows is highly variable (see Figure 4.3). Both flows are socially regulated. Income from the farm is for 'small expenses' (daily expenses for the household, clothes, *quanxi*,[21] etc.) whilst the income from elsewhere is for the 'big expenses', such as the education of the son, building a house for him and financing his wedding. These are the big and important expenses that delineate and ensure the standing, welfare and prospects of the family. This defines, in a seemingly self-evident way, the identity of men: they are important because their labour allows for the big expenses that help the family to move ahead. Women's identity, on the contrary, is at best supplementary. They *only* contribute to the 'small spending'. Interestingly, this is quite close to the dominant framing in, say, the Netherlands. In large-scale, specialized farms, women's role is also perceived (and materially structured) as supplementary: they are to be standby, available to assist the men (see above). This equates to and reinforces stereotypical gender identities of the men being 'in charge' and the women giving 'a helping hand', or of the 'boss' and his 'secretary'.[22]

There is yet another and illustrative exception. That is the one of large-scale, specialized, high-tech farms in the Netherlands (or, more generally, in northwestern Europe) where the man is considered to be the *entrepreneur* (also according to his self-definition) and the wife operates, as a well-paid professional, in another economic sector. In regards to the farm she seems to be absent, she is not there, she is elsewhere. Only when the children return from school can you find her in the farm *house*. There she is seen, above anything else, as mother and just as mother. What is typically left out in the social construction of these engendered identities is that this apparently contradictory whole can only function because the *household* is completely financed with the earnings of the women, whilst the earnings of the farm (if any) are used for re-investment in the farm. Here some people comment that the farm is just 'a very expensive hobby of the farm men'.

Notes

1. There is a notable difference with what is called the farming systems approach which seeks to tie together the different 'subsystems' (arable production, animal production, the household, the context, etc.) through their functional connections. The notion of strategic ordering is basically absent here. Other research traditions classify farms according to categories defined by agricultural policies (e.g., large, medium and small farms).

2. See Bolhuis and Ploeg (1985) and Ploeg (1990:151–203). The cited comment echoes long-time experiences that go back to the *Conquista*.

3. In this way, farmers are responding to the abruptly increased market-induced risk that comes with substantial amounts of formal credit.

4. Laurent et al. (1999) applied this approach to South African agriculture.

5. *Domus* is where you live and work together with the previous and the next generation (i.e., the extended family) which together own the place. *Domus* is the place to which you belong. It protects you but also requires your solidarity (see, among others, Le Roy Ladurie, 1980).

6. Similar findings are shown in Hu (2014) and Meng (2014).

7. A high cattle density means having many animals per hectare. This implies high levels of fertilization and high levels of bought-in feed and fodder.

8. In the past there had been many such activities, but having a farm involved in a range of activities (instead of specializing in the production of only one agricultural product) was considered, from the 1960s onwards, as an historical anachronism that definitely did not fit in with the modernization agenda. Farmers with an additional job were pejoratively called 'postman farmers'. Their land, production quota, etc., should, according to the hegemony of the time, have been reallocated to large, quickly developing farms. It did not occur to the scientists and politicians that in the late 1980s this phenomenon would re-emerge as an increasingly popular strategy.

9. See, for example, Broekhuizen and Renting (1994), Ettema et al. (1995), Bruin et al. (1997), Broekhuizen et al. (1997), Scettri (2001), Joannides et al. (2001) and Ploeg et al. (2002). In a way Ye (2002) fits in the same tradition, giving detailed descriptions of novel farming practices.

10. 'Hidden' because they were left unnoticed, if not neglected, by hegemonic science and policies. It took another epistemic community to explore, understand and represent the newly emerging realities. Fine examples are the Ph.D. theses of Lola Domínguez García (2007), Pierluigi Milone (2004) and Line Louah (2020). See also Flaminia Ventura (1995) and Jim Kinsella et al. (2002).

11. For example, Altieri (1990), Reijntjes et al. (1992), Gliesmann (1997), Alonso Mielgo et al. (2008), Rosset et al. (2011), Rosset and Altieri (2017), Sevilla Guzman (2007), Petersen (2015, 2017) and Petersen and Silveira (2017).

12. This is a typical example of the institutional blindness of agricultural sciences (see Figure 1.7). The distance between novel practices and science is further increased by the 'not-invented-here' syndrome: things are only valid if invented by science.

13. This model is synthesized in the following equation: $VA/GVP * GVP/LU = VA/LU$. This is an adaptation of the model developed by Hayami and Ruttan (1985) that was discussed in Chapter 2. The advantage of this model is that it reflects the effects of cost reductions brought by an agroecological approach. The higher the cost reductions, the higher the VA. See Ploeg et al. (2019).

14. In graphical terms, this notion comes down to the difference between the upper left flow and the upper right one in Figure 3.1. The 'clean part' is the difference between what goes in and what goes out (both flows expressed in amounts of money). It is very close to labour income as specified by Chayanov (1925/1966).

15. In dairy farming, the notion of grassland-based denotes a well-adjusted equilibrium between the available pasture land and the number of cattle. The animals are (largely) fed by the grasslands of the farm. There is no (or minimal) necessity to acquire additional feed and fodder. The manure produced is used to fertilize the meadows, creating a closed cycle.
16. This breed originally stems from the river basins of the Meuse, the Rhine and the Yssel: hence MRY. It mostly is red-and-white coloured.
17. Time spent with and for the children is not included here.
18. The nearly classical example is the hooking up of machinery to tractors. This is nearly impossible to do alone. Thus, men call the women (whenever they are needed according to the male schedule) in order to help them with this particular task. Typically, in farms where there is no woman, other solutions are created, such as using several small, and probably older but well-maintained, tractors, each of which is permanently hooked to a specific implement. This makes 'hooking up', a continuously reappearing job, no longer necessary.
19. In many parts of the world, farm women are responsible for the domain of reproduction, or at least for considerable parts of it. In the Andes, for instance, taking care of the potato seeds for the cycle to come, the safeguarding of genotypical diversity and *knowing* all the different varieties (and their properties) all are the responsibilities of women.
20. This is sometimes referred to as a 'feminization' of farming – a characterization that is, at best, only partly true.
21. Relations of reciprocity and patronage.
22. As a matter of fact, one of the daily realities most hated by farming women (especially since the second feminist wave) is when advisors, traders, veterinary doctors, extensionists (whatever exponent of the *nomenklatura*) come to the farm and ask them 'whether the boss is around'.

Bibliography

Adesugba, M., E. Oughton, and S. Shortall. (2020), 'Farm household livelihood strategies', in *Handbook of Gender and Agriculture*, edited by C. E. Sachs, L. Jensen, P. Castellanos, and K. Sexsmith, 315–325, Routledge, London.

Alonso Mielgo, A. M., G. I. Guzman Casado, L. Foraster Pulido, and R. González Lera. (2008), 'Impacto socioeconómico y ambiental de la agricultura ecológica en el desarrollo rural', in *Producción Ecológica. Influencia en el Desarrollo Rural*, edited by G. I. Guzmán Casado, A. R. García Martínez, A. M. Alonso Mielgo, and J. M. Perea Munoz, 72–267, Ministerio de Medio Ambiente y Medio Rural y Marino, Madrid.

Altieri, M. A. (1990), *Agroecology and Small Farm Development*, CRC Press, Ann Arbor, MI.

Bock, B. B. (2004), 'Fitting in and multitasking: Dutch farm women's strategies in rural entrepreneurship', *Sociologia Ruralis* **44** (3), pp. 245–261.

Bock, B. B. (2006), 'Rurality and gender identity: An overview', in *Rural Gender Relations: Issues and Case Studies*, edited by B. Bock and S. Shortall, 279–298, CAB International, Wallingford, UK.

Bolhuis, E. E., and J. D. van der Ploeg. (1985), *Boerenarbeid en Stijlen van Landbouwbeoefening*, Ph.D. thesis, Leiden University, LIDESCO, Leiden, the Netherlands.

Brandth, B. (2002), 'Gender identity in European family farming: A literature review', *Sociologia Ruralis* **22** (3), pp. 227–244.

Brandth, B., and M. S. Haugen. (2010), 'Doing farm tourism: The intertwining practices of gender and work', *Signs: Journal of Women in Culture and Society* **35**, pp. 425–446.

Broekhuizen, R. van, L. Klep, H. Oostindie, and J. D. van der Ploeg. (1997), *Renewing the Countryside: An Atlas with Two Hundred Examples from Dutch Rural Society*, Misset/Elsevier, Doetinchem, the Netherlands.

Broekhuizen, R. van, and H. Renting. (1994), *Pioniers op het Platteland: Boeren en Tuinders op Zoek naar Nieuwe Overlevingsmogelijkheden*, CLO pers, Den Haag, the Netherlands.

Bruin, R. de, R. van Broekhuizen, and J. D. van der Ploeg. (1997), *Ondernemen van Onderop, Plattelandsvernieuwing in Gelderland*, Circle for Rural European Studies, LUW, Wageningen, the Netherlands.

Chayanov, A. V. (1925/1966), *The Theory of Peasant Economy*, edited by D. Thorner et al., Manchester University Press, Manchester, UK.

Commandeur, M. (2003), *Styles of Pig Farming: A Techno-sociological Inquiry of Processes and Constructions in Twente and the Achterhoek*, Ph.D. thesis, Wageningen University, Wageningen, the Netherlands.

Cush, P., Á. Macken-Walsh, and A. Byrne. (2018), 'Joint farming ventures in Ireland: Gender identities of the self and the social', *Journal of Rural Studies* **57**, pp. 55–64.

Dominguez García, L. (2007), *The Way You Do, It Matters. A Case Study: Farming Economically in Galician Dairy Agroecosystems in the Context of a Cooperative*, Ph.D. thesis, Wageningen University, Wageningen, the Netherlands.

Ettema, M., A. Nooij, G. van Dijk, J. D. van der Ploeg, and R. van Broekhuizen. (1995), *De Toekomst, een Bespreking van de Derde Boerderij-Enquete voor het Nationaal Landbouwdebat*, Misset Uitgeverij bv, Doetinchem, the Netherlands.

Gliessman, S. R. (1997), *Agroecology. Ecological Processes in Sustainable Agriculture*, Ann Arbor Press, Chelsea, MI.

Groen, Ab F., K. de Groot, J. D. van der Ploeg, and D. Roep. (1993), *Stijlvolvol Fokken, een Oriënterende Studie naar de Relatie tussen Sociaal-Economische Verscheidenheid en Bedrijfsspecifieke Fokdoeldefinitie*, Bedrijfsstijlenstudie 9, Landbouwuniversiteit, Wageningen, the Netherlands.

Hayami, Y., and V. Ruttan. (1985), *Agricultural Development: An International Perspective*, John Hopkins University Press, Baltimore, MD.

Hu, Z. (2014), *Socio-economic Drivers of Agricultural Production in a Transition Economy, a Case-Study of Hu Village, Sichuan Province, China*, Ph.D. thesis, Plymouth University, Plymouth, UK.

Joannides, J., S. Bergan, M. Ritchie, B. Waterhouse, and O. Ukaga. (2001), *Renewing the Countryside*, Institute for Agriculture and Trade Policy, Minneapolis, MN.

Kelly, R., and S. Shortall. (2002), '"Farmers' wives": Women who are off-farm breadwinners and the implications for on-farm gender relations', *Journal of Sociology* **38** (4), pp. 327–343.

Kinsella, J., P. Bogue, J. Mannion, and S. Wilson. (2002), 'Cost reduction for small-scale dairy farms in County Clare', in *Living Countrysides: Rural Development Processes in Europe – The State of Art*, edited by J. D. van der Ploeg, A. Long, and J. Banks, 149–161, Elsevier, Doetinchem, the Netherlands.

Langthaler, E. (2012), 'Balancing between autonomy and dependence. Family farming and agrarian change in Lower Austria, 1945–1980', *Contemporary Austrian Studies* **21**, pp. 385–404.

Laurent, C., S. Cartier, C. Fabre, P. Mundler, D. Ponchelet, and J. Remy. (1998), 'L'activité agricole des ménages ruraux et la cohesion économique et sociale', *Économie Rurale* **224**, pp. 12–21.

Laurent, C., and J. Remy. (1998), 'Agricultural holdings: Hindsight and foresight', *Etudes et Recherches des Systemes Agraires et Developpement* **31**, pp. 415–430.

Laurent, C., J. C. van Rooyen, P. Madikizela, P. Bonnal, and J. Carstens. (1999), 'Household typology for relating social diversity and technical change. The example of rural households in the Khambashe area of the Eastern Cape province of South Africa', *Agrekon* **38**, pp. 190–206.

Leder, S., and C. E. Sachs. (2019), 'Intersectionality at the gender-agriculture nexus. Relational life histories and additive sex-disaggregated indices', in *Gender, Agriculture and Agrarian Transformations. Changing Relationships in Africa, Latin America and Asia*, edited by C. E. Sachs, 75–92, Routledge, Abingdon, UK.

Leeuwis, C. (1993), *Of Computers, Myths and Modelling: The Social Construction of Diversity, Knowledge, Information and Communication Technologies in Dutch Horticulture and Agricultural Extension*, LUW, Wageningen, the Netherlands.

Le Roy Ladurie, E. (1980), *Montaillou: Cathars and Catholics in a French Village, 1294–1324*, Penguin, Harmondsworth, UK.

Louah, L. (2020), *The Nature of Farming, Agricultural Efficiency Revisited through the Lens of Diverging Survival Strategies of Farms within the same Micro-Territory, Wallonia, Belgium*, Ph.D. thesis, Université Libre de Bruxelles, Brussels.

Meng, X. (2014), *Feminization of Agricultural Production in Rural China. A Sociological Analysis*, Ph.D. thesis, WASS, Wageningen University, Waginengen, the Netherlands.

Milone, P. (2004), *Agricoltura in Transizione: La Forza dei Piccoli Passi, un Analisi Neo-istituzionale della Innovativitá Contadina*, Ph.D. thesis, Wageningen University, Wageningen, the Netherlands.

Monllor, N., and T. Fuller. (2016), 'Newcomers to farming: Towards a new rurality in Europe', *Documents d'Anàlisi Geogràfica* **62** (3), pp. 531–551. http://dx.doi.org/10.5565/rev/dag.376

Morel, K., M. San Cristobal, and F. Léger. (2017), 'Small can be beautiful for organic market gardens: An exploration of the economic viability of French microfarms using MERLIN', *Agricultural Systems* **158**, pp. 39–49.

Morel, K., M. San Cristobal, and F. Léger. (2018), 'Simulating incomes of radical organic farms with MERLIN: A grounded modelling approach for French microfarms', *Agricultural Systems* **161**, pp. 89–101.

Paredes, M. (2010), *Peasants, Potatoes and Pesticides: Heterogeneity in the Context of Agricultural Modernization in the Highland Andes of Ecuador*, Ph.D. thesis, Wageningen University, Wageningen, the Netherlands.

Petersen, P. (2015), 'Hidden treasures: Reconnecting culture and nature in rural development dynamics', in *Constructing a New Framework for Rural Development (Research in Rural Sociology and Development*, Vol. 22), edited by P. Milone, F. Ventura, and J. Ye, 157–194, Emerald, Bingley, UK.

Petersen, P. (2017), *Arreglos Institucionales para la Intensificación Agroecológica; Una Mirada al Caso Brasileño desde la Agroecología Política*, Ph.D. thesis, Universidad Pablo de Olavide, Sevilla, Spain.

Petersen, P., and L. M. Silveira. (2017), 'Agroecology, public policies and labor-driven intensification: Alternative development trajectories in the Brazilian semi-arid region', *Sustainability* **9** (4): pp. 543–569. doi: 10.3390/su9040535.

Ploeg, J. D. van der. (1990), *Labour, Markets and Agricultural Production*, Westview Press, Boulder, CO.

Ploeg, J. D. van der. (2003), *The Virtual Farmer: Past, Present and Future of the Dutch Peasantry*, Royal van Gorcum, Assen, the Netherlands.

Ploeg, J. D. van der, D. Barjolle, J. Bruil, G. Brunori, L. M. Costa Madureira, J. Dessein, Z. Drąg, et al. (2019), 'The economic potential of agroecology: Empirical evidence from Europe' *Journal of Rural Studies* **71**, pp. 6–61. doi:10.1016/j.jrurstud.2019.09.003.

Ploeg, J. D. van der, A. Long, and J. Banks. (2002), *Living Countrysides: Rural Development Processes in Europe: The State of Art*, Elsevier, Doetinchem, the Netherlands.

Ploeg, J. D. van der, S. Miedema, D. Roep, and R. van Broekhuizen. (1992), *Boer Bliuwe Blinder ... Bedrijfsstijlen, Ondernemerschap en Toekomstperspectieven*, Bedrijfsstijlenstudies 6, AVM/CCLB and Vakgroep Ontwikkelingssociologie LUW, Leeuwarden/Wageningen, the Netherlands.

Ploeg, J. D. van der, and J. Ye. (2010), 'Multiple job holding in rural villages and the Chinese road to development', *The Journal of Peasant Studies* **37** (3), pp 513–530.

Reijntjes, C., B. Haverkort, and A. Waters-Bayer. (1992), *Farming for the Future: An Introduction to Low-External Input and Sustainable Agriculture*, Macmillan/ETC ILEIA, London/Leusden.

Roep, D. (2000), *Vernieuwend Werken, Sporen van Vermogen en Onvermogen (een Socio-materiele Studie over Vernieuwing in de Landbouw Uitgewerkt voor de Westelijke Veenweidegebieden*, Ph.D. thesis, Landbouw Universiteit Wageningen, Wageningen, the Netherlands.

Rooij, S. J. G. de. (1992), *Werk van de Tweede Soort, Boerinnen in de Melkveehouderij*, Van Gorcum, Assen, the Netherlands.

Rooij, S. J. G. de, E. Brouwer, and R. van Broekhuizen. (1995), *Agrarische Vrouwen en Bedrijfsontwikkeling*, Studies van Landbouw en Platteland 18, Landbouwuniversiteit, Wageningen, the Netherlands.

Rosset, P. M., and M. Altieri. (2017), *Agroecology: Science and Politics*, ICAS Small Book Series, Fernwood Publisher, Vancouver. http://practical-action.org/10U-588QE-B869K6V111/cr.aspx (international edition).

Rosset, P. M., B. M. Sosa, A. M. Roque Jaime, and D. R. Avila Lozano. (2011), 'The campesino-to-campesino agroecology movement of ANAP in Cuba: Social process methodology in the construction of sustainable peasant agriculture and food sovereignty', *Journal of Peasant Studies* **38** (1), pp. 161–191.

Sachs, C. E. (2019), 'Gender, agriculture and agrarian transformations', in *Gender, Agriculture and Agrarian Transformations. Changing Relationships in Africa, Latin America and Asia*, edited by C. E. Sachs, 3–7, Routledge, Abingdon, UK.

Sachs, C. E., and E. Garner. (2017), 'Gender transitions in agriculture and food systems', in *Gender and Rural Globalization: International Perspectives on Gender and Rural Development*, edited by B. B. Bock and S. Shortall, 253–271, CAB International, London.

Scetri, R., ed. (2001), *Novitá in Campagna: Innovatori Agricoli nel Sud*, ACLI Terra/IREF, Rome.

Seuneke, P., and B. B. Bock. (2015), 'Exploring the roles of women in the development of multifunctional entrepreneurship on family farms: An entrepreneurial learning approach', *NJAS - Wageningen Journal of Life Sciences* **74–75**, pp. 41–50.

Sevilla Guzman, E. (2007), *De la Sociologia Rural a la Agroecologia: Perspectivas Agroecologicas*, ICARIA Editorial, Barcelona, Spain.

Sherwood, S. G. (2009), *Learning from Carchi, Agricultural Modernisation and the Production of Decline*, Ph.D. thesis, Wageningen University, Wageningen, the Netherlands.

Shortall, S. (2006), 'Gender and farming: An overview', in *Rural Gender Relations: Issues and Case Studies*, edited by B. B. Bock and S. Shortall, 19–26, CAB International, London.

Shortall, S. (2017), 'Rurality and gender identity', in *Gendered Rural Globalization: International Perspectives on Gender and Rural Development*, edited by B. B. Bock and S. Shortall, 162–169, CAB International, London.

Teather, E. (1996), 'Farm women in Canada, New Zealand and Australia redefine their rurality', *Journal of Rural Studies* **12** (1), pp. 1–14.

Ventura, F. (1995), 'Styles of beef cattle breeding and resource use efficiency in Umbria', in *Beyond Modernization: The Impact of Endogenous Rural Development*, edited by J. D. van der Ploeg and G. van Dijk, 219–232, Van Gorcum, Assen, the Netherlands.

Whitley, H., and K. Brasier. (2020), 'Women farmers and women farmer's identities', in *Routledge Handbook of Gender and Agriculture*, edited by C. E. Sachs, L. Jensen, P. Castellanos, and K. Sexsmith, 360–369, Routledge, London.

Wiskerke, H. (1997), *Zeeuwse Akkerbouw Tussen Verandering en Continuïteit: Een Sociologische Studie naar Diversiteit in Landbouwbeoefening, Technologieontwikkeling en Plattelandsvernieuwing*, Ph.D. thesis, Landbouwuniversiteit Wageningen, Wageningen, the Netherlands.

Ye, J. (2002), *Processes of Enlightenment: Farm Initiatives in Rural Development in China*, Ph.D. thesis, Wageningen University, Wageningen, the Netherlands.

5 Farm development trajectories and agricultural growth

Overview: Main concepts discussed in Chapter 5

Growth patterns: direction, rhythm, mode and nature
Proportionate/disproportionate growth
Labour-driven/technology-driven growth
Inclusive growth/selective growth
Malleability
Differential outcomes (or macro effects)
Dualism
Policy scenarios
Entrance barriers
Constant data sets
Inflow
Outflow
Throughflow
Social logic
Economic logic
Vanguard approach to agricultural development
Redistributive and egalitarian reform
Storytelling
Differentiation
Institutional mediation
Balances governing the organization and development of the farm
Drudgery
Utility
Re-assessing balances
Reversal of curves
The power of sheer numbers

Farming styles are not static entities. Instead, they are specific *flows of activities through time*. Each farming style has its own particular development trajectory that is characterized by a style-specific *direction*, *rhythm*, *mode* and *nature*. Beyond that, all styles are confronted with a context that is subject to change, sometimes dramatically: institutions change, markets can be volatile, new technologies emerge,

DOI: 10.4324/9781003313274-5

every now and then the climate and other ecological conditions bring unexpected surprises, patterns of consumption vary – and farmers need to respond to all of these contextual changes. Their ways of doing so, however, will often sharply differ from style to style. That is to say, the responses to contextual changes are style specific.

The variations in *direction* can be grasped via the concepts introduced in Chapters 1 and 2: *intensification* and *scale enlargement*. Intensification means that levels of production per object of labour are increased. This shows up as yields going upwards (or, in the case of intercropping: as total calorific or monetary production/ha going upwards). Scale enlargement means that the quantity of labour objects per unit of labour force is augmented. Both intensification and scale enlargement are the outcome of a strategic ordering of the farm labour process. This means that one cannot change overnight from the one to the other or vice versa. Co-production is moulded in contrasting ways, specific skills are generated and specific technical artefacts are incorporated into the farm; thus, a kind of path dependency is generated.

As shown in Figure 1.2, intensification and scale enlargement are trajectories that can co-exist in one and the same institutional, economic and ecological setting; at the same time, there might be trajectories that combine the two.

The *rhythm* of farm development can also vary considerably. In farmers' languages (whether in Italy, Peru, the Netherlands, Brazil or China) this is grasped by the notion of growing in a *step-by-step* way versus the notion of growing through *jumps*. This runs more or less parallel to scientific categories that distinguish between proportionate growth (levels of growth that are proportional to the magnitude of the farm) and disproportionate growth. Here, again, different modalities of growth are rooted in the farm labour process, just as they impact very strongly on them. Step-by-step growth is typically rooted in, and limited by, the capacity of farms to generate the savings that are invested in further development (or to dedicate labour into labour investments; see Figure 1.6 for a schematic overview). Setting the balances within and between farm and family is also crucial (see previous chapter). The prevention of risks and the willingness to maintain autonomy and preserve the patrimony ('keeping the name on the land' as it was called in a classical study from Ireland) are important drivers (Arensberg and Kimball, 1948). Disproportional growth, which often occurs 'as a jump', typically relates to the implementation of new, and expensive, technological profiles. Because the latter involve high financial costs (which often require access to credit), enlargement of the farm is needed in order to be able to make the investment profitable. But in many instances it is as much the conviction of being engaged in a battle for the future ('only a few farms are going to survive, and these are the large farms') that makes farmers opt for such 'jumps forward'. In the short term, the financial costs coming with such a growth pattern can put the family income under considerable pressure; the hope is that it will improve in the medium or long term.

The *mode* of farm development also refers to the farm labour process. Depending on the already established routines, farm development can occur as *labour driven* or as *technology driven* (this applies especially to the nature of the process of intensification at farm level, but it also pertains to issues of scale enlargement).

The difference between the two can be found in the relative weight of variable and fixed costs (of bought inputs and the depreciation of technological artefacts). Through labour-driven development trajectories the labour income can be maintained (or even improved), whilst technology-driven solutions often come with a decrease of labour incomes (and therefore with the need for further scale enlargement). These differences have gained again enormous relevance with the rise of (proto) agroecological movements (see Table 4.3, which presents data on grassland-based farms in Brittany).

As an aggregate phenomenon (that is, by taking all farm development processes together), the *nature* of agricultural growth can be structured in sharply differing ways. Growth can unfold as an *inclusive* process that gives room to all farms and all farming people to play a role in it – or it can follow a so-called vanguard approach. In the latter case, the overall process of agricultural growth is selective: specific categories of farms and farmers and often specific geographical areas are excluded and driven to the wall. The role of the state is often decisive in this selective process.

The search (or struggle) for adequate incomes in farming can follow, and will materialize in, different trajectories – the proverbial multitude of roads leading to Rome. Malleability not only applies to the farm labour process (and the resulting emergence of a specific process of production) but it also and equally applies to the *flow of activities through time.*

Taking the theoretically possible differences *in direction, rhythm, nature* and *mode* of farm development (at the level of the single units of production) together allows for a wide array of possibilities. Empirical research mostly encounters a specific range (see Figure 1.2). This range not only confirms the malleability of co-production and co-evolution but also reveals the room for manoeuvre that exists at both the micro and macro levels – the latter because these differential farm development patterns might (explicitly or implicitly) also become the object of agricultural policies and social movements.

Recently, the repertoire of possible farm development trajectories has been enlarged by the emergence of many different farm-based rural development activities that offer new possibilities to build on already developed and available mechanisms but also bring new mechanisms and potentialities, sometimes with considerable impacts. I will discuss these new activities in Chapter 7. In this chapter, I will focus on the malleability of farm development (and food production) *tout court.*

Differential outcomes

Due to the varying structuration of the farm labour process, the different styles of farming come with style-specific outcomes. This is illustrated in Table 5.1, which gives some macro outcomes at the level of the Dutch province of Friesland, assuming that the entire provincial milk quota (at that time around 1.8 billion kg of milk per year) were to be produced by one particular style. What would the total amount of agricultural land needed be *if* all farms produced according to the style of farming economically? The answer to this question was then compared to the result *if* other styles were universally adopted. In a situation, for instance,

Table 5.1 Differential macro effects of different farming styles (Friesland 1990)

	Amount of land needed (in ha)	Number of dairy cows needed	Labour input needed	Total number of farms	Total N surplus (in millions of kilograms)
High cattle density	117,000	213,000	4,754	2,515	51.1
'Cow men'	132,000	216,000	4,114	2,154	43.6
'Economical farmers'	144,000	232,000	5,065	3,107	59.0
'Large farmers'	145,000	246,000	4.377	2,200	51.3
'Machine men'	154,000	240,000	4,255	2,880	61.6
'Breeders'	167,000	235,000	6,440	4,000	53.6

Data from Vakgroep Rurale Sociologie (1993:13).

in which much land was required for nature conservation (whilst the total production of milk were to remain the same), such a comparison would suggest that the style of farming that puts cattle density centre stage would be the best option: this style would only make a spatial claim of 117,000 hectares. This is a considerable contrast with the claim on space inherent to the style of the breeders (167,000 hectares). The contrasting macro effects of different styles also come to the fore if we focus on the total number of labour units (full-time equivalents, FTEs) needed (ranging from just over 4,000 to almost 6,500), the total number of farms (ranging from 2,154 to 4,000) and environmental pressures. The style of farming mechanically (the one of 'machine men') would render an overall N surplus (at the level of the province) of 61.6 million kg of nitrogen per year. If, however, the style of the 'cow men' (in which fine-tuning is central) were to be generalized, this surplus of N would be 29% lower (43.6 instead of 61.6 million kg of pure N).

Thus, the malleability of agriculture as a whole comes to the fore – at least potentially. That is to say, different styles of farming help co-shape the macro structure of agriculture. If the relative weight of the different styles vis-à-vis each other shifts, these macro outcomes will also change.

It is possible, of course, to widen the range on which the comparisons are based. An interesting example is the comparison between dairy farming in Friesland (oriented towards delivering milk for industrial transformation) and dairy farming in Emilia Romagna (oriented towards the production of Parmesan cheese). This comparison (summarized in Table 5.2) has played, at certain moments, an

Table 5.2 Widening the comparison of agricultural systems (Broekhuizen and Ploeg, 1999)

1.8 billion kg of milk	
Bulk trajectory Friesland	*Quality trajectory Emilia Romagna*
3,500 dairy farms	8,000 dairy farms
70 milking cows per farm	30 milking cows per farm
1.6 labour units per farm	3.1 labour units per farm

Incomes per labour unit are more or less equal

important role in discussions within the European Commission, especially when it came to issues of employment and generation of incomes in peripheral economies (Broekhuizen and Ploeg, 1999; Roest, 2000; see also Ploeg and Saccomandi, 1995).

Evidently, the scope of comparisons such as those included in Tables 5.1 and 5.2 cannot be taken to suggest that particular styles of farming can easily be generalized over larger areas or even be 'transplanted' from one particular area to another. There are far too many linkages with institutional context, history, ecology, skills and identities. But these comparisons make it clear that there is *malleability*. Farming can be moulded and remoulded in different ways. There is not just one trajectory but a wide range of possibilities, some of which will materialize whilst others will get blocked.

Such insights into the malleability of farming are a crucial pre-condition for the elaboration of agricultural policies and also for the setting of agendas of social movements operating in and around agriculture. It is precisely this *malleability* – that is, *not* being solely determined by markets and technology – that opens up the needed socio-political debate: What is farming meant for? What are the objectives? How should it relate to nature and wider society? Who are the ones supposed to do farming? How should farming be done, and how should the efforts of those engaged in it be remunerated? Where is it to take place? Why is it that in many situations farming differs from what it could potentially be? Why does it so often run counter to nature (and the required sustainability) as well as to the needs and prospects of the actors involved and societal expectations? These questions are even more pertinent in areas (such as the European Union, EU) where much farming is heavily dependent on subventions and where its trajectory should be subject to broader societal values and aspirations. In Chapter 6 I will probe deeper into such questions or, more precisely, suggest some concepts and methods that might help to tackle these issues.

Agricultural growth as aggregate phenomenon

Agricultural growth in aggregate terms (whether measured in terms of total production, total income generated, total contribution to the development of gross domestic product [GDP], etc.) is a seemingly simple but, in reality, a highly complex outcome of the many different, and partially contradictory, development trajectories that exist at the micro level. Discussing the interrelations between farm development trajectories (at the micro level) and agrarian growth (at the macro level) brings us face to face (once again) with the complexities of micro-macro relations in agriculture (see Chapter 1).

It is possible to use more dynamic approaches (than the simple comparisons shown in Tables 5.1 and 5.2) to grasp these differential development trajectories. Linear programming (LP) is a much used technique for forecasting (for technical details, see Dorfman et al., 1987), but other, more compound forms of modelling and/or so-called fuzzy logic can be used as well (Hennen, 1995).

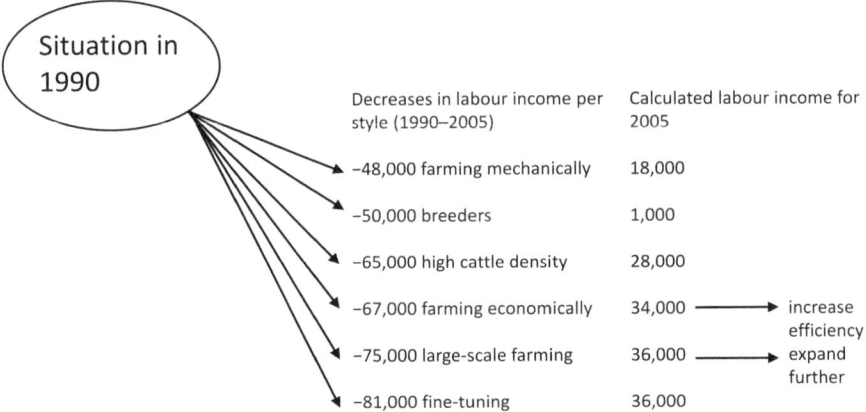

Figure 5.1 The likely development of labour income in different styles of farming, under the assumption of a policy trend scenario (1990–2005) (Antuma et al., 1993).

Figure 5.1 summarizes an application of LP to a set of 300 dairy farms in the province of Friesland, covering the period 1990–2005. This study (realized in 1992–1993) was requested by the province of Friesland and the three provincial farmers' organizations with the aim of providing empirically grounded forecasting for the Second Agricultural Development Plan of Friesland. The study was done by a team of economists, sociologists, agronomists and animal scientists from Wageningen University, together with experts from the National Information and Knowledge Centre (IKC) and leaders of farmers' organizations (see Antuma et al., 1993; Vakgroep Rurale Sociologie et al., 1993).

The trends shown in Figure 5.1 express the calculated development of labour incomes under different farming styles (using LP) over the 15 years following 1990. This was done under the assumption that existing policies would be continued and develop according to the logics that were already built into them. Thus, a further 10% reduction of milk quota was assumed together with a tightening of different environmental restrictions (primarily regarding N emissions).

The information summarized in Figure 5.1 firstly highlights the *differential* effects induced by this policy trend scenario. Labour incomes are *unequally* affected (ranging from minus 48,000 to minus 80,000 Dutch guilders[1]) just as the final outcomes (the labour incomes in 2005) were highly differentiated (and running from just 1,000 guilders/year to 36,000).

Secondly, it is somewhat surprising that the calculated labour income for the farms operated in an economical way (low external input levels, highly efficient use of internal resources, modest and proportionate growth levels) would be (in 2005) more or less the same as that of the far larger farms operated according to the logic of ongoing expansion and intensive farms in which fine-tuning is the central strategic beacon. The average (economic) size of the style of farming economically was (at that time) 513,000 kg of milk per year, whilst it was nearly 800,000 kg of milk in the style centred around ongoing expansion. Nonetheless, the calculated labour incomes for 2005 were nearly identical (this strongly reflects the comparison of low-cost and hi-tech farms discussed in Chapter 3).

Thirdly, the calculations indicated that the negative tendencies could be countered, to a degree, through style-specific responses. In the style of farming economically, further improvements in the use efficiency of internal resources could provide (relative) relief,[2] just as further expansion would allow for some improvement in the style of the large (and growing) farmers.

The calculations were repeated for scenarios that differed from the policy trend scenario. These were a 'free trade scenario' and a 'diversity scenario'. The free trade scenario assumed a modification of the quota system (a protected quota for supplying the European market and liberty to go beyond this in order to export dairy products to areas outside the EU), a decrease in the average milk price (protected prices for the internal market, fluctuating price for exports) and an associated decrease in the prices for meat, calves and heifers – in short: a 'mild' form of free trade. The diversity scenario focussed on partly recombining dairy and arable production and modest levels of multi-functionality (adding complementary economic activities to the farm such as landscape management; see Chapter 7).

Again, a range of interesting outcomes was generated. For the sake of brevity, I limit myself to discussing two opposite poles. Table 5.3 summarizes the labour incomes that might be obtained in 2005, under different agro-political regimes, for two contrasting styles of farming: low-cost and ongoing expansion.

These outcomes reveal the sensitivity of farming to (changes in) agricultural policy. Agro-political regimes (McMichael, 2009)[3] do matter: they impact directly on farms and farming families. That is why they are so frequently contested (and re-negotiated). The outcomes also reveal that this impact is differential: it differs for different farming styles (and that is why farmers so often *disagree among themselves* about agro-political changes). Finally, Table 5.3 once again highlights the enigmatic fact that it is *not* the large and expanding farms that fare the best under free trade conditions. It is, instead, the style of farming economically that is better equipped to face the deteriorating circumstances implied by free trade conditions.

As a matter of fact, this may be the central politico-economic contradiction of our times (I am talking about the early 2020s). Under the previous conditions of a protected market and a stable price regime, the style of large-scale and highly intensive farming could develop and prosper. Stable prices allowed for long-term

Table 5.3 Differential outcomes of two contrasting farming styles under different policy scenarios (predicted levels of labour income in 2005, in guilders) (Vakgroep Rurale Sociologie, 1993)

	Style of farming economically (low cost)	Style of large-scale farming focussed on ongoing expansion
Policy trend scenario	34,000	36,000
Free trade scenario	31,000	10,000
Diversity scenario	75,000	42,000

planning, high investments, associated debts and ongoing expansion. Such farms were not only 'pampered' by, and through, the main regulatory schemes; they also became the proud beacons (the logo) of the reigning agro-political regimes: they were, indeed, the 'vanguard farms' that were thought to be the main promise for the future. As the largest farms (also in economic size), they were thought to be the most competitive ones and far more able than others to face the harsh conditions of the world market. The involved farmers themselves were also convinced of such prospects. However, once deregulation and globalization became facts of life (and the agro-political regime took on definite neo-liberal contours), it showed – especially at moments of crisis – that the combination of high debts and high levels of external input use, on the one hand, and price volatility and a tightened squeeze, on the other, created a highly fragile and vulnerable constellation. The 'vanguard farms' frequently entered into negative cash flows and 'the best pupils of the class' (this expression is from the French rural sociologist Nicole Eizner, 1985) became highly frustrated and tended towards right-wing populist denials of the specific reality that they themselves had helped to co-create. The irony now is that this fragile component of the farming system is maintained by costly public interventions (hectare-based income payments from the European Union are an important part of the sustaining system) that generate, taken all together, few or no public benefits at all. As early as the 1970s this system was referred to as *sub-ventionierte unvernunft* (subsidized madness; Priebe, 1985). At the interface of the previous modernization-centred policy regime and the current neo-liberal regime, a model has been parachuted into our times that fits neither with the economic nor with the social and ecological conditions of today nor of the decades to come.

In the meantime, the peasant-like components (illustrated here with the style of farming economically) show their potential to face the harsh and awkward situations of today in convincing ways and to provide goods and services (marketed and un-marketed) that are increasingly in demand. However, the main agro-political institutions are mostly unable and unwilling to see this potential.

The most important methodological point here is that the possibility of different policy approaches to agriculture combined with the differential responses to policy that come with different farming styles provide considerable room for manoeuvre and change (at least potentially) at the macro level. This combines with the finding that this variability will result in different outcomes. Such outcomes (at the macro level) are not the simple addition of effects produced at the micro level (i.e., farm level). They are, instead, critically affected by changes between styles.

Let me illustrate this with the dualistic farm structures as we know them in, for example, Latin America with one segment of small-scale but intensive farms (the *minifundia*) and another of large-scale but extensive farms (*latifundia*; see Figure 5.2). If the goal is to increase total production but this is accompanied by a shift of resources from the *minifundia* to the *latifundia* segment, the increases in production might well be neutralized. The combination of changes will result in stagnation at the macro level. The compounded micro-macro interaction thus brings adverse, and often unexpected, effects.

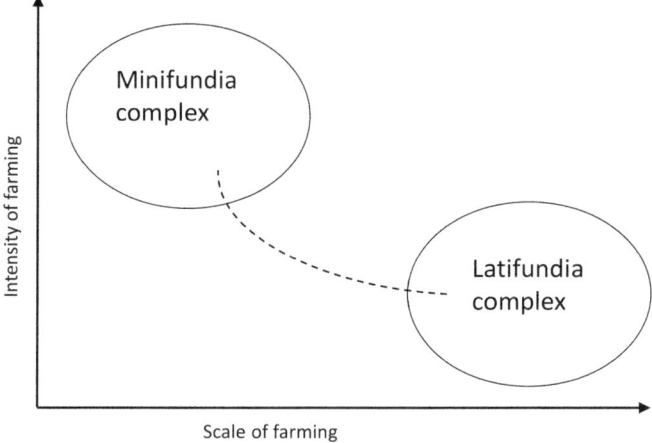

Figure 5.2 The *latifundia/minifundia* complex (author's own elaboration).

Adverse effects can also be noted in the Frisian example. At that time, the increase in total production was excluded by the quota system (although in the free trade scenario a partial increase of production could be *assumed*). In addition, environmental limitations equally hampered production increases at the macro level.

However, there were serious issues in relation to the other dimensions. The different policy scenarios significantly affected the total number of farms in the province, total direct and indirect employment and the total generated labour income. This would bring far-reaching consequences for the strength of the rural economy and the total number of farms. Table 5.4 sums up some of the outcomes at the provincial level (in the next chapter I will discuss such outcomes, intended or not, in terms of externalities).

Table 5.4 The macro outcomes of different policy scenarios for Friesland as a whole (Vakgroep Rurale Sociologie, 1993)

	Situation in 1990	Policy trend scenario 2005	Free trade scenario 2005	Diversity scenario 2005
Total number of farms	5,500	3,550	1,940	4,700
Direct employment (FTE)	9,300	5,300	3,100	7,600
Indirect employment (FTE)	12,000	9,000	5,000	11,000
Labour income (in millions of guilders)	426.2	216.3	114.4	378.5
Agricultural area (in ha)	200,000	155,000	160,000	190,000

Inclusive or exclusive agricultural development?

As Table 5.4 shows, with data on different (potential) development trajectories for Frisian agriculture, the number of farms that would be able to continue their operations under different policy regimes would vary considerably: a range of 1,940 to 4,700 dairy farms is far from negligible (the more so because this figure is associated with employment levels, the strength of the rural economy, multiplier effects, etc.). Some growth trajectories will bring considerable inclusion, but there equally might be considerable exclusion, with up to 60% of farms being closed down and farmers being forced out of agriculture. In such a scenario, 'agricultural development' would inevitably lead to a (further) rural exodus.

Currently, there is a widespread consensus (i.e., an institutionally supported view; see Figure 1.7) that agricultural development occurs as, and through, exclusion: small and supposedly inefficient farms are eliminated, after which their resources are re-allocated into large, efficient and growing farms. It is thought that, in this way, total agricultural production will grow (the more so because the large farms are more likely and able to adopt new, 'more productive' technologies that are too expensive for small farms) and the total number of farms will decrease whilst the average farm size increases. In such a scenario (as observed by Chinese scholars), 'capital flows in, labour goes out' (Ye and LeGates, 2013:45). Agriculture becomes more capital intensive and labour is increasingly substituted by capital. This substitution is, according to standard theories, part and parcel of 'modernization', which is also equated to 'structural development' (meaning that development *necessarily* occurs this way because it is structurally determined).

There are many episodes, though, that show that agricultural development can occur in ways that differ strongly from the model assumed in modernization theories.[4] Apart from that, the pace of expulsion and the degree of exclusion can also differ considerably between different countries during the same epochs. Having said this, it remains true that, especially since the 1980s, in most countries, agricultural growth and exclusion have de facto occurred simultaneously – and there are clear reasons for this, some of which I will discuss in Chapter 6. Nonetheless, the hegemonic idea that growing exclusion is seemingly unavoidable does not exclude the (theoretical and practical) possibility of agricultural development occurring as an inclusive process – the more so because the higher the levels of exclusion that we witness, the more need there is for agricultural growth to be transformed into an inclusive process.

Alongside the issue of inclusiveness or exclusiveness, there is another dimension: the openness and accessibility of the agricultural sector for newcomers, displaced persons and/or victims of segregation policies. This openness can be contrasted with the closure that results from the high entry barriers that make it impossible for persons with a non-agricultural background to enter the sector and build a livelihood as farmers. These barriers may lie in high prices for land, the difficulty in linking with the networks needed for commercialization, the hostility of settled farmers towards newcomers, a lack of skills, etc. All of these factors can turn the agricultural sector into a 'closed shop'.

Who contributes what?

The compound nature of agricultural processes of growth and development is highlighted further if we ask 'who contributes what' to the overall process of growth (note that this again relates to the complex and often contradictory dynamics of micro-macro relations). We have to be aware, though, that such a question is seldom asked. It has been, as it were, *already set out* in the main modernization theories (and, for that matter, in critical theories inspired by Marxist analysis as well). Within such theories it is self-evident that:

a small farms disappear, whilst large farms further grow and develop, and
b large and growing farms are the engine of agricultural growth; they contribute the most to it, whilst the contribution of small farms is neglectable, nil or even negative.

Empirical data, however, show that, in reality, the dynamics of agricultural development are not that simple and unilinear. The general impression, mostly based on comparing census data, is not untrue as far as the overall trend is concerned (fewer farms whose average size is increasing) but does not inform us about the underlying logics and dynamics. To grasp this, different methods are needed (see Methods Box #10).

Methods Box #10   **Grasping the dynamics of agricultural growth and development**

Empirical development in agriculture is mostly reconstructed through, and represented by, moving average data that mostly cover 10-year periods. They are based on census data (most agricultural censuses are organized every 10 years and entail making a detailed inventory of each farm). The results of these censuses are nearly always reported in terms of different farm size categories. The changes in these categories, noted between the beginning and the end of such a 10-year period, are then interpreted as reflecting the real developmental tendencies. However, what is not known is what *actually* happens during this period, either *within* or *between* the different size categories. It is like reconstructing the course of a river by measuring the flow at two points (A and B) separated, say, 10 km from each other. Doing so provides no information about whether the river meandered strongly or followed a more or less direct trajectory. More important, we still cannot derive, from the difference in flow rate (the volume of water passing per unit of time) between A and B, what really happened between those two points of observation. Maybe a lower flow rate at B is the result of leakages (or a consequence of high levels of water abstraction between the two points). But it is equally possible that between A and B there is a side branch of the river that takes the water flow elsewhere. There might even be a side branch feeding the river and another one diverting the water.

This problem can only be resolved by using so-called constant data sets (that allow us to follow single units during the whole of the considered time period) and adequate methods of statistical analysis that reflect the different, and possibly differential, development trajectories.

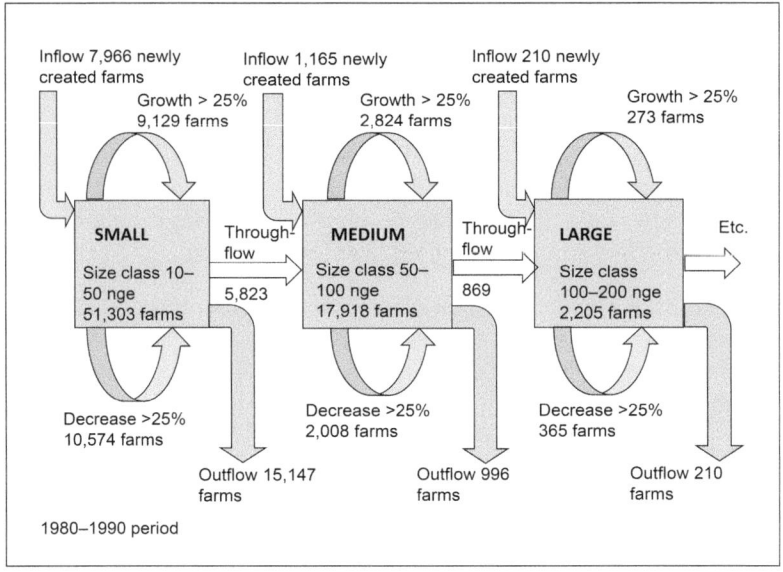

Figure 5.3 Farm development in the period 1980–1990 (constant database) (Ploeg, 2018).

Figure 5.3 reveals some of the mechanisms that, taken together, shape the agricultural growth process. The micro-macro interactions are crucial here. The figure is based on an analysis of a constant data set[5] that comprises all Dutch farms with grazing animals (mostly dairy cows but also beef cows, goats, sheep, etc.). In 1980 these numbered nearly 72,500 farms. Out of these, 51,503 were classified as small (between 10 and 50 NGE),[6] 17,918 as medium sized (50–100 NGE), 2,205 as large (100–200) and the remainder as very large. The constant data set covers the period 1980–1990. *Constant* means, in this context (see also Methods Box #10), that *every single* farm can be followed from year to year. So we are not limited here to comparing a 'moving average', but we can probe into the *micro-level* mechanisms that together make up for the changing averages (that appear at the *macro level*).

We can make several observations about the data contained in Figure 5.3. First, it shows that there is, indeed, a considerable *outflow* among small farms. In the 10-year period considered, 15,147 small farms closed down.[7] But, at the same time, there were also outflows among the medium and large size categories. Farm closure is not limited to, nor unique for, small farms. It also occurs in the categories of medium and large farms, although farm closure is, in relative terms, less prevalent in the latter two categories (5% and 10% respectively as opposed to 29.5% among small farms).

Secondly, the data show that alongside outflow there are *inflows* as well (apart from inter-generational takeovers of the farm). These are *newly* created farms (i.e., *not* farms passing, within the family, to the next generation but farms created *anew*). Inflow occurs in all size categories, but it is especially remarkable that alongside the 15,147 small farms that were closed, another 7,966 were created anew. Thus, there is a kind of *balance* of disappearing and newly created farms (over the decades, this balance has deteriorated, especially as it becomes more and more difficult to establish a farm of whatever size) – and this raises interesting questions about the driving forces behind the *outflow* and *inflow* (or, in theoretical terms, about the actor-structure relations, for here it is one and the same structure that simultaneously generates *differential* effects; i.e., inflow *as well as* outflow).

Thirdly, the figure gives, for each size category, the number of farms that increase their economic size by 25% or more as well as the number of those that see their economic size decrease by 25% or more. Again, this reveals multiple dynamics. Small (and, to a lesser degree, medium) farms are not only, or mainly, stagnating (and/or decreasing their economic size as a prelude to the closure to come): *considerable segments are growing by 25% or more*. Out of all small farms, 9,129 (18%) grew by more than 25%. Notably, in the category of large farms, only 12% grew by 25% or more.

This shows that there are multiple dynamics among all the size categories. Growth is not a unique property of (already) large farms. Nor is closure an exclusive property of small farms. There is, instead, an intriguing combination of trends and categories, and this raises questions as to why, how, when and by whom. The sociology of farming has developed important answers throughout its rich history. I will discuss some of these in the following section on social and economic logics. But first I will discuss another feature shown in Figure 5.3.

It appears, in the fourth place, that there is also through-flow between the different size categories. Some small farms grew so much in the 10-year period that at the end (i.e., in 1990) they belonged to the category of medium farms and some even to large farms. This occurred, in total, on 5,823 farms. This also occurs in other size categories.

Text Box 5.1   Differential but combined processes of growth

The analysis summarized in Figure 5.3 was repeated for the decade 1990–2000, for the period 2000–2004, as well as for the whole period 1980–2004. It was also done for arable farming. The overall conclusion is that there is not such a thing as one single, linear and selective growth process that occurs with smaller farms being closed down and large farms growing further (as hypothesized in modernization theories where such growth is represented as 'structural development').

Indeed, many farms disappear (notably small ones) and the average size of farms is increasing. But this cannot simply be extrapolated to one single process of unavoidable 'structural development'. There are different processes unfolding at the same time here.

A first, and permanent, process, at least in the period analyzed here, is the disappearance of small farms. This process was very much spurred by agricultural policies, agro-industries and banks dealing selectively with farms. The decisive moment mostly, though not uniquely, occurs at the point of intergenerational change: young people consider continuation of farming as too burdensome (also because there are *no* institutional patterns that allow for a smooth passage of the farm to the next generation). The resources that thus come available (land, environmental space, quota, sometimes the farmhouse and barns) are acquired (bought or leased) by other farmers, who can then expand their farms. I refer to this process as *farm closure* (and the subsequent *transfer* of resources to other farms).

A second process, which occurs alongside the first, is the one of substantial and continued farm growth. I refer to this process as *expansion* (especially when it is disproportionate growth). Through hard work (and sometimes due to 'luck'), small farmers are able to develop their farms into medium farms and sometimes even into large or very large farms. To do so, they need to take out considerable loans, thus raising their indebtedness, to induce new jumps in their growth process. The resulting process of growth brings considerable stress, risks, frustrations and uncertainty and can at a certain point lead to farm closure. This expansion is, as such, a Sisyphean process: a lot of drudgery but hardly any enduring satisfaction. 'Everything grew, apart from our income' (as French farmers who followed this route said; INOSYS Réseaux d'Elevage, 2016). The specific causes can differ, but at a certain moment the involved actors lose their faith in the project and sell the farm (the more so because its value is very high). The resources that come available are absorbed by small and medium farms that start their own uphill, Sisyphus-type, battle. So while many large farms disappear, new small farms are created, some of which will again develop into large farms.

Thus we meet another balance (at the meso level): on the one hand, there is a set of small and disappearing farms and, on the other, a set of equally small, but nonetheless growing farms. This particular balance is governed by the encounter and interplay of agrarian policies and emancipatory aspirations. Although the combination of the two processes leads to a reduction in the number of small farms, this combination also makes clear that being small cannot be equated to being doomed to disappear.

Next to *farm closure* and an *expansion* that, in the end, runs counter to its own in-built limits, there is *transition*. This occurs in, and with, farms that are able to escape from the grow-or-disappear dilemma. They remain in the same size category (or only grow or contract modestly) but adapt their price-cost structure and/or develop additional remunerative activities in their farms. Consequently, they improve their incomes whilst the economic size of their farms remains the same. In Chapter 7 I will further outline this process which, over the last two decades, has gained considerable momentum.

Fourthly, there is *inflow*: non-agricultural people becoming farmers. They use the resources that become available through closure of existing farms. Currently, inflow depends much on young people and nearly always goes together with transition (because it is the only possible way forward).

Finally, there is the construction of mega-farms. This occurs as a massive take-over of existing farms and the building of very large farms that represent a deep *rupture* vis-à-vis existing styles of farming.

There are evident interactions between these different processes. It also applies that the balance between them will be affected very much by agricultural policies just as they depend on what society at large is willing to accept.

It is also clear that the mathematical *average* of these processes (ongoing reduction in the number of farms, ongoing increases in farm size) gives a far from satisfactory representation of the different sub-processes. The *average* trend conceals rather than reveals what is really happening.

On social and economic logics

Underlying the complex, if not contradictory, trends and the intriguing relations between trends and size categories, there is the interplay between social and economic logics. These concepts were introduced by Teodor Shanin (1990), building on the work of Alexander Chayanov. From this perspective, we can understand farming styles as distinctive combinations of social and economic logics. The economic logic resides in the domain of markets and the social logic in the domain of family and community (see Figure 1.5 in Chapter 1).

Economic logic refers to economic relations and their dynamics. This logic comes to the fore if and when the relations, tendencies and expectations reigning in the markets are actively projected to the farm and translated into a plan for action. The organization and development logic of the farm are defined by the markets. If, for instance, farm products are relatively well priced (especially when compared to cost levels and their development over time) and the markets are able to absorb growing levels of production, economic logic both allows for, and spurs, growth at the farm level (resulting in, e.g., the through-flow shown in Figure 5.3). But, of course, economic logic can also operate the other way around: low and decreasing prices, rising cost levels and the impossibility of obtaining the additional resources that would allow the farmer to react to these adverse conditions may together induce a decrease in farm income, a marginalization of the farm and a de-activation, firstly decreasing the economic size and then closing the farm altogether (the outflow shown in Figure 5.3). In short: economic logic can push a farm in different directions.

The *social logic* refers to non-economic circumstances and drivers. The aspiration to construct, in a step-by-step manner, one's own and independent farm might be a very strong driver that translates first as inflow and then as further growth and finally possibly as through-flow, just as the willingness to develop an attractive farm for the next generation to take over can be a main driver for farm development. In such cases drudgery is taken for granted or seen as justified. But, just as in the case of economic logic, the social logic can operate in other directions as well. Running a large farm can bring considerable stress (not only in the farm but in the farming family as well) and may inspire a process of contraction that implies downsizing the farm (or even its closure).

Social and economic logics can combine and intertwine in a myriad of ways and induce a (re-)consideration of the farm and the potentials it entails (for better or worse), and this applies to all farms regardless of the size category to which they belong or farming style they pursue. This does not mean that the study of farm development comes down to methodological individualism. It means that complex balances and the way they relate to farm sizes (and farming styles) need to be put centre stage.

Figure 5.4 (inspired by the work of Teodor Shanin, 1990:218) compares the conventional view of the dynamics in the agricultural sector (summarized above as the modernization view) with the far more complex tendencies observed in empirical reality (shown in Figure 5.3). In both small and large farms, the interplay

Conventional view

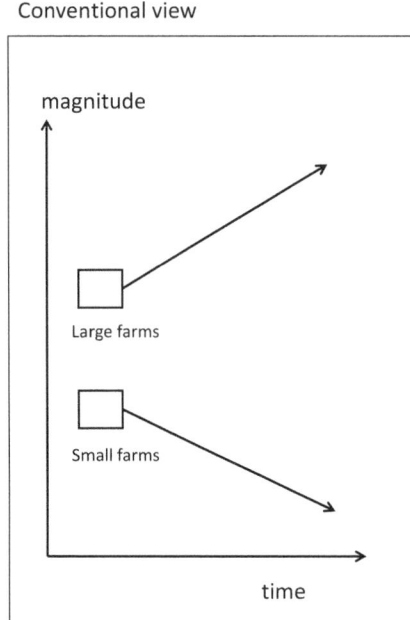

Empirical reality

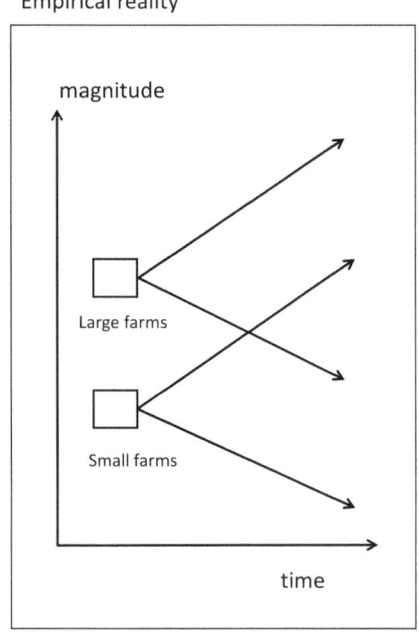

Figure 5.4 From the conventional view to empirical reality (author's own elaboration).

of social and economic logics produces processes of growth and contraction, and it does so simultaneously. Now, by adding growth and contraction (per size category), an *average* growth level will emerge. *Differences* in these size-specific averages can be understood as the outcome of *differently structured balances* between positive and/or negative loadings on social and economic logics. Agricultural policies, emancipatory aspirations, town-countryside relations, intra-family relations and poverty and wellbeing – they all play a role in setting these balances.

As a sideline I note that, over the recent decades, modern agronomy (see also Chapter 1) has developed a 'production logic'. It has taken the potential yields (and the technological means for realizing them) as the ordering principle for the organization of farms (combining it, at best, with economic logics). The specification of a 'reproduction logic' (that centres on the continuity of co-production and the living nature entailed in it) – preferably in combination with 'social logics' – is one of the new, and very urgent, challenges for agrarian sciences in the years to come.

Inclusive agricultural growth and development revisited

The data synthesized in Figure 5.3 formed part of a wider study that embraced a 24-year period: 1980–2004. This wider study included a retrospective question: how much did the differently sized farms contribute to the *overall* growth of Dutch agriculture in the period considered? In order to delineate this contribution,

Table 5.5 Net contribution of different size categories to overall agricultural growth 1980–2004 (Ploeg, 2018)

Size category departing from the 1989 situation (in NGE)	Net contribution to total growth in the 1980–2004 period (in NGE)
<50	175,416
50–100	258,913
100–200	37,979
200–400	3,237
>400	119

a subset was created of all farms that operated during this period as a whole, after which the net effect of growth, degrowth and closure was calculated. The results of this exercise are presented in Table 5.5.

It shows that, taking everything together, the small and medium-sized farms contributed far more to agricultural growth than the large, very large and extremely large (or mega-) farms. At first sight, such a finding is counter-intuitive, for small farms only grow modestly, whilst the large farms show the impressive jumps ahead. However, multiplied by their *sheer number*, the overall contribution of the small farms greatly exceeds that of the large farms: 175,000 NGE versus 38,000 – more than four times as much.

The findings summarized in Table 5.5 are highly relevant, I think, for a debate that has already raged for decades in the Global South. In most so-called developing countries there is the evident and urgent need to augment total agricultural production. In more general terms, the world as a whole needs to produce more food to meet the growing needs of an ever-growing population, and this applies especially to the Global South. The central question, then, is how is this to be achieved?

There are basically two contrasting approaches for making agriculture grow. One is *inclusive growth*; the other might be referred to as the *vanguard approach* (which shows up, in practice, as I have argued, as a process of exclusion). Inclusive growth builds on the mathematics of the 'sheer numbers'. It involves all rural producers, especially the peasant population, and builds on their strengths and potentials. Inclusive development introduces measures and mechanisms that unlock the productive potentials entailed in peasant farming (in this respect, it is also an endogenous process) and thus triggers processes of growth that unfold in a step-by-step way, basically through labour-driven intensification. Inclusive growth proceeds as an essentially democratic process and is socially and politically supported and driven by the peasantry. It needs a state that chooses, in its policies, to support, first of all, *the rural, the poor and the peasants*. These are the key words that explain most of the success of agriculture in several Asian countries, such as Japan, Korea, China and Vietnam (as convincingly shown by Henley and van Donge, 2013).

Inclusive growth is often preceded by a redistributive and egalitarian land reform: the available land is equally shared by all rural dwellers (or by all people actively engaged in farming), and the same applies to other important resources: they are evenly distributed. The egalitarian redistribution of land (and other resources) and an associated process of inclusive development are options that are especially relevant for countries with large but poor rural populations and levels of agricultural production that chronically fail to meet the needs of the nation. However, during the last decades, the reigning power relations have moved the pendulum away from options such as inclusive development and egalitarian distribution which are nowadays represented as being dystopias: unworkable concepts that only lead to disaster. However, we should not forget, as the example of China has made crystal clear, that the two can work well and bring unprecedented agricultural growth and a simultaneous alleviation of rural poverty. Secondly, we have to take into account that the more agricultural policies in the Global South follow the vanguard approach, the more exclusion, poverty and food shortages at national level will be seen. Thus, egalitarian land reform and inclusive growth will inevitably reappear on the agenda.

The vanguard approach is grounded on the proposition that peasant agriculture (or the segment of small-scale family farms) is unable to generate growth and that new technologies and organizational models need to be brought in. Growth is thought to be necessarily an exogenous process. It needs to focus on, and start from, those farms (a minority) that are more able than others able to take over and operate the new technologies and to link with the markets. This focus also introduces a spatial bias that shows some clearly delineated growth poles and an ever-growing, *marginalized hinterland*. Nonetheless, the approach appeals strongly to politicians. They can distribute the 'benefits of development' to targeted followers, and the turnkey (or demonstration) projects that are central in this approach seem to suggest that impressive jumps in development will be made. The vanguard approach is strongly associated with 'seeing like a state' (as argued by James Scott, 1998).

The way smallholders are perceived is a kind of a boundary marker. In narratives that propose an inclusive development, the smallholder is not seen as someone to blame (or, as it goes in policy jargon, as someone 'sitting on the fence')[8] but as 'people that are able to produce, even on small pieces of land, a living for themselves and food for others', as Frei Sergio Antonio Görgen (2017), a Brazilian priest strongly linked to rural social movements, likes to argue. It also applies that inclusive development is people centred and decentralized, whilst the vanguard approach is capital centred and centrally directed. The vanguard approach aims at large-scale farming and sees the associated expulsions as unavoidable. It argues that the new employment opportunities will be created in the cities and metropoles – a prospect that seems to be unlikely for the time to come.

Although the two modes of agricultural growth are, at least theoretically, mutually exclusive, in many areas of the world they co-exist, albeit in an uncomfortable way. This is the case in, for example, China (as shown by Donaldson [2011], who compared 'small works' with large-scale development), Latin America (during, e.g., the land reform process in Peru, but currently also in Brazil) and Europe (where agroecology is unfolding alongside continuing scale enlargement).

The amazing thing here is that in one and the same societal (or, if you want: the same politico-economic) setting there are these two strikingly different processes of growth, and this occurs all over the world. This points again, and very clearly now, to the fact that farming and agricultural growth are *not* completely determined by structural factors such as markets, technologies and (agricultural) policies. There definitely are *different* socio-economic drivers just as there are *different* sets of relevant social relations of production.

Differentiation

The notion of differentiation refers to the growing differences between farms. These differences might relate to the magnitude of the farms,[9] the size of farm families, the relations they maintain with food industries (contract farming or not), the class position of farmers[10] (and, consequently, class relations within farms and between farmers) and/or their position in the local community as reflected in storytelling (see Methods Box #11). The imposition of a classification scheme by

---

Methods Box #11   The importance of storytelling

National statistics mostly document moving averages (especially when census-based). They are not organized as *constant* databases; they do not allow for tracing the specific farm biographies that reveal differential movements through time. Sometimes there are notable exceptions as the 'mutation data' from the Netherlands (used in the analysis underlying Figure 5.3 and Table 5.5) and the farm fiches from Austria (Langthaler, 2012). Where such data are lacking, research into this important field needs to be grounded on other data. The use of cadastral data is one possibility (as shown by Marc Edelman and Seligson, 1994). Oral history and, more specifically, the stories about the trajectory of farms is another. I myself came to know several such stories that are mostly considered to be (even by the storytellers themselves) merely anecdotal but which together very well can reveal a (hypothetical) underlying pattern that usually can be checked with other means. Sometimes specific trajectories are condensed in the denominations used to refer to particular trends. Where I was born and raised, the notion of *wrotter* (literally: those who are toiling and moiling, scratching a life from bare earth) was everyday language to refer to those we met above (see again Figure 5.3) as 'inflow': people who have obtained a small farm and then set out to develop it further with all their energy – indeed, regardless of all drudgery, dreaming of the satisfaction that a well-developed farm will render them in the future. Next to that there was another term: *gezeten boeren* (literally 'the well settled farmers'), which referred to large farmers who did not feel any urge for further development; they were happy with what they had and preferred to dedicate their time to pursuits such as horse racing. When we passed a closed farm, the comment often was 'there used to be a *settled farmer* over there, now you see what it led to').

In terms of methods, this means that it is very important to engage in colloquial talk about the territory, the farms located there and the people working and living in these farms and to be very attentive to the wording used. This wording (such as *wrotter* and *gezeten boer*) can give insights into meaningful and probably significant deviations and can be also used as a starting point for further inquiries (targeted interviewing, using old maps and reconstructing the destiny of the farms indicated on them with local experts, reading tombstones, etc.).

Sometimes novels or popular theatre play scripts can render useful insights as well. They sometimes reveal things that remain concealed in the hegemonic narrative. I will illustrate this with a short fragment of an article that reviewed *Stepmother Earth*, a novel by Theun de Vries (who knew peasant life in Friesland thoroughly):

> The two main storylines in *Stepmother Earth* (represented by Jarig and Tjalling respectively) come down to the interaction of economic and social logic. Pride and horses on the one hand, and the economic crisis of the 1880s on the other, make for the collapse of the Wiarda *domus*. [...] Jarig, the eldest son, large farmer and outspoken 'horseman' goes to jail. [...] Tjalling, his younger brother, becomes a rural worker, but through marriage acquires a small farm that he is able to develop. [Thus], against all orthodoxy it is the large farmer who disappears and the rural worker, who is able to develop a small farm into a more prosperous one. [...] It is precisely the unity of these two contrasting trajectories [...] that gives *Stepmother Earth* its narrative splendour. (Ploeg, 2021:1131)

state agencies for agricultural policy may reify (or even induce) a differentiation as well.[11] Differentiation (as a process) may have its roots within agrarian society as such; it may equally well result from the interactions between the farming population on the one hand and the state and capital on the other. Sometimes differentiation is actively co-shaped by parts of the farming population, especially when they opt for the design and implementation of novel types of farm development (see Chapter 7). Other important theoretical distinctions are the ones between class-based differentiation, market- and policy-driven differentiation and demographic differentiation (grounded in the internal composition of the farming family and its changes over time). Different types of differentiation may go together and mutually strengthen the final results.

Institutional mediation

This chapter has discussed farm development as a process that can take different *directions* and *rhythms*, just as it can be grounded in different *modes* and impact differently on the farming population (bringing about, or interacting with, specific processes of differentiation). The chapter has also discussed how, at an aggregate level, agricultural growth is either shaped as an *inclusive process* or follows a *vanguard approach*. The chapter has equally argued that growth is not only attributable to an *economic logic* but that there is a *social logic* as well. Each can operate in different directions (encouraging growth, slowing it down or even reversing it). Thus, differently structured processes of agricultural growth emerge which are definitely *not* situated on different planets but often unfold within one and the same time-and-space location (assumedly governed by one set of so-called structural – i.e., explanatory – factors).

In this last section I will introduce the concepts and approaches that help us to explore and explain agricultural growth as a differential phenomenon – both at

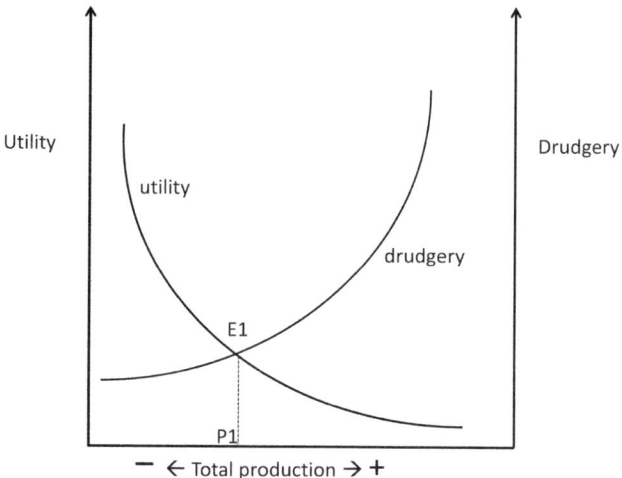

Figure 5.5 Levels of on-farm production as governed by the balance of utility and drudgery (Ploeg, 2013).

the farm level and at an aggregate level. I will do so by taking Figure 5.5 (derived from the work of Alexander Chayanov, 1923, 1925/1966) as a point of departure.

Utility and drudgery are two incommensurable phenomena whose equilibrium determines the functioning of the peasant farm. In Figure 5.5, *drudgery* refers to the extra efforts required to increase total production (or total farm income) by an additional unit. Drudgery is associated with hardship, long working days, sweating under a burning sun, pre-dawn starts and/or working under freezing or sodden conditions. Agricultural work can sometimes be experienced as a joy and as a meaningful activity. However, it also involves physical exertion, often in uncomfortable conditions, and when the work to be done increases, its strenuous nature will be more pronounced. This is what the notion of drudgery captures. Utility represents the mirror side. Utility is the extra benefits (of whatever nature) that come through (additional) increases in production. Utility might emerge as food for the family, as marketable surplus or as a combination of the two, but it may also take the form of monetary income (in Chayanovian theory, this is referred to as *labour income*; that is, all of the monetary benefits minus all monetary expenses related to the production of these benefits). The precise nature of utility will differ according to time and space and individual households. However, in general terms, it equates to the benefits that result from the production of the peasant farm and which satisfy the needs of the peasant family *in both the short and long term.*

The central point of Chayanov's approach is that there needs to be a balance between drudgery and utility. Peasants expand their production to the point where the extra drudgery (needed to increase production by one more unit) equals the extra utility obtained through this drudgery. Going beyond this point would be pointless: it would require more drudgery for less extra utility. A peasant household's assessment of this equilibrium (E1) creates a specific level of production (P1 in Figure 5.5).

Clearly, subjectivity enters into assessing this equilibrium, but this subjectivity does not come down to mere individualism. The active and ongoing assessment of the equilibrium requires *agency*: active, goal-oriented and knowledgeable decision making (Long, 2001) through which the development of the farm is linked with surrounding market opportunities, the wider power structures, class relations, cultural repertoires and local history. The specification of both utility and drudgery is also critically shaped by the composition of, and relations within, the peasant family. Relations between men and women and between different generations are strategic here (for an illustration from China, see Ploeg and Ye, 2016:45–65). The same balance is also crucial for understanding the *dynamics* of the peasant farm, for production in the peasant farm does not necessarily stay at level P1 (in Figure 5.5). *Changes* in the balance of drudgery and utility might lead peasants to actively develop production or to diminish it.

Both the drudgery and the utility curves may move upwards, move downwards or remain 'fixed'. External circumstances, relations within the peasant family and community and the politico-economic, social and cultural relations prevailing in wider society will all have an influence here. The interaction between 'external' forces (located in the wider society) and 'internal' drives (located in the family, village and community) are especially decisive in this respect.

Throughout history, the emancipatory aspirations of different peasantries have been (and still are) very strong drivers of agricultural growth. They were and are a crucial part of what was discussed previously as social logic. If the actors involved in farming want to actively develop their farms (to obtain, in the medium and longer run, an improved income, a higher status and/or better prospects for their children), they seek to shift, as it were, the utility curve upwards (see Figure 5.6).

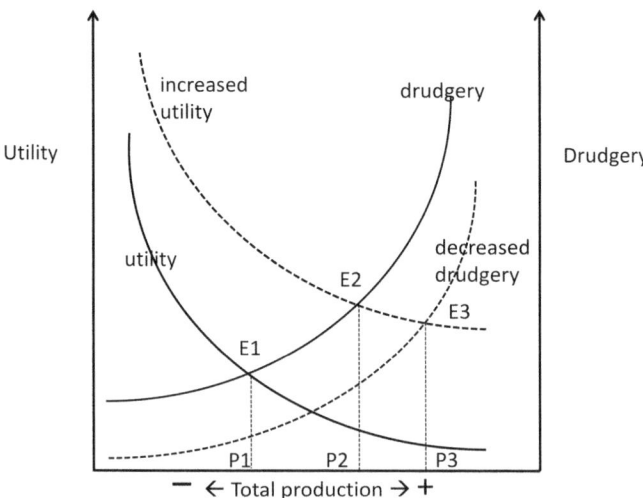

Figure 5.6 Re-assessing the balance of drudgery and utility (Ploeg, 2013).

They are not only producing food and income for today: the definition of utility now critically embraces the future as well (this is often codified in the local cultural repertoires and *calculi* operated by farmers). This orientation towards the future allows the acceptance of more drudgery: working harder, making additional labour investments. The practice of farming is perceived as a struggle: a struggle to move ahead, to improve one's life and (especially if this struggle is linked to the enlargement of the self-controlled resource base) as a struggle for autonomy.

Whether this emancipatory orientation toward the future can be sustained and materialized in building a 'beautiful farm' (see, e.g., Figure 4.1) depends on many societal conditions. In this respect, remunerative and stable prices for agricultural products will help very much. The availability of institutional mechanisms, such as peasant communities that protect against land grabbing, allow for access to, for example, additional land and/or offer security insofar as water supply is concerned can also contribute very strongly. Heritage patterns can help to secure the availability of land as non-commodity for the next generation (but they can equally fail to do so – this depends on their concrete nature). Institutions (such as SAFER in France) that regulate the market for land may equally help to sustain and support the emancipatory aspirations of young farmers and new entrants (or, again, fail to do so). SAFER can pre-emptively buy land (when it becomes available) and then sell it to selected (prospective) farmers who meet specific criteria.[12] Text Box 5.2 summarizes the main criteria used for such 'pre-sales'. In short: the institutional environment can both strengthen the social logic and mediate (at least some of) the economic logic. All of this helps to shift the utility curve upwards (see Figure 5.6).

Text Box 5.2   SAFER: Objectives of pre-sales

The criteria under which SAFER can effectuate a presale are determined by law.

- *Settling*, resettling or helping farmers. This is one of the SAFER's fundamental objectives.
- *Enlargement of existing farms*. This is also a fundamental principle of SAFER, given the context of its creation, when the majority of French farms were 'too small'.
- *Helping farms survive* if some of their land is expropriated for a public works project.
- *Preserving family farms*.
- *Fighting against real estate speculation*.
- *Assisting when a viable farm is endangered* because its buildings and land are separated.
- *Using and protecting forest areas*.

This also applies to the hardship of farming. Cooperation, labour exchange, study groups, etc., can all greatly ease the operation of the farm, and this also applies to institutes for applied agricultural research, the presence of people interested in cooperating with farmers and gaining an (additional) income from it,[13] as well as religious and social movements. They all help push the drudgery line downwards (see Figure 5.6).

Agricultural policies occupy a prominent place among the institutions that impact the direction, rhythm, mode and nature of agrarian growth processes. Agricultural policies are both allocative (meaning that they distribute resources in particular ways) and authoritative (meaning that they 'authorize' particular practices and prospects as being allowed or recommendable, whilst others are sanctioned or proscribed). Finally, agricultural policies are the object and outcome of socio-political struggles: Where is agriculture heading? Who is to benefit? What are the legitimate interests? Thus, agricultural policies strongly impact who can join the process of growth and who will be excluded. One way in which they do so is by unevenly distributing the benefits and costs that come with such policies (see Figure 6.13 in the next chapter) and/or explicitly outlining how agricultural development is to proceed.[14] These two mechanisms (uneven allocation and selective authorization) are often combined. Take, for instance, the oft-used policy instrument of interest subsidies.[15] These lower the relative price of capital vis-à-vis the factor price of labour, and this favours technology-driven processes of growth whilst disfavouring labour-driven intensification. Such mechanisms nearly always come under the umbrella of modernization, defined, at the farm level, as scale enlargement, specialization and a high-tech type of intensification. The specific instruments (and their selective effects) are legitimized by the specific project-for-the-future (nearly always represented as unavoidable) and the latter is, in the end, confirmed by the effects brought by these instruments. Thus, agricultural policies become self-fulfilling prophecies and will be supported by the (select group of) beneficiaries. In this way, (re-)allocation, authorization, and bargaining come together in an unholy alliance.

Institutional patterns matter very much. They impact the satisfaction and prospects that come with farming as much as they impact the difficulty and hardship that seem to be, at least at first sight, inherent to it. Drudgery and utility are institutionally mediated. Therefore, institutions and institutional patterns can substantially and significantly help to shift (as shown in Figure 5.6) the respective curves in such a way that production at farm level is increased, from level P1 to P2 and then to P3.

The opposite can occur very easily as well: if the markets only offer low and disappointing prices and if volatility is so high that it excludes any planning, then the satisfaction curve will probably move downwards (especially when the squeeze is tightened so much that one cannot think of oneself as anything but an idiot working for others without any decent remuneration). In the same vein, the drudgery curve might go upwards: if, for example, there is no or slow access to the internet, farming will become harder (for young people especially). The same occurs with the impression that there is hardly any future in farming: it will move the drudgery curve upwards (note that here drudgery is no longer just about physical hardship but, above all, a question of prospects, the availability of means and attractiveness). With such changes, the equilibrium of satisfaction and drudgery moves and the levels of production go down. In this way, farming is institutionally driven to *de-activation* and *farm closure*. Phrased differently: drudgery and satisfaction are affected (and reshaped) in such a way that in some cases deactivation is made quite probable, whilst in other cases growth of a specific kind is spurred.

Entrepreneurial farming, which is mainly, though not exclusively, located in the Global North might be seen as a special case (see Figure 5.7). It can be hypothesized that the combination of industrialized farming and entrepreneurial reasoning has brought a *reversal of the satisfaction (or utility) curve.* Instead of diminishing utility, the expansion of production will bring increased utility – precisely because it is thought that it is only the large (or very large) farm that offers the prospect of surviving the rough competition that increasingly governs the globalizing markets. Thus, 'survival' is the main utility in the entrepreneurial style of farming.[16] Other forms of utility (as an adequate income) are secondary (and understood as derivative). These farmers view small farms as worthless, because they have no prospect of sharing in the future.

At the same time, though, the drudgery curve is also changed. There is a tendential reduction due to new technologies such as robotization and automation. The possibility of externalization (or 'outsourcing' of considerable parts of the processes of production, notably the most risky and/or labour-intensive parts) also contributes to this reduction. At the same time, though, there is a substantial change: whilst physical fatigue goes down, psychological stress goes up. Small mistakes, suddenly occurring animal diseases, contaminated inputs, unexpected

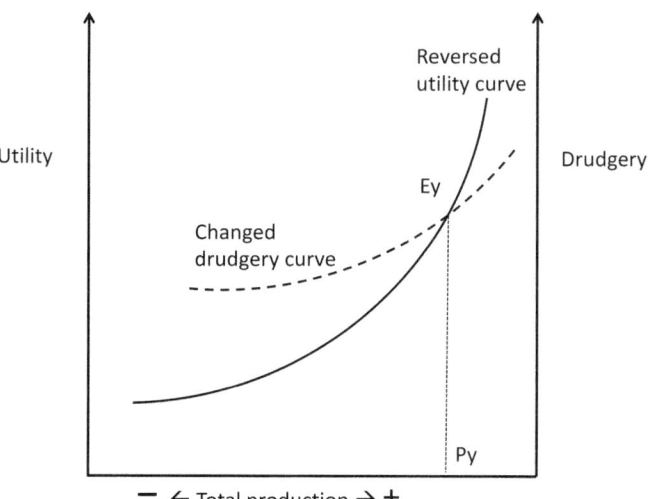

Figure 5.7 Utility and drudgery in large, high-tech, entrepreneurial farms (newly elaborated).

changes in the markets, malfunctioning software and high levels of indebtedness can all have huge consequences due to the magnitude of the farm. The stakes are high and the possibly growing levels of production (resulting from the new, albeit precarious, equilibrium) are ironically combined with higher levels of stress and incertitude. And this is precisely what explains the disintegration of many large farms discussed earlier in this chapter.

Some conclusions

Firstly, agricultural growth is a socially regulated process. It is not a simple derivate of technological development and the 'dull compulsion of the market' (as Karl Marx would have it). The direction, rhythm and modes of agricultural growth are all co-defined by the direct producers and how they perceive the (changing) context and the (wide range of) opportunities it entails. The review of agricultural patterns of growth reveals that there is a social as well as economic logic and that both can operate in different directions.

Secondly, there is, at the aggregate level, a fourth important dimension: the *nature* of agrarian growth (i.e., its degree of inclusiveness). As opposed to the currently dominant vanguard approaches, an inclusive type of development is equally possible, but it brings completely different outcomes. The vanguard approach (centred on the application of the newest technologies in large-scale and expanding farms) is not the only possible engine for high levels of agrarian growth. Inclusive development can render an even higher and more robust increase in food production (in situations where this is most needed) and incomes.

Thirdly, decision making at the farm level (especially when it comes to growth, stagnation or de-activation) can be studied by using the (adapted) balance of utility and drudgery as developed by Chayanov. A redefinition of utility and/or drudgery directly impacts the level of agricultural production realized in and by the farm. The changes in balances are institutionally mediated.

Fourth, the different curves discussed so far (see Figures 5.6 and 5.7) are, of course, 'constructs' (like the 'functions of production' discussed in Chapter 2). They reflect, synthesize and represent the complex interplay of emancipatory drives and institutional patterns and how they impact the decision-making processes taking place within the farming families.

This implies, in the fifth place, that both emancipatory aspirations (especially those that are rooted in local agrarian history and in the specificities of the local politico-economic situation) and the institutional context are integral parts of the social relations of production: they co-shape the processes of production and the way they develop, just as they impact, directly or indirectly, the social distribution of the produced wealth.

Sixth: if the institutions of rural society and the many interfaces between the rural and the urban matter very much (in the sense that they help to facilitate, hinder or block agricultural development), this means that these institutions and interfaces are an important arena for socio-political struggles that define the prospects of farming.

For the sociology of farming, this implies, in seventh place, that it continues to be important to feed these institutions with critical information on the impact they have both on the decisions taken within farming families and, therefore, on the concrete development of agriculture as a whole.

Notes

1. 2.20 Dutch guilders at that time equaled 1 euro.
2. Interestingly, in 1990, the style of farming economically already showed the highest efficiency (calculated as labour income/monetary costs). This ratio was 3.07 for economical farmers and as low as 2.12 for those with high levels of cattle density.
3. I use this somewhat broader concept because it is not only prices that matter – a range of other agro-political instruments (subsidies, spatial reorganization, quality controls, etc.) and fiscal and social policies (regarding, e.g., stimuli for investments or pensions) are also important.
4. This applies to, for example, the agrarian history of the Lowlands (Bieleman, 2010) and China in the post–World War II period (Ploeg and Ye, 2016). It also applies to the family farming sector in Brazil during the 1990s and 2000s.
5. The technical details and institutional background of this exceptional database are given in Ploeg (2018). The analysis of these data was done on request of the European Parliament. Similar findings were presented by Ernst Langthaler (2012) and Marc Edelman and Seligson (1994).
6. NGE (*Nederlandse Grootte Eenheid*, Netherlands Size Unit) is a measure for the economic size of farms. It is a composite figure that includes land, animals and crops, bringing them together into one indicator of economic size. At the time, 50 NGE was equal to some 10 hectares and 15 dairy cows.
7. Technically speaking, not all of these farms really close down. Some decrease and 'diminish' below the threshold level of 10 NGE. Others stop having grazing animals and might change to horticulture or intensive husbandry.
8. 'Farmers for the Future' (EUR 30464 EN), a Science for Policy Report, prepared by the Joint Research Centre (JRC) of the European Commission.
9. Expressed in hectares, or number of animals or in terms of economic size. This mostly comes in categories such as 'small', 'medium' and 'large' farmers (or 'poor', 'medium' and 'rich' farmers).
10. Basically capitalist farmers, independent farmers and farmer-workers (i.e., farmers with few resources who work part of their time elsewhere as a wage labourer).
11. Such classification schemes include categories such as 'viable farms', 'non-viable farms' and 'intermediate farms' (that can still go both ways). Everyday language terms associated with such schemes are, for example, 'stayers'. 'losers' and 'those who are hanging on'.
12. New institutions such as Terre de Liens in France (Rioufol and Wartena, 2011) function in comparable ways but arose from civil society. They are creating newly constructed commons. The 'Conference Report on Access to Land for Farmers in Europe' gives an overview of similar initiatives all over Europe. See 2017_access_to_land_2017_conference_report.pdf (agrarbuendnis.de)
13. Such as villagers participating in the harvests (a phenomenon noted on all continents).
14. A well-known example, in this respect, is the Mansholt Plan that proposed, more than 50 years ago, the outlines of agricultural development in what was then the Common Market. The reduction of the number of farms and farmers by more than 50%, an accelerated growth of the remaining ones and decreased prices were among the main cornerstones. This provoked massive protests (including what is still today the largest ever post–World War II demonstration in Brussels).

15. If an investment project for a farm (mostly elaborated by, e.g., extension services) is approved and accepted by the state, the farm will receive a subsidy on the interests to be paid for the loan that is to finance the investment (sometimes this is combined with fiscal policies making the redemptions of the loan deductible from the income, which implies that less taxes will be paid).
16. With this probably comes as well the pride of being seen by others and by oneself as 'survivor' (if not as a 'winner').

Bibliography

Antuma, S. J., P. Berentsen, and G. Giessen. (1993), *Friese Melkveehouderij, Waarheen? Een Verkenning van de Friese Melkveehouderij in 2005; Modelberekeningen voor Diverse Bedrijfsstijlen onder Uiteenlopende Scenario's*, Vakgroep Afrarische Bedrijfseconomie, LUW Wageningen, Waginengen, the Netherlands.

Arensberg, C. M., and S. T. Kimball. (1948), *Family and Community in Ireland*, 2nd ed., Harvard University Press, Cambridge, MA.

Bieleman, J. (2010), *Five Centuries of Farming: A Short History of Dutch Agriculture 1500–2000*, Mansholt Publication Series 8, Wageningen Academic Publishers, Wageningen, the Netherlands.

Broekhuizen, R. van, and J. D. van der Ploeg. (1999), 'The malleability of agrarian and rural employment – The political challenges ahead', paper presented at the EU Seminar on the Prevention of Depopulation in Rural Areas, Joensuu, Finland, 2 October.

Chayanov/Tschajanow, A. (1923), *Die Lehre von der Bäuerlichen Wirtschaft, Versuch einer Theorie der Familienwirtschaft im Landbau*, mit einem Vorwort vor Dr. Otto Auhagen, professor an der Landwirtschaflichen Hochschule zu Berlin, Verlagsbuchhandlung Paul Parey, Berlin.

Chayanov, A. V. (1926/1966), *The Theory of Peasant Economy*, edited by D. Thorner et al., Manchester University Press, Manchester, UK.

Donaldson, J. A. (2011), *Small Works: Poverty and Economic Development in South-western China*, Cornell University Press, Ithaca and London.

Dorfman, R., P. A. Samuelson, and R. Solow. (1987), *Linear Programming and Economic Analysis*, Dover Publications, New York.

Edelman, M., and M. A. Seligson. (1994), 'Land inequality: A comparison of census data and property records in twentieth-century southern Costa Rica', *The Hispanic American Historical Review* **74** (3), pp. 445–491. doi: 10.2307/2517892.

Eizner, N. (1985), *Les Paradoxes de l'Agriculture Française: Essai d'analyse a partir des Etats Generaux de Developpement Agricole: Avril 1982–Fevrier 1983*, Harmattan, Paris.

Görgen, F. S. A. (2017), *Trincheiras da Resistência Camponesa: Sob o Pacto de Poder do Agronegócio*, Instituto Cultural Padre Josimo, Candiota, RS, Brazil.

Henley, D., and J. K. van Donge. (2013), 'Diverging paths: Explanations and implication', in *Asian Tigers, African Lions: Comparing the Development Performance of Southeast Asia and Africa*, edited by B. Berendsen, T. Dietz, H. Schulte Nordholt, and R. van Veen, 27–50, Brill, Leiden, the Netherlands.

Hennen, W. H. G. J. (1995), *Detector: Knowledge-Based Systems for Dairy Farm Management Support and Policy Analysis: Methods and Applications*, Agricultural Economics Research Institute (LEI-DLO), The Hague, the Netherlands.

INOSYS Réseaux d'Elevage. (2016), *Vaches, Surfaces, Charges ... Tout Augmente sauf le Revenu (Quinze ans de Suivi en Bretagne, Pays de la Loire et Deux-Sévres*, L'Institut de l'Elevage, Paris.

Langthaler, E. (2012), 'Balancing between autonomy and dependence. Family farming and agrarian change in Lower Austria, 1945–1980', *Contemporary Austrian Studies* **21**, pp. 385–404.

Long, N. (2001), *Development Sociology: Actor Perspectives*, Routledge, London.

McMichael, P. (2009), 'A food regime genealogy', *The Journal of Peasant Studies* **36** (1), pp. 139–169. https://doi.org/10.1080/03066150902820354

Ploeg, J. D. van der. (2018), 'Differentiation: Old controversies, new insights', *Journal of Peasant Studies* **45** (3), pp. 489–524.

Ploeg, J. D. van der. (2021), '*Stiefmoeder Aarde* [*Stepmother Earth*] by Theun de Vries (1936)', *Journal of Peasant Studies* **48** (5), pp. 1124–1139. https://doi.org/10.1080/030 66150.2020.1774701.

Ploeg, J. D. van der, and V. Saccomandi. (1995), 'On the impact of endogenous development in agriculture', in *Beyond Modernization, the Impact of Endogenous Rural Development*, edited by J. D. van der Ploeg and G. van Dijk, 10–27, Van Gorcum, Assen, the Netherlands.

Ploeg, J. D. van der, and J. Ye. (2016), *China's Peasant Agriculture and Rural Society: Changing Paradigms of Farming*, Routledge, London and New York.

Priebe, H. (1985), *Die Subventionierte Unvernunft: Landwirtschaft und Naturhaushalt*, Siedler, Bonn, Germany.

Rioufol, V., and S. Wartena. (2011), *Terre de Liens: Removing Land from the Commodity Market and Enabling Organic and Peasant Farmers to Settle in Good Conditions*, https:// www.terredeliens.org (accessed 10 October 2021).

Roest, K. de. (2000), *The Production of Parmigiano-Reggiano Cheese: The Force of an Artisanal System in an Industrialised World*, Royal van Gorcum, Assen, the Netherlands.

Scott, J. C. (1998), *Seeing Like a State: How Certain Schemes to Improve the Human Condition Have Failed*, Yale University Press, New Haven and London.

Shanin, T. (1990), *Defining Peasants, Essays Concerning Rural Societies, Expolary Economies, and Learning from Them in the Contemporary World*, Basil Blackwell, Oxford.

Vakgroep Rurale Sociologie, Vakgroep Agrarische Bedrijfseconomie, Vakgroep Ruimtelijke Planvorming, AVM/CCLB and IKC-Veehouderij. (1993), *It Kearpunt Foarby: Bouwstenen voor het Agrarische Ontwikkelingsplan Friesland*, Landbouwuniversiteit, Wageningen/Leeuwarden, the Netherlands.

Ye, Y., and R. LeGates. (2013), *Coordinating Urban and Rural Development in China: Learning from Chengdu*, Edward Elgar, Cheltenham, UK.

6 Farming, society and capital

Overview: Main concepts discussed in Chapter 6

Externalities (positive and negative)
Market failures
Water use
Water use efficiency
Balance of water availability and demand
Energy use
Use efficiency
I/O balance
Landscapes
Interconnecting landscape elements
Biodiversity
Substitution curves (for labour and machine use)
Agrarian questions
The classical agrarian question
Agricultural production and its triple articulation
Squeeze on agriculture
Skimming
Enlarging material dependency on capital
The agricultural 'treadmill'
Shifting the locus of decision-making
Takeovers of resources
Land and water grabbing
Mega-farms
Biased agricultural policies
Framing technologies

How farming is organized affects wider society and has a strong impact on society at large – in both positive and negative ways. At the same time, societal norms and expectations and legislation strongly condition farming and the ways in which it develops. In this chapter, I shall briefly examine some of the interrelations between the agricultural sector and wider society, paying special

DOI: 10.4324/9781003313274-6

attention to the most important concepts and methods that help to unravel and explain these interrelations. The interplay of farming and wider society covers several dimensions, and there is considerable diversity within each of them. It is important to thoroughly explore the resulting heterogeneity because this can highlight the possibilities of endogenous change at the regional level and/or the level of the agricultural sector as a whole. In this respect, the huge heterogeneity of agriculture represents a substantial part of the symbolic and material wealth of a nation or region. This heterogeneity shows (implicitly or explicitly) the wide array of empirically visible effects that can be generated by, and through, different ways of organizing and developing farming. Thus, it carries a series of potential solutions for problems and opportunities (of whatever nature) that exist at the interfaces between agriculture and society. At the end of Chapter 3 I discussed the multi-year research programme that systematically compared the operation of a low-cost farm and a high-tech farm in the Netherlands. This research programme clearly showed that within the existing economic and institutional context there was considerable space for manoeuvre: the production of the national milk quota could involve between 12,500 and 25,000 viable farms, whilst total sector income could vary from 500 to 1,000 million euros. This shows, in a nutshell, the interplay between society and agriculture in general. Dimensions that are important to society as a whole (e.g., employment, income generated, fossil fuels used, biodiversity, the quality of food, etc.) are critically affected by the way in which agriculture as a whole is organized (which, in turn, is affected by the relative prevalence of different farming styles). The external effects can be both positive and negative (in scientific jargon, these are often referred to as positive and negative externalities). By reorganizing farming, such effects can be re-shuffled, for better or worse. If specific goals are desired, they need not necessarily be 'added' from the outside – they might also be pursued through goal-oriented adaptations *within* the practice of farming. Together such adaptations provide a considerable potential for endogenously changing the interplay between agriculture and society.

The structure of the following sections is simple. In each section the *range* of possible effects on particular dimensions will be discussed. This range is presented as highlighting the boundaries within which agriculture can switch, thus realizing significantly different effects for the wider society. Because employment and income have been discussed already, I will limit myself here to dimensions such as sweet water use, climate change, the landscape and biodiversity, all of which are generally regarded as non-economic aspects of farming.

In terms of transition theory, the presentation of a *range of possible effects* means that particular localities and farms (especially if they engage in forms of action research mentioned in Methods Box #12) might be understood as *field laboratories*: experimenting with ways of doing that might be generalized to the agricultural sector as a whole – thus making the effects that were initially limited to a small sub-sector of agricultural producers more widely available (Stuiver et al., 2003). I hope to show that such effects (which come from internal shifts) can have a considerable impact.

Methods Box #12   Looking for endogenous potential

When problems are recognized within the agricultural sector as a whole, the standard reaction is to look for new technological fixes that might help to remedy them (if needed in combination with new regulatory schemes and/or economic interventions). Thus, exogenous solutions are sought, whilst endogenous solutions receive far less attention or are even ignored. Typical expressions of this tendency can be found in the debates about, and policies for, sustainability in agriculture. Dutch agriculture produces a large surplus of nitrogen. The solutions suggested have generally centred on the introduction of new technologies (closed reservoirs for slurry, the injection of slurry in the subsoil, air-washing machines to reduce ammonia emission, new floors in stables, etc.). Often such technologies become mandatory for all farmers, and for many the associated costs have been a reason to quit farming.

This approach has been contested from the beginning. It was argued, in the first place, that not all farms contributed in the same way to the nitrogen surpluses (and associated ammonia emissions). Secondly, growing numbers of farmers claimed that there were far better endogenous solutions available. A *reorganization* of farming practices (see Figure 6.1) could bring reductions equal to, if not greater than, those promised by the new technologies: by altering cattle feeding, a different, 'improved' slurry could be produced (this, interestingly, reflects farmers' knowledge that centres on feedback and feedforward relations and effects; see Chapter 2). Thirdly, a kind of *reconstruction* of the new technologies was suggested in order to make them far more effective and cheaper (this could be done, for instance, by combining the flow of slurry with a flow of water when spreading slurry over the fields).

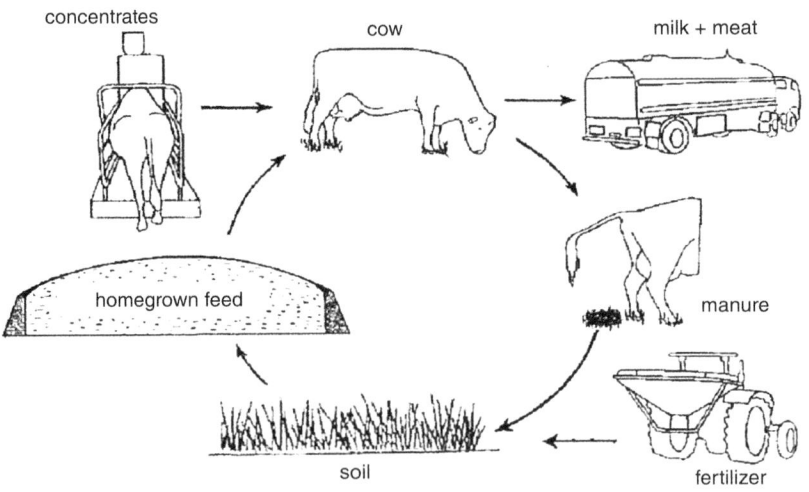

Figure 6.1 Improving the animal-manure-soil-fodder cycle (author's own elaboration).

These alternatives were (with a few exceptions) neglected by the state, because they did not allow for control at a distance. With satellites one can control (or at least have the illusion of control) whether slurry is injected or not (with the required heavy machinery). An internal reorganization of farm practices is far more difficult to control – at least from a distance. In cases like this, the notion of *control* strongly aligns technological developments, capital interests, science and generic state policies. The interests of large and rapidly expanding farms also became part of this unholy coalition. Large farmers generally welcome the demise of small farms (because this frees up resources that they can 'take over' and because they are also far better positioned to integrate the new technologies into their farms).

However, when confronted with such 'sector-wide problems' (as they are called in the dominant discourse), it is important to bear in mind the following steps:

1 Identify the most important indicators related to the suggested problem and check whether the values around them show considerable deviance from the average level. This can help us identify whether one is facing a generic problem or a differentiated one.

2 Relate, if possible, the encountered variability to other variables concerning the farms, in order to explore underlying patterns of cohesion that might explain the variations of the main indicators. The identification of such patterns might help identify levers that will allow for changes in these indicators. It is very important to discuss the derived insights with groups of farmers to check whether the indicators and possible levers are seen as possibly relevant and worthy of joint action. From then on, different steps (or combinations of different steps) are possible.

3 Organize a wider sample (and extend the number of variables used) in order to consolidate the findings of (2).

4 Design a project for action research that aims to systematically induce, try out and assess the effects of the changes suggested in (2), involving a sufficient number of farmers and farms and obtaining, if needed, legal permission to start the action research.

5 Organize feedback mechanisms, right from the start of (4), that make the findings of stages 2, 3 and 4 publicly known. Ensure media attention. Get support from farmers' movements (and, if possible, other social movements) and from representatives in parliament. Involve, if possible, interested authorities and civil servants who are sympathetic to the proposed solutions.

6 Organize a *scientific support group* that will help to address the inevitable backlash from those actors who designed the generic approach that is challenged by, and through, the previous steps.

7 If (4) is successfully realized, try to obtain room within the regulatory system(s) to allow a more generalized application of the newly developed and tested approach (that is, try to construct and consolidate a *strategic niche*).

The experience of the Northern Frisian Woodlands (NFW, a territorial cooperative discussed below) offers an interesting example of such an approach. They railed against the generic prescription of slurry injection, arguing that by combining on-surface application with the production of improved manure (which, technically speaking, has a higher carbon/nitrogen ratio [C/N] and contains less ammoniacal nitrogen) they could achieve similar, if not superior, results. This *improved manure* was obtained by feeding the cattle differently, with silage containing an increased C/N ratio (or, as farmers say: 'with less protein and more structure'). In sum, these farmers proposed a reorganization of one of the main agronomic cycles within the dairy farm: the animal-manure-soil-fodder cycle, represented in Figure 6.1. An improved soil biology (which results from the timely application of appropriate quantities of improved manure) is a key in this process. By improving the cycle and simultaneously reducing the use of industrial concentrates and chemical fertilizers, the N surpluses and ammonia emissions could be (and were) reduced significantly.

The same approach had important additional (and often unforeseen) effects: the improved soil biology positively contributed to biodiversity as well as to grassland yields (see also the discussion in Chapters 9 and 10).

For further reading, see the special issue of *Netherlands Journal of Agricultural Sciences* (NJAS), Vol. 51, nos. 1–2 from 2003.

Water

Adri van den Dries' (2002) study on farmer-managed irrigation systems in northern Portugal,[1] which interestingly refers to these irrigation systems as 'skill-oriented technologies' (4), describes two styles of farming that operate alongside each other. The first one, firmly rooted in history, centres on the breeding of *Barrosã* cattle: high-quality meat that once was even exported to the UK, where it was a favourite dish of the royal family. This breeding is grounded on a complex system for food supply. Each farm has irrigated *lameiros* that are used for harvesting roughage and hay, as well as meadows for pasturing. All farms use the *baldios*: commons that are situated in the mountains. Here farmers also harvest heather and broom that are used in the stables and turned, with the dung, into high-quality manure, which is used to fertilize the fields for summer crops (such as potatoes, cabbages and other vegetables). Irrigation is a key component and helps to optimally tune this style to the local ecological conditions. This is shown in Figure 6.2, which shows how the use of irrigation water, the associated water allocation, the use of *lameiros* and cattle feeding are carefully coordinated during the year and thus allow for a precise match of water availability and water use. In the months of December, January and February, the abundant rains are 'harvested' in the area as a whole and then 'routed', through the irrigation systems, to the fields where the water is 'stored' in the subsoils, allowing for abundant grass growth in spring and a good hay harvest in July and August. Hardly any water gets lost.

Alongside this agro-pastoral style of farming a new style emerged as a result of state interventions. The core of this newly built style is large-scale milk production (using cubicle stalls for the dairy cows) and an associated increase in maize cultivation in order to provide sufficient maize silage to feed the cattle.[2] Participating farmers received credit facilities and technical assistance through an EU-funded programme. The newly introduced technological model represented a rupture with local ecological conditions, summarized in Figure 6.3, which shows

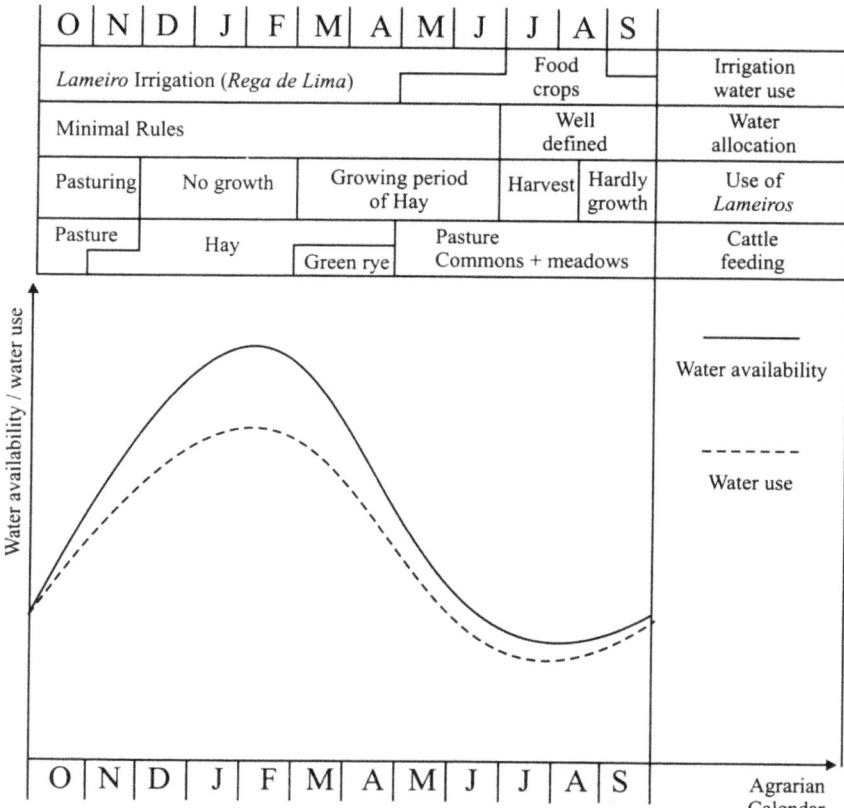

O	N	D	J	F	M	A	M	J	J	A	S	
Lameiro Irrigation (*Rega de Lima*)									Food crops			Irrigation water use
Minimal Rules									Well defined			Water allocation
Pasturing	No growth		Growing period of Hay				Harvest		Hardly growth			Use of *Lameiros*
Pasture	Hay		Green rye		Pasture Commons + meadows							Cattle feeding

Figure 6.2 Irrigation water use and the agrarian calendar in the Tras-os-Montes agro-pastoral style of farming (Dries, 2002:171).

how the requirements for irrigation water became disconnected from the water supply due to the widespread cultivation of maize with a high water requirement in the hot summer period. As Adri van den Dries observed: 'Here, *water scarcity appears as a consequence of a specific type of farm development and not as a technical constraint "an sich"*'(2002:179; italics in original). The newly created water shortage induced competition between farmers for the scarce water sources and local pressure for the construction of new, large water reservoirs to resolve the problems in the dry, hot summers.

But the regression did not stop there. By combining hydrological and farm accountancy data, Adri van den Dries showed that gross income per cubic metre of scarce summer water was 600 escudos[3] for the agropastoral breeders compared to 210 escudos for the modernized milk producers and the net income per cubic metre of scarce summer water varied from 500 for the former to just 110 for the latter.

This means that, in this case, with a given (and limited) amount of water, nearly five times as much income (and employment) could be generated by the local

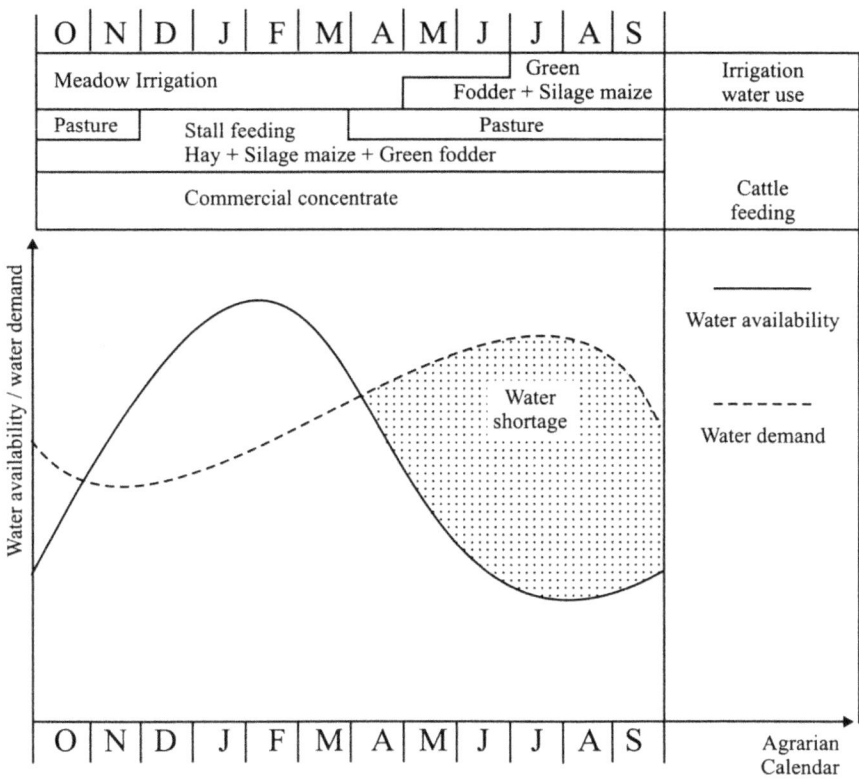

O	N	D	J	F	M	A	M	J	J	A	S	

Figure 6.3 The balance of water availability and demand in the newly introduced style of farming (Tras-os-Montes, 1989) (Dries, 2002:179).

style of agro-pastoral breeding compared to the style of high-tech milk production. Water use is not just a fixed or purely technical issue. It is malleable and subject to change. The re-organization of farming, with a shift from traditional to modern high-tech styles, had significant impacts on water use patterns (and potentially upon the ecology of the region by creating a demand for large scale reservoirs). Similarly, a return from high-tech milk production to locally adapted forms of agro-pastoral breeding might raise employment and the total income produced.

This malleability applies to many other dimensions of farming practices. There are a range of positions that can be taken in relation to these dimensions, all of which can have significant socio-political consequences. Agriculture is never fixed in a given moment in time and space. The potential for change is omnipresent, as is repeatedly reaffirmed by empirical research that reveals this variability. It is also true that from any given point in time and space there are multiple possible developmental trajectories that can be pursued. One of the central tasks and duties of the sociology of farming is to empirically explore these possibilities and contrast them with claims that there is just one 'optimal' way of farming and just one 'optimal' way forward. Consolidating knowledge on actual and potential heterogeneity helps to avoid claims about the singularity and unavoidability of just one track becoming hegemonic. In this respect, the sociology of farming becomes 'subversive', undermining the view held in certain politico-economic constellations, where it is often thought that agricultural development must inevitably follow one trajectory.

The malleability of agriculture is not a random phenomenon; rather, it is intrinsic to farming as co-production. Co-production is identical to the moulding and ongoing re-shaping of both the social and the natural and their interactions, which implies that there are different ways of unfolding over time. The specific forms that will emerge out of this are not known beforehand but depend on the reigning institutions, their interactions and the social relations of production implied by the interplay between institutions. What is certainly true is that particular ways of farming can be changed, just as their effects on particular dimensions can be altered. This will occur if and when the interinstitutional interplay (or socio-political political struggle in the broadest sense of the word) allows for new institutional patterns that facilitate new ways of farming (or the re-emergence of old ones).

Energy

In a meticulous study, Meino Smit (2018) reconstructed the development of fossil energy use in Dutch agriculture between 1950 and 2015 (see Figure 6.4). The total energy input increased far more than the output (i.e., the energy content of all of the agricultural produce). In 1950 the output/input (I/O) ratio was 1.20 (implying

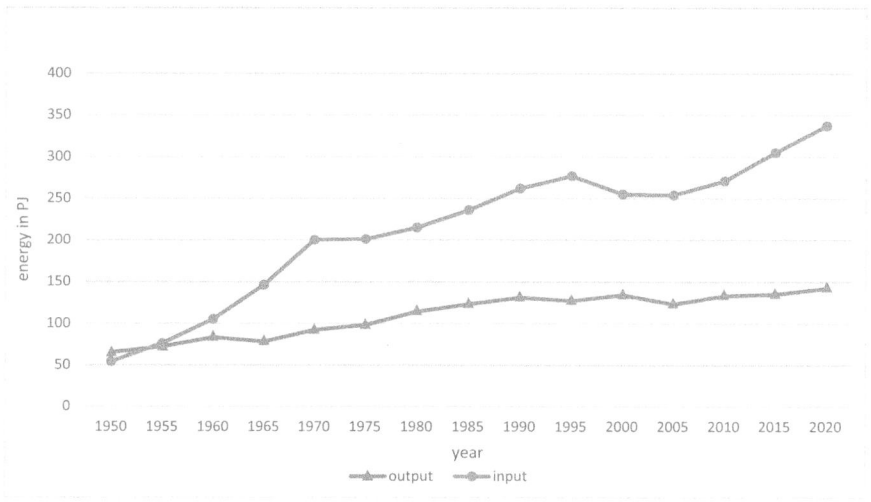

Figure 6.4 The input and output of Dutch agriculture in energetic terms (in petajoules) (Smit, 2018).

that agriculture was a net producer of energy); in 2015 this ratio had decreased to 0.41. Agriculture had become a net consumer of energy. Far more energy was used to produce food than the energy contained in the food.

At the same time, the composition of the energy used also changed. The direct energy input (in the form of electricity, diesel, gas, coal) went from 27.4 peta-joules (PJ) in 1950 to 135.6 PJ in 2015, whilst the indirect energy use (the energy embodied in fertilizers, buildings, technologies, etc.) rose from 26.4 PJ to 169.4 PJ (Smit, 2018).

The ongoing increase in consumption of fossil energy (and the associated decrease in use efficiency) closely relates with the change of Dutch agriculture from being mainly a peasant agriculture (that largely depended on the quantity and quality of peasant labour)[4] to a mainly entrepreneurial agriculture (although styles of farming that are very much peasant-like have remained and are re-emerging again). Being strongly grounded on an elevated use of external inputs (many of them embodying high quantities of energy) and the application of mechanical technologies (which consume equally high quantities of fossil energy), entrepreneurial agriculture has introduced a massive dependency on fossil energy into the production of food. The interdependency between the nature of farming and the level of energy use also comes to the fore in contemporary comparisons between different countries. Yu Tian (2019) compared current levels of energy use in Chinese and Dutch agricul-ture.[5] Figure 6.5 gives an impression of some of his findings.

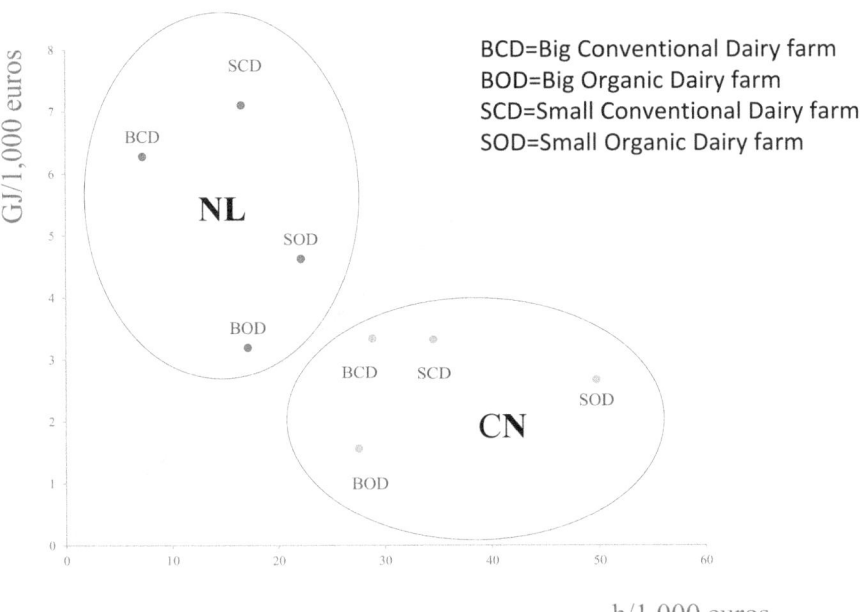

Figure 6.5 Comparison of energy and labour use in dairy farming in the Netherlands and China (Yu Tian, 2019).

To produce a similar amount of milk, Dutch farms have, on average, a far higher usage of fossil energy than Chinese dairy farms (roughly 5.5 GJ/1,000 euros versus 2.5 GJ/1,000 euros).[6] On the other hand, Chinese farms use far more labour than the Dutch farms (35 hours/1,000 euros versus 15 hours/1,000 euros).

Thus, again, we see a considerable empirical range, this time for indicators that refer to the use of fossil energy and labour (see also Figure 6.10 later). This range reveals, once again, that there is considerable potential for changes.

Landscapes and biodiversity

Farming is an activity that critically requires and affects *space*. The co-production of man and nature is, exceptions apart, rooted in space. Farming needs space, sometimes constructs space and nearly always moulds and re-moulds space into the particular spatial patterns which we know as landscapes. Landscapes are an unintended outcome of co-production (although their maintenance can become an explicit objective), and because co-production unfolds in different ways, there is an overwhelming variety of landscapes that co-exist alongside each other, often impressively beautiful.

The Sierra of Manantlán in Mexico is a richly chequered, colourful and densely populated landscape that is full of contrasts and contains an amazing amount of resources. In addition, there is an impressive biodiversity (which is strongly rooted in the resource diversity that results from the co-production of man and nature). Peter Gerritsen (2002) worked with peasant producers from the Cuzalapa community to construct a map of the 'Cuzalapa landscape as perceived through farmers' eyes' (151; see Figure 6.6). For the farmers working here (they apply a kind

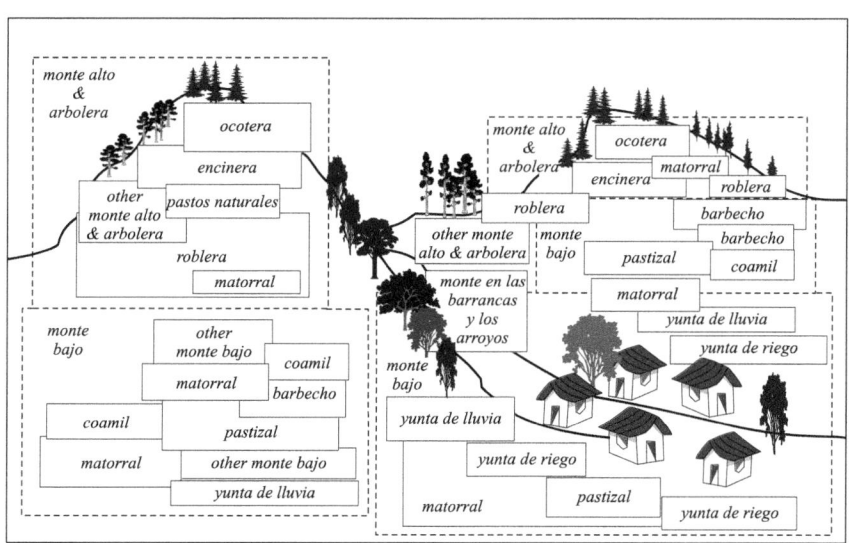

Figure 6.6 Resource diversity in the Cuzalapa landscape through farmers' eyes (Gerritsen, 2002).

of shifting cultivation),[7] the landscape is far from monotonous. It is nature that has been transformed by peasants who, by doing so, became intimately familiar with nature, the landscape and biodiversity.

Farmers perceive the many differences (that are meaningful and relevant to them) and categorize the different parts by applying a classification scheme that is grounded on vegetation type, altitude and previous use (see table 5.4 in Gerritsen, 2002). Using this classification scheme, they clearly observe (just as they actively construct) 'an organization of space that is dependent on time horizons underlying the different units, the area's specific ecological characteristics and farmers' decision making' (Gerritsen, 2002:154). With telling interview fragments (gathered when discussing the landscape with the producers), Peter Gerritsen reveals the many interactions and flows within this landscape and how famers use them:

> The *barbechos* [fields that are recovering] are always at a lower altitude than the *robleras* [oak forests]. That is why they can collect the *pudrición* [the flow of decomposing leaves, topsoil, etc.] of the *roblera* and why the *milpa* [field dedicated to maize cultivation] can grow beautifully. Some of the decomposition also goes to the irrigation fields, through the rivers. ... (Gerritsen, 2002:154)

Through farming practices, the different landscape elements (located at different altitudes and each having a specific vegetational pattern) are interconnected through a *rotation scheme* that specifies the sequence of use of the different fields. Figure 6.7 summarizes the rotation scheme used in Cuzalapa.[8] In many places throughout the world farmers use such schemes: sometimes simple ones, sometimes highly sophisticated ones that cover very long time spans. In some places

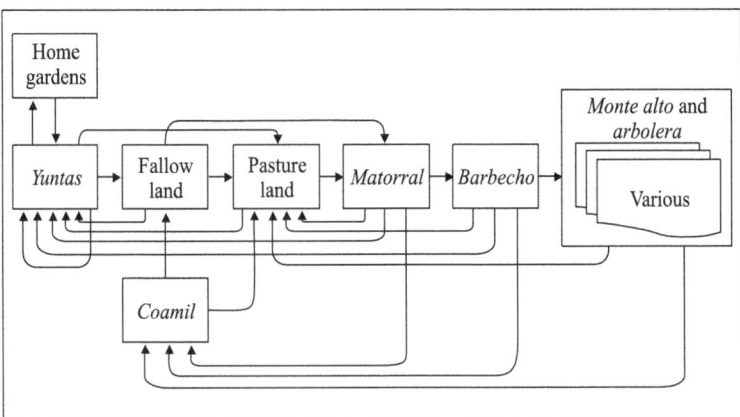

Figure 6.7 Succession management of diverse units of resources (rotation scheme) in Cuzalapa (Mexico) (Gerritsen, 2002).

the rotation is communally regulated in order to prevent pests and diseases and to avoid the intrusion of animals, whether wild or domesticated, that could damage the crops.

The rotational use of the land implies that in a given field (a fixed space) there is an ongoing seasonal and yearly alteration: from year to year different things are done, just as in a single year there are, everywhere, different activities. This double diversity (plus the frequent ruptures created through the application of controlled bush fires) creates the optimal conditions for biodiversity (see also Text Box 6.1).

Text Box 6.1   A blossoming mountain

In spring, the top of the Monte Subasio (located near Assisi in Umbria, Italy) is completely covered with orchids, among them the *Orchis purpurea*, also known as the Lady Orchid, and the *Orchis italica* Mediterannean. The number and diversity of orchids is almost unbelievable. They make Monte Subasio look like a Garden of Eden, untouched by mankind. However, the presence of orchids here is critically due to ruptures induced by farmers. Chianina cattle graze here (especially the heifers), and this causes the transport of nutrients from the top of the Subasio to the lower valleys (where the farms are located), which causes the systematic impoverishment of the soil, which is ideal for orchid growth (see Figure 6.8). The trampling of the topsoil and controlling the growth of shrubs equally support the reproduction of the beautiful flora and impressive biodiversity. Without grazing, this wealth would disappear.

Figure 6.8 Monte Subasio (Hans Dijkstra/GAW).

Figure 6.9 Different rural landscapes (late 1970s; the Netherlands) (Redaktiekollektief, 1980).

Whenever and wherever agriculture develops into a set of contrasting styles, landscapes are moulded into a diversified whole with different types of landscape co-existing next to each other. This was beautifully demonstrated by Jeroen Bosch and Johan Meeus, two gifted landscape architects who, in the late 1980s, made a set of drawings of different Dutch landscapes,[9] some of which are shown in Figure 6.9 (Redaktiekollektief, 1980).

Drawing #1 in Figure 6.9 shows a small-scale dairy farm in the wet peat areas (note the little bridge connecting the farm to the main road). At that time, the milk was still collected in churns – nowadays it probably would be organic milk that goes to a small organic dairy. The landscape is multi-faceted and colourful: there is grazing, hay production and abundant meadow bird life. Drawing #2 shows what, at that time, was considered as a bit of a monstrosity. It is a multi-functional farm *avant la lettre*. One can see horticultural production, dairy production, fruits and energy production as well (the mill). There is a kind of a *tipi*: maybe for agro-tourists? This type of combination of activities turns space (at least this part) into an idyllic agro-pastoral landscape, full of diversity.

Drawing #3 shows another landscape. As opposed to the first two, it is open and wide and contains long sightlines. It is the typical landscape made by the relatively large-scale, specialized, but extensive dairy farm of that time,[10] illustrated here by the typical 'head-neck-body' architecture of the farm buildings. Drawing #4 shows an industrialized farm, which started to emerge at that time for the first time. New architectural designs, monotonous grasslands (merely 'parking lots' for the cows) and large silos for storing concentrates dominate the scene.

Landscapes are not only an outcome of decades, if not centuries, of co-production; they also become, over time, one of the structural elements that condition and mould the farm labour process. In a way, landscapes, at least specific landscapes, operate as a social relation of production (see Chapter 1), because they shape the farm labour process and (indirectly) help to regulate the distribution of the wealth produced.

The Northern Frisian Woodlands are home to one of the most beautiful hedgerow landscapes in northwestern Europe. They were constructed by the peasants of the area in the early part of the 19th century. Following Napoleonic rule, the commons were divided into individual plots and hedgerows were constructed to delineate the different, individually owned, fields. In addition, the hedgerows provided protection from wind erosion, offered shelter (and dietary supplements) to grazing animals, provided a supply of timber and firewood and allowed for hiding rifles used for poaching. Nowadays the hedgerows are highly valued due to the rich biodiversity (both flora and fauna) they harbour.

Methods Box #13   Slotting in landscape and biodiversity

Following Ruiz and Gonzalez-Bernaldez (1982), Rene de Bruin developed an elegant method to explore the interrelations between farming styles and landscapes. He carefully chose pairs of photographs that show, at different levels, contrasting landscapes. The first pair showed a contrasting pair of grasslands: a standardized, productive, monocrop and a more diverse field with different grass varieties and the visible presence of cow dung. The second pair showed farm buildings in relation to their immediate surroundings: again two opposite images. The third pair related to the wider landscape: many small fields separated by hedgerows and an open landscape with large fields. De Bruin asked farmers which, out of each pair, they thought best suited for farming, after which he assembled the different answers into landscapes most favoured by different farmers. The underlying assumption was that, over the years (through the many decisions to be taken on, for instance, eliminating or maintaining hedgerows), the real landscape will be moulded, step by step, towards the most favoured one.

In discussing the role of living nature (at different levels, including the landscape), Paul Swagemakers (2008) developed a technique that is a bit similar to the one discussed in Methods Box #8. He showed his respondents two contrasting images of the relation between farming and nature (see below). In one of these, nature and farming are in two separate boxes; in the other there are two, partly overlapping, boxes. As in Methods Box #8, these are only very rough and schematic approaches to a very complicated issue. But then, these images, boxes or models are not meant to be precise descriptions of reality: they are, instead, *invitations* to describe how these elements, or variables, appear, relate, interact and develop in real life and why. It provides an occasion to discuss *cultural repertoire* and to explain how things can and should be done in practice.

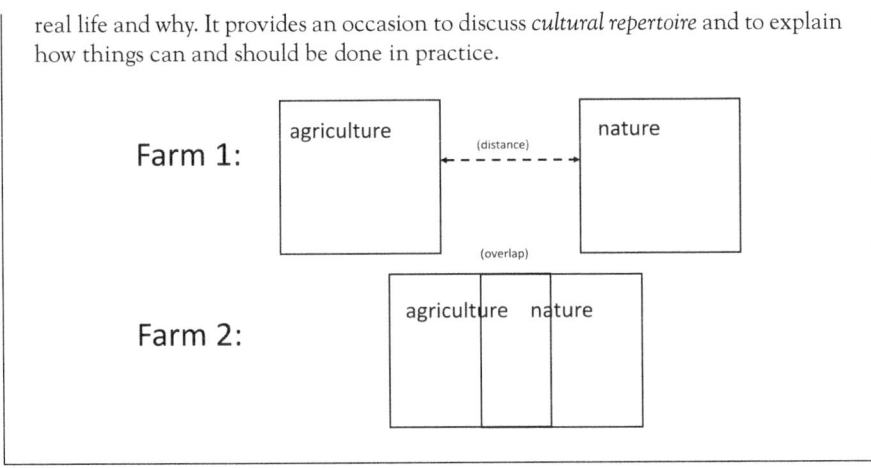

If one compares old and new topographical maps of the NFW (see Ploeg, 2003:134–136), one can see that the pattern of hedgerows has been remarkably stable over time. This is due to tradition, local cultural repertoire (which highly values the shelter, both materially and symbolically, offered by the hedges), the high costs of uprooting them and, more recently, regulatory schemes. Thus, farmers here have had to deal with small fields and meadows delineated by the many hedgerows – and have done so quite successfully. One way in which they did so was by developing specific combinations of machinery and labour input that allowed them to operate in small fields while keeping the costs down. I will discuss the modality developed, in the Northern Frisian Woodlands, with reference to Figure 6.10.

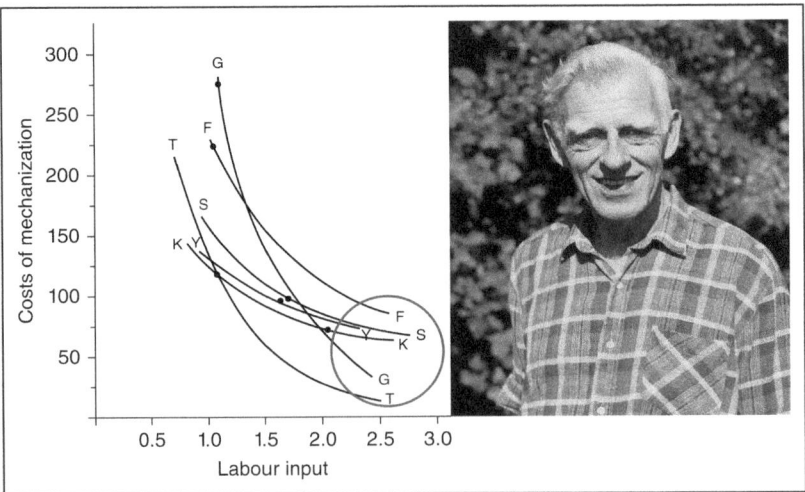

Figure 6.10 Different substitution curves representing different style-specific modalities for combining labour-input and mechanization (Ploeg, 2003).

The graph shown in Figure 6.10 refers to the amount of machine costs (depreciation plus running costs) and labour input needed to produce 500,000 kg of milk per year. The different lines show the substitution curves created through different farming styles within the province of Friesland. The diversity underlines, once again, the malleability inherent within agriculture. Farms can move along a curve and thus reduce the direct and indirect use of fossil fuels and increase the labour input (see the discussion on energy use earlier in this chapter). They can equally try to move slowly towards another style that contains a different curve.

The farmers in the Northern Frisian Woodlands mostly operate within (or close to) the circle drawn in the graph. They combine low mechanization costs (using relatively small, often secondhand and well-maintained machinery and equipment) with a high labour input. This allows them to keep operating in the typical hedgerow landscape of the area. One of the most outspoken pioneers of this way of farming, Taeke Hoeksma, is shown in the photograph in Figure 6.10. He was convinced that farming could be well attuned to the specificities of nature and the landscape and became one of the advocates for farmers' managed maintenance of nature and the landscape. Taeke Hoeksma ran the farm with his two sons (a total labour supply of, say, 2.5 FTE), and their machine costs were considerably lower than those of comparable farms. In other words, Taeke Hoeksma successfully rebuilt his farm's process of production to better align it with the hedgerow landscape. At the same time, he obtained a labour income per 100 kg of milk that was almost twice as high as a group of comparable farms (14.80 euros versus 8.53 euros/100 kg milk; Ploeg, 2000, table 1). So, alongside adapting the processes of labour and production to attune them with the surrounding landscape, he increased the amount of value added retained.

There is, of course, considerable irony in this story. Here the landscape (as a materialized social relation of production) blocked the full-fledged application of the modernization project (large-scale, high-tech farming and continuous expansion; I will come back to the theoretical and socio-political implications of this finding in Chapter 10). Farms in the Northern Frisian Woodlands are smaller than those in the rest of the province of Friesland. Because farm expansion is less prevalent here than elsewhere (the landscape does not lend itself to the construction of extended fields that allow for greater mechanization and an associated reduction of labour input), farmers became less dependent on banks than those elsewhere. The low debts and low costs for mechanization (among other factors) helped them fare better through the difficult economic years that came in the first two decades of the 21st century – far better than farmers elsewhere. Thus, the landscape not only (partially) blocked modernization – it also functioned as a line of defence against the negative and unforeseen consequences of the same modernization project. The hedgerows protected, as it were, the farmers against the cold winds of neoliberalism (more so because the farmers succeeded in obtaining public payments for the maintenance of nature and the landscape; for further discussion, see Chapters 9 and 10).

Agrarian questions

Farming articulates with a wide range of other issues: the generation of employment and income, water use, energy use (and the associated contribution to climate change), scenic landscapes, biodiversity, inclusion/exclusion, contribution to overall economic development, the maintenance of cultural heritage, etc., Of course, not all of these dimensions are relevant in all places and at all times. Each specific time-space location comes with a selection of dimensions considered to be relevant – and each time agriculture has a highly variable effect on them.

When there are substantial and enduring negative impacts (which may imply considerable external costs that need to be paid by others and/or in the future), these effects are referred to as *negative externalities*. In the opposite case, we speak of *positive externalities*.[11] These are the positive effects that come with a particular way of farming. These positive effects might be unintended or intentionally produced. The key is that they do not show up (only) within the farm – they are external to the farm (at least partially). Positive externalities come to the fore as contributions to public goods. This is associated with another key feature: they are not paid for. There is no market that remunerates the production of positive externalities and/or sanctions the production of negative externalities. The market for agricultural products does not take these externalities into account. Negative externalities are also referred to as *market failures* (I will come back to this concept in Chapters 7 and 8).

If there are too many persistent, and highly damaging, negative externalities, there is an *agrarian question*. There is not a single agrarian question – there are many – and their nature and morphology impact wider society and the ways to solve them differ very much according to the specificities of time and space.

The first and now classic version of the agrarian question was formulated by Marxists. It centred on agriculture not contributing enough to overall economic development (more specifically: not sufficiently feeding the process of capital accumulation) and simultaneously turning the countryside into a space of misery and poverty. Both negative effects were thought to be structural outcomes of the predominance of peasant ways of farming that could only be remedied by and through a transformation of the peasantries into a small class of capitalist farmers and large masses of landless people, most of whom would go to strengthen the ranks and files of urban industrial workers. This would spur accumulation in several ways. Capitalist farmers would generate profits to be reinvested in industry, trade and the banking circuit. At the same time there would be a large supply of cheap labour to work in those sectors. In this context, commoditization was seen as a strategically important trigger. Separating peasants from their land would create large markets for both land and labour.

This type of agrarian question very much reflected the reality of England during the industrial revolution in the 18th and 19th centuries and was theoretically elaborated (especially by Lenin, 1914/1972) as '*the* agrarian question' (which was thought to be universal). Later on, several Marxists started to distinguish different modalities of the agrarian question (Byres, 1991; Moyo et al., 2013), whilst

Haroon Akram-Lohdi and Cris Kay (2010a,b), who built on previous debates in revolutionary Russia, argued that providing cheap labour, land and capital (to feed the process of industrialization)[12] was far from being the only possible way for agriculture to contribute to overall economic development. By developing agriculture and increasing the income levels of peasant families, an extensive *domestic market* can be build that represents considerable purchasing power, which, in turn, can strongly contribute to the process of industrialization.[13]

Other politically engaged scholars (many of them Marxists) argued that, contrary to the classic view, the low contribution to economic development and the widespread poverty in the countryside did not result from the *absence* of capital as an organizing principle but was due to *the direct and indirect exploitation of agriculture by capital*. It was only through the massive expropriation of large capitalist farms and properties, the subsequent redistribution of land to peasants, the creation of cooperatives and the development of productive forces in the countryside that progress could be wrought. Within this process, peasant communities could become a main driving force and thus play a progressive role in the transformation of society as a whole. This latter position was advocated by, among many others, Alexandre Chayanov in Russia (1920/1976, 1924/1966, 1927/1991), José Carlos Mariátegui in Latin America (1925) and Fei Xiao-Tung in China (Fei and Chang, 1945).

I will not go here into the many details of the debate between the different views (the debate and the different positions positions re-emerged in Latin America in the late 1960s and 1970s as the polemics between *campesinistas* and *descampesinistas*; see Feder, 1977). Instead, I will jump to the many-sided *agrarian questions* (in the plural) as they existed at the end of the second millennium and the beginning of the third one.

Wherever it is located, in time and space, farming always articulates with nature, society and, specifically, the farming population (see Figure 6.11). Each of these dimensions can be well balanced, which means, among other things, that agriculture meets society's needs and expectations, reproduces (and further enhances) nature and provides the actors involved with adequate incomes.

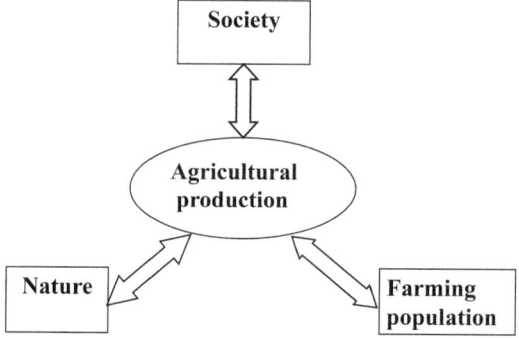

Figure 6.11 The triple articulation of agriculture (author's own elaboration).

Yet there might also be enduring frictions. This occurs when, and if, agriculture is organized and developed in such a way that substantial and long-lasting negative externalities are created.[14]

Figure 6.11 helps us to understand the agrarian crises that we are witnessing in the first decades of the 21st century. Many countries, especially in the Global South, are experiencing shortages of food (partly because considerable parts of agricultural production are exported), whilst other countries face serious problems with the quality of food. Nature is being harmed nearly everywhere (to the degree that even agricultural production itself is endangered) and some 70% of all poor people in the world (living on less than $1 a day) are rural: agricultural production is not able to provide them with adequate employment and incomes. Of course, the specific mix of the negative externalities that come with agricultural production varies from country to country, and there often are important variations even within countries. But on the whole, it applies that the agrarian questions of today articulate *on all three dimensions* distinguished in Figure 6.11. It also applies that these different agrarian questions are tightly *interlinked*: the destruction of the Amazon rainforest to make way for soy production is closely tied to factory-like pig and poultry breeding in, for example, the Netherlands and Belgium, which in turn links to the out-competition of local producers in West Africa through massive and cheap imports of the less desirable parts of pigs and poultry from northwest Europe. The high interdependency also implies that alleviation of one partial problem easily translates into an aggravation of problems elsewhere. Together the agrarian crises of today compose a *Gordian knot* that is nearly impossible to disentangle.

At a higher level of aggregation, these different agrarian questions (*different* because their morphology differs from country to country) make up one single and global agrarian question. From a theoretical point of view, this *global* agrarian question is new: new in that it differs structurally from the classical agrarian question. Central to the classical agrarian question (see above) was the *failure of agricultural production to contribute to* the process of capital accumulation. Currently, however, it is no longer about the levels of exploitation and accumulation – it is, instead, the unprecedented domination of capital that characterizes and causes the agrarian questions of today. Wherever located, farming is subordinated to capital through a range of mechanisms, specified in the next section. This subordination induces a range of negative externalities that together create the agrarian questions. In some places, parts of the farming population are able to resist (albeit only partially) this subordination and develop alternative models for agricultural production. By doing so, they show ways to resolve the agrarian questions – or, at least, develop some elements that might contribute to the construction of promising ways forward.

Farming and capital

The first series of mechanisms through which the subordination of agriculture to capital occurs might be grouped under the heading of 'skimming'. It involves mechanisms that facilitate the shift of value from farming to capital groups located in trading, processing and banking. Unequal exchange is central to these mechanisms.[15] In this respect, the ongoing deterioration of the terms of exchange (between agriculture and other economic sectors) is often central: the prices paid to farmers are stagnating whilst the prices for the industrial goods they need are rising. Sometimes this tendency is enforced by the state: low food prices allow for salaries and wages in the non-agricultural sectors to remain relatively low, and this enhances the profitability of the industrial and service sectors.[16] Dependency on the banking circuit (or money lenders) also contributes to skimming. Payments of interest generate a permanent value flow from the agricultural sector to financial capital. In the Netherlands, for instance, total debts (loans within the family excluded) amount to some 33 billion euros. Even with low interest rates (say 3% on average) this amounts to an income flow of 1 billion euros from agriculture to the main banks – *equal to half of the total family income generated by farming*. But value can be drained in other ways as well, such as the provision of cheap labour to other economic sectors. This is typically the case in colonial, post-colonial and racist regimes, but it also seems to be inherent to highly dualistic agricultural regimes (as the Latin America *minifundia/latifundia* complex) where small farmers have to earn parts of their income outside of their farm and thus offer their labour cheaply. Finally, the payment of taxes (to the state) completes this first series of mechanisms. It goes without saying that the relative weight of each mechanism (vis-à-vis the others) varies between time and place.

These mechanisms provoke a permanent and downward pressure on farming incomes. Theoretically speaking, this pressure is due to the shift of substantial parts of the value produced by farmers towards non-agricultural sectors and actors. As argued, unequal exchange is decisive here: generally speaking, power relations in the food markets (many suppliers, few buyers and perishable products) are detrimental to farmers. In today's food markets, food empires are able to operate as obligatory passage points. Food is mostly only able to travel from producers to consumers through the networks (and associated points of inlet, conversion and outlet) controlled by large capital groups. When it comes to the capital market, rural labour markets and relations with the state there is, again, unequal exchange, due to skewed power relations. In the 1970s, Colombian peasants defined themselves sometimes as *patasucias*: having 'dirty feet' meant that one was not allowed to enter banks (with their beautiful marble floors) and thus they became dependent on usurers and the like. This auto-denomination also came with the connotation that one had no recourse whatsoever to denounce injustices.

A second series of mechanisms centres on enlarging the material dependency of farming on goods and services provided by capital groups (agribusiness and banks). Technological development (especially when it occurs as externalization and the increased use of mechanical technologies; see Chapters 2 and 3) implies the acquisition of expensive technological artefacts and increased spending on external inputs. This increases the material basis for 'skimming' (see above).

Together, the trend towards increased material dependency and the compound skimming that comes with it constitute a trend that is known as *the squeeze on*

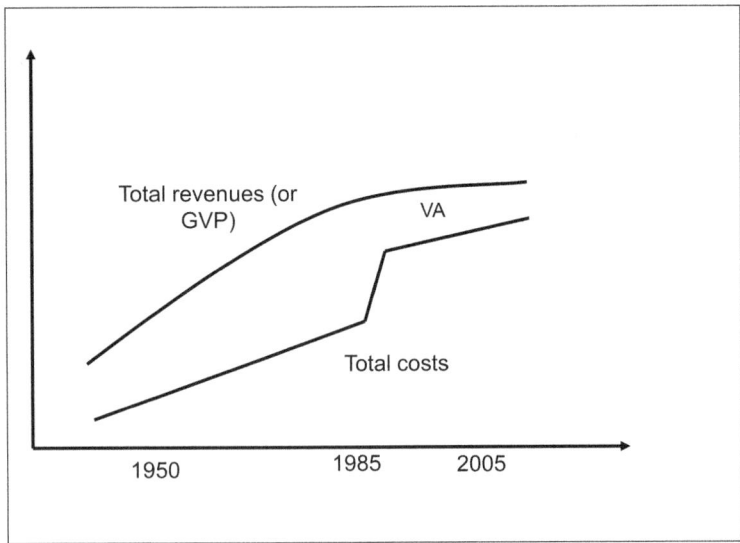

Figure 6.12 A schematic representation of the squeeze on agriculture (Ploeg, 2003).

agriculture (see Figure 6.12).[17] Whilst the gross value of production (GVP) for the agricultural sector as a whole is growing slowly or even stagnating, the total costs for generating this GVP (expenses for external inputs plus depreciation of investments) keep growing. Consequently, the value added (VA) of the agricultural sector suffers downward pressures (at least in real terms). Thus, value is, indeed, *squeezed* out of agriculture. It is re-allocated elsewhere and significantly contributes to processes of accumulation in non-agricultural sectors.

Figure 6.12 also explains (especially if it is projected onto a single farm) the so-called treadmill: Farmers face declining incomes (due to different forms of skimming) and try to change the trend by enlarging their farm (or, analytically, enlarging the scale of production). This implies huge investments (not only in land, quota, etc.) but also, and especially, in new technological artefacts that allow for greater production per unit of labour force. Thus, the material dependency on capital increases, and this brings extra costs, which, in the end, require further expansion. The more farmers invest in escaping the squeeze, the more this very squeeze is intensified.

Translated to the level of the agricultural sector as a whole, the scale enlargement this generates means that the available income (VA) is shared by fewer people (thus allowing for an increased income per person). However, the same scale enlargement further narrows the gap between revenues and costs, and this prompts further scale enlargement. In a way, this is the basic drama of modernized farming: farmers themselves are fuelling the process that in the end will eat them all (a 'race to the bottom', as Terry Marsden [2012] argued). As in a true treadmill, the faster you run (in order to not fall down), the higher the velocity of the mill.

It is important to signal here that these processes are not just driven by external capital groups. They are not simply imposed from the outside but are equally driven by farmers who opt for ongoing scale increases and high-tech ways of farming. To *control* an extended and permanently growing resource base, they need the *material dependency* on external capital groups (and the inputs and technologies they provide). In this context, the agricultural sciences operate as an important vehicle, designing the models that allow for the desired control over living nature after which agro-industries turn such models into specific technologies to be sold to farmers. The state, especially when it is interested in a positive and substantial contribution of agriculture to the trade balance, will equally encourage these compound processes. To do so, it has a wide range of instruments that include technical assistance, subsidized credit, favourable tax regimes, spatial reorganization, infrastructural projects and applied research. Hegemony, condensed in the widely accepted notion that scale enlargement is self-evident, unrivalled and the only possible way forward, is strategic here.

The subordination of farming to capital not only concerns the 'material' side (discussed here as value shifts). It also occurs through *shifting the decision-making processes* from the farm and farming family towards external agencies. This shift is as important as the material expropriation. And, again, it is grounded in different mechanisms. One of these is the regulatory schemes imposed by the state and/or agri-business groups. Such schemes specify what is to be done, by whom, under which conditions and with which particular technological artefacts. Some schemes directly prescribe (parts of) the farm labour process; others do so indirectly (by specifying, e.g., environmental measures). When the state and agribusiness align themselves in the definition and operation of such schemes, the farm labour process becomes regimented (and guided by interests other than those of the involved farmers). Such regimentation can also occur as a consequence of a high dependency on banks (which will only refinance if certain requirements are met) and high degrees of externalization (when farming is increasingly reduced to a mere assembling of building blocks elaborated elsewhere). Sometimes, the growing involvement of contract workers (such as enterprises providing machine services to farms), accountants and/or consultants can have similar effects.

A fourth series of mechanisms for linking capital and farming in unequal ways centres on the *takeover* of resources, linkages and/or symbols. The currently most discussed expression of such takeovers is 'land grabbing'. The grabbing of land (or usurpation or clearance as it was called in other times and places) occurs through market transactions and/or coercion. A combination of the two is frequently encountered, even to the degree that coercion dominates and the market transaction only follows in order to suggest legality. But there are many other forms and mechanisms. Diverting water flows from peasant agriculture towards other areas and scopes (for instance, to production areas controlled by capital groups and/or crops for export) is another widely used mechanism (see Text Box 6.2). Monopolizing the right to use seeds is yet another, and particularly shameful, example. Establishing (monopolistic) control over marketing channels can equally be understood as takeover. And, finally, I want to highlight the battles that are

---

Text Box 6.2   'Underground' struggles for water in Morocco

Moroccan agriculture largely depends on irrigation. Since ancient times the country has had ingenious irrigation systems: 'shallow wells', which can go as deep as 40 metres and which are connected by extended underground tunnels that (a) tap aquifers and (b) connect the shallow wells. In this way, scarce water could be collected over wide areas and concentrated at specific points for agricultural use. Currently, there are several different systems in use. Single shallow wells used in peasant-like agriculture; boreholes (sometimes constructed with percussion rigs) that have a depth of up to 90 to 100 metres and which are used in newly emerging entrepreneurial agriculture (see Lisa Bossenbroek, 2016) and, finally, boreholes constructed with sophisticated machinery that go as deep as 200 metres and more. The latter are exclusively used by large-scale, export-oriented agricultural enterprises that are controlled by international food empires such as the former Van Oers Group (see Text Box 6.3).

These three systems do not simply co-exist alongside each other. They enter (and are meant to enter) into an unequal type of competition. The second type of boreholes often catch the water from aquifers that used to deliver to the shallow wells used in peasant-like farming. In turn, the very deep boreholes grab the water of the less deep wells and boreholes. Thus, a kind of underground struggle for scarce water is going on that, in the end, produces drought in peasant agriculture whilst concentrating water in entrepreneurial, and especially corporate, agriculture.

---

going on between food empires and the sections of agriculture they control, on the one hand, and the new constellations being constructed by farmers' and peasant movements (which centre on new ways of farming, processing and marketing), on the other. A considerable part of these battles is about symbols; that is, about which food products are thought to be healthy, high quality, artisanal, sustainably produced, fair, animal friendly, low in fossil fuel use, local, etc. As soon as alternative constellations gain some momentum, food industries and large retail (and sometimes state agencies as well) jump in to try to appropriate the new symbols (and the economic opportunities that come with them).

Another form of takeover (sometimes combined with the mechanisms mentioned above) resides in capital directly organizing parts of agricultural production. Thus, the 'mega-farms' of today are constructed (see Text Box 6.3) in the same way as the plantations, *haciendas*, large estates, etc., from the past (and sometimes the present). This implies that the opportunities that otherwise could have been used for strengthening family farms (or peasant agriculture) are directly taken over by capital groups and brazenly used for capital accumulation. Such newly constructed capitalist farms often push family farmers and/or peasants directly into the margins (or even completely out of agriculture). With the possibilities of long-distance trade, technologies that keep food products 'fresh', even for extended periods; very cheap land; and political regimes that favour 'investments in land', an extended corridor has been created around Europe (a corridor

Text Box 6.3   Mega-farms

The most salient features of mega-farms are the large stretches of land, the extensive herds and/or the exclusive access to abundant water resources they control and use. Mega-farms exist as junctions between capital groups and the *direct* control over these resources. Mega-farms monopolize extra-economic opportunities to directly (1) connect place of poverty to places of richness,(2) obtain large amounts of cheap credit and/or extensive public funding, (3) play a key role in financial engineering (including speculation, tax evasion and the creation of safe havens), (4) grasp market shares that allow for considerable control over these markets and/or (5) accumulate resources in quantities that go far beyond the possibilities of family farms. Below are a few telling examples.

Ekosem-Agrar GmbH is a German holding company that owns the EkoNiva group: a huge dairy farm that operates in Russia. In 2019 this mega-farm covered 631,000 hectares and had 182,000 head of cattle, including 97,640 dairy cows (an eightfold increase since 2012). Total milk production is 759,000 tonnes, part of which is now processed within the company itself. The total turnover of the EkoNiva groups is 564 million euros (in 2019) with a net profit of 36 million euros (also in 2019). It is the largest dairy farm in Russia and, indeed, in all of Europe. Ironically, it greatly benefits from the EU boycott of Russia that followed the military invasion of Ukraine: its position is no longer threatened by dairy imports from the EU.

Van Oers United BV is a chain of connected large horticultural enterprises that operate in France, Spain, Morocco, Senegal, Ethiopia, Kenya, Zambia, Zimbabwe, Rwanda, Egypt, Peru and Guatemala. By producing in different climatic zones, the company can provide a year-round supply of vegetables. The site in Morocco has 800 hectares of irrigated land for vegetable production (which is extraordinary when compared to the average vegetable-producing family farm in Europe). The produce is sent directly (with the company's own lorries) via Tangiers to Rotterdam (where the freight arrives within 24 hours), from where it is distributed further. Van Oers mostly delivers to large supermarket chains (thereby bypassing the cooperative auctions meant to protect the horticultural family farms of northwestern Europe). The Van Oers group was recently taken over by the French AGRIAL group.

Emiliana-West Rom is a Romanian mega-farm fuelled by Italian capital from Unigra. It cultivates nearly 11,000 hectares of cereals – maize, wheat, sunflower, rapeseed, rye and barley – employs 99 workers and also raises 1,200 Limousin beef cattle. It was able to obtain the ownership of the land for only 100 to 150 euros per hectare. In 2013 it received 1.3 million euros from the EU of 'income support' (and an additional 365,000 euros from the Romanian government).

Currently, the enterprise is one of the partners of the Danube Soy Project: a concerted action of different capital groups (among which Cargill is prominently present) to organize a European market for non-GMO (genetically modified organisms) soy produced in the Danube basin. In addition to tapping into EU hectare payments, it will access European funds meant to decrease the 'protein dependency' on the USA and Brazil.

Similar cases can be found in Africa (e.g., Senhuil, with 20,000 hectares of bioethanol production in Senegal, and Zambeef, the largest beef producer in Zambia and beyond). One can also find several mega-farms in China (such as the 'King Bull' meat producing enterprise).

that includes North Africa, Eastern Europe and Russia). This corridor harbours growing numbers of mega-farms that produce for the European food market (and sometimes for the USA as well). They effectively link areas of poverty (with cheap land, cheap labour, a cheap environment,[18] etc.) to areas of richness, thus making windfall profits.

Alongside the value flows that go from primary agricultural production towards capital groups located elsewhere in the economy, there is a reverse flow in which public resources (mostly channelled through the state) are used in, and as part of, agricultural policies. Like all policies, agricultural policies *authorize* (they prescribe specific activities whilst sanctioning others) and *allocate* (they distribute resources, mostly financial ones). At the junction of the two, agricultural policies can become a powerful tool that strongly influences the organization and development of agricultural production.

Precisely at this particular junction, a fifth set of mechanisms can be discerned that concerns the deliberate, or unintended, construction of *biases* in agricultural policies (Vecchio et al., 2022). These biases discriminate against family farming and peasant agriculture and favour types and styles of farming that are smoothly aligned with dominant interests and views. Such biases also function (albeit mostly in an indirect way) as mechanisms that help to further subordinate farming to capital.

Agricultural policies normally include a range of different instruments, each of which assumes specific conditions that are to be met for the instrument to be applied. For instance, in the period 1960 to 1990, interest subsidies were given for building cubicle stalls (as opposed to renovating the traditional type of Frisian or Dutch stable). The cubicle stalls allowed for considerable scale increases (allowing one person to control a much larger herd), whilst the other types were considered to be 'traditional' (although farmers loved them because they allowed far better care). Thus, a kind of bias was introduced. Support, in the form of an interest subsidy, was only given for cubicle stalls. In contrast, the farmers who wanted to continue with another type of architecture for their stables were put at a disadvantage. In Peru, credit programmes for small farmers only financed the acquisition of new, scientifically designed potato seedlings (see Text Box 3.5) and the associated package of inputs. No credit was provided for working with 'traditional' varieties. The European Union does not give Pillar 2 support (see Chapter 7) to part-time farmers, even though pluri-activity (or multiple job holding) is a widespread phenomenon throughout Europe. In this way, agrarian policies become a selective mechanism: certain styles of farming are included and supported while others are excluded. And just as the benefits that come with agrarian policies are unequally divided, so are the costs associated with them. Imposing an environmental policy that requires investments in expensive machinery disfavours small farms, because these investments are far easier for large farms to make.

Thus, a systematic mismatch (summarized in Figure 6.13) is created and prolonged. Certain styles are favoured and others are kept at arm's length from the opportunities entailed in the reigning policies. At the same time, these policies induce new styles that are thought to hold promise for the future. In policy circles

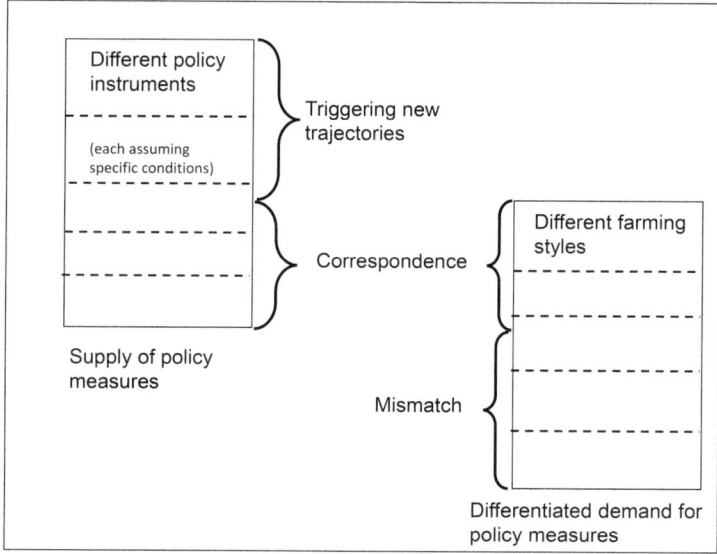

Figure 6.13 Biased agricultural policies (author's own elaboration).

this is often legitimized by using an explicit classification scheme that usually distinguishes between small, medium and large farms. The large farms are thought to be able to develop towards being viable farm enterprises; small farms are, by contrast, considered to be too small to be able to develop, and medium farms are kind of in between. Subsequently, it is argued that it makes no sense to give public money to farms that are thought to be unable to develop further because this would be tantamount to squandering. 'Don't throw good money after bad', as state agents like to say.

The same occurs in, and through, agricultural sciences. As all applied sciences, they use an often explicit and hardly disputed *horizon of relevance* (see Figure 1.7). Solving the particular problems of small farms is thought to be irrelevant, for these farms will disappear anyway (it would be tantamount to 'dedicating good research money to useless research'). This horizon also influences the practices, crop varieties, animal breeds, production areas, etc., that are thought to merit research and those that don't. Finally, it specifies what is to be perceived as successful and what as failure. Thus, current research (and the supply of technical and biological artefacts it produces) represents a remarkable contrast with agricultural research in the immediate post–World War II period. At that time there was an urgent need to increase food production, and small farms self-evidently had an important role to play in meeting this challenge. There was even a special directorate for small farms in the Netherlands (*dienst kleine boerenbedrijven*) that, among other things, designed innovations especially aimed at small farms. But once food production had risen enough (by the second half of the 1950s), the horizon of relevance was redefined, and this affected the 'biography' of specific

Figure 6.14 Fence rack (author's archive).

technical artefacts in amusing, but also telling, ways. Fence racks for drying hay (see Figure 6.14) were very much promoted in the first decade after World War II. These racks allowed farmers to obtain a higher quantity and better quality of hay (compared with hay that was just dried on the surface of the land). Using these artefacts required a considerable labour input, but at that time this was not seen as a problem: especially in the smaller farms there was enough labour available. However, after the mid-1950s, when the contours of the big modernization project became slowly visible, the use of fence racks was strongly discouraged because the same instrument was said to require too much labour. It had become an expression of 'old-fashioned farming'.

A permanent and many-sided disarray

The many-sided exploitation of farming by capital generally, and by particular capital groups (trading companies, food industries, large retail organizations, agribusiness, banks) specifically, introduces three interrelated and more or less permanent problems in farming. Firstly, there is the nearly chronic tendency towards decreasing incomes and the fear of poverty. This triggers, in the second place, the many endeavours to squeeze more out of co-production, which can be detrimental for the quality of resources and for sustainability in general. And thirdly, it impacts on agriculture's relations with wider society. With the reorganization of spatial divisions of labour (such as moving vegetable production to Morocco) and the artificialization of food by food industries, all kinds of new problems emerge at the interface of society and agriculture (the quality and nutritional value of food is threatened, scenic landscapes are turned into monotonous plains, etc.). On all three axes shown in Figure 6.11 there is the threat of systematic disarticulation. In terms of previous polemics: the agrarian question does not arise because agriculture is at odds with capital. It is, nowadays, exactly the other way around. Contemporary agrarian questions exist precisely because agriculture is subordinated to, and reshaped in multiple ways by, capital. Of course, the specificities of the agrarian question in countries such as Peru and Brazil differ from those of Europe and China (see Chapter 7). What these apparently different agrarian questions have in common, though, is that they are deeply rooted in the subordination of farming to capital. Whether this subordination is the result of direct interventions of capital, or resulting from state policies, or emerging out of the activities of segments of the farming population looking for integration and control does not matter in as far as the end result is concerned: farming becomes subjected to a permanent draining of value that is rooted in a multitude of (interchangeable) mechanisms and which is therefore very difficult (but not impossible) to resist.

The standard response to the appropriation of value has been and still is further technological development, thus enlarging the production per unit of labour force. However, apart from the fact that this response is more or less exhausted (current modalities of technological development come with such an increase in material dependency and increased skimming that the negative effects on income mostly outweigh the positive effects),[19] it creates exclusion, accelerates the rural exodus and often endangers the quality of natural resources. In short, the current modalities of technological development have become mechanisms that are pivotal to the further subordination of farming to capital. They are part and parcel of the agrarian question.

Nonetheless, the subordination to capital can be addressed (and redressed) in a variety of other ways, some of which will be discussed in the chapters to come.

Notes

1. A synthesis is given in Dries and Portela (1995).
2. There already was considerable small-scale milk production. These small producers owned collective milking parlours, which helped to considerably to raise the quality of milk and reduce the drudgery experienced.
3. One hundred sixty-five escudos was equal to US$1 in 1989.
4. In this respect, it is telling that the absolute number of people working in agriculture (expressed in FTE = full-time equivalents) kept growing until 1956 when it reached its peak of some 570,000 people (in FTE) directly involved in agricultural production. After 1956 it started to decrease. By 2015 it had dropped to 113,099 FTE.
5. This required a complicated process of making the data comparable. For the relevant details, see Yu Tian (2019).

6. Calculated using the purchasing power parity exchange rate (see Yu Tian, 2019:51).
7. Shifting cultivation involves a time sequence that specifies the use of different fields, and it nearly always includes a more or less lengthy period to allow the land to rest (and regain fertility). Often shifting cultivation uses the controlled application of fire to clear, after a sufficient rest, the fields in order to restart cultivation.
8. In a later publication, Peter Gerritsen (2018) showed how this scheme changed between the late 1960s and the late 1990s as a consequence of neoliberal policies.
9. See also Meeus et al. (1988) for similar drawings for Europe as a whole. Sabine de Rooij (2018) gives a synthesis and links the discussion to spatial policies. See also Bruin and Ploeg (1992) for an application to the hedgerow landscapes of Friesland, the area where the first territorial cooperative of the Netherlands, NFW, was formed (see also Chapters 9 and 10 in this book).
10. These farms were relatively extensive because they did not use high levels of concentrate and fertilizer. Long periods of grazing were the rule. All of this clearly benefitted the meadows and pasture lands, which contained considerable biodiversity.
11. By re-adjusting the internal animal-manure-soil-fodder cycle (see Methods Box #12), the farmers of the Northern Frisian Woodlands also created a range of positive externalities: cleaner natural resources (air, water, soil), more biodiversity, less pressure on the small-scale hedgerow landscape, improved milk quality, less stress and more longevity in the herds and a strengthening of the regional rural economy. This is further discussed in Chapter 10; see especially Figure 10.4.
12. Especially heavy industry.
13. Here the industries producing consumer goods would be pivotal.
14. The 'classical' agrarian question fits very well in the general scheme shown in Figure 6.11. On the axis between society and agriculture it is impossible for agriculture to contribute to overall economic development (or, more specifically, to the process of capital accumulation). Indeed, it is even unable to move rural producers out of chronic poverty and despair. In addition, the capitalist organization of agricultural production often leads to ecological destruction. In this respect, see Karl Marx on social metabolism (González de Molina and Toledo, 2014).
15. Unequal exchange occurs as a consequence of extra-economic factors such as skewed power relations in the markets.
16. This mechanism was actively used in most Western countries in the post–World War II period. It was also central in China during the first decades of the People's Republic.
17. Figure 6.12 is based on Dutch data. Revenues and costs are corrected for inflation. The 'hiccup' in the cost line (in the mid-1980s) was caused by the introduction of the quota system (this made growth at the farm enterprise level far more expensive) and obligatory investments in environmental protection measures.
18. Meaning that there are hardly any regulations to protect the environment or that prevailing regulations are barely applied.
19. This is beautifully synthesized in the title of a publication of the INOSYS Network of French Farmers (INOSYS Réseaux d'Elevage, 2016): *Cows, Areas, Taxes, Everything Is Increased Except Our Incomes*.

Bibliography

Akram-Lodhi, A. H., and C. Kay. (2010a), 'Surveying the agrarian question (part 1): Unearthing foundations, exploring diversity', *The Journal of Peasant Studies* **37** (1), pp. 177–202. https://doi.org/10.1080/03066150903498838.

Akram-Lodhi, A. H., and C. Kay. (2010b), 'Surveying the agrarian question (part 2): Current debates and beyond', *The Journal of Peasant Studies* **37** (2), pp. 255–284. https://doi.org/10.1080/03066151003594906.

Bossenbroek, L. (2016), *Behind the Veil of Agricultural Modernization: Gendered Dynamics of Rural Change in the Saïss, Morocco*, Ph.D. thesis, Wageningen University, Wageningen, the Netherlands.

Bruin, R. de, and J. D. van der Ploeg. (1992), *Maat Houden: Bedrijfsstijlen en het Beheer van Landschap en Natuur in de Noordelijke Friese Wouden en het Zuidelijk Wester Kwartier*, BLB, Ministerie van Landbouw en Landbouwuniversiteit Wageningen, Wageningen, the Netherlands.

Byres, T. (1991), 'The agrarian question and differing forms of capitalist agrarian transition: An essay with reference to Asia', in *Rural Transformation in Asia*, edited by J. Breman and S. Mundle, 3–76, Oxford University Press, Oxford.

Chayanov, A. V., and I. Kremnev. (1920/1976), 'The journey of my brother Alexis to the land of peasant utopia', *Journal of Peasant Studies* **4**.

Chayanov, A. V. (1924/1966), 'On the theory of non-capitalist economic systems', in *The Theory of Peasant Economy*, edited by D. Thorner et al., 1–28, Manchester University Press, Manchester, UK.

Chayanov, A. (1927/1991), *The Theory of Peasant Co-operatives*, Ohio State University Press, Columbus.

Dries, A. van den. (2002), *The Art of Irrigation: The Development, Stagnation and Redesign of Farmer-Managed Irrigation Systems in Northern Portugal*, Ph.D. thesis, Wageningen University/Wageningen Studies on Heterogeneity and Relocalisation nr. 5, Wageningen, the Netherlands.

Dries, A. van den, and J. Portela. (1995), 'Irrigation in two contrasting agrarian development patterns in the northern Portuguese mountains', in *Beyond Modernization: The Impact of Endogenous Rural Development*, edited by J. D. van der Ploeg and G. van Dijk, 191–218, Van Gorcum, Assen, the Netherlands.

Feder, E. (1977), 'Campesinistas y descampesinistas: Tres enfoques divergentes (no incompatibles) sobre la destrucción del campesinado', *Comercio Exterior* **27** (12), pp. 1439–1446.

Fei, H.-T., and C.-I. Chang. (1945), *Earthbound China, a Study of Rural Economy in Yunnan*, University of Chicago Press, Chicago.

Gerritsen, P. R. W. (2002), *Diversity at Stake: A Farmers' Perspective on Biodiversity and Conservation in Western Mexico*, Ph.D. thesis, Wageningen University, Wageningen, the Netherlands.

Gerritsen, P. R. W. (2018), 'Rural landscapes in dispute: On coproduction, farming styles and resource diversity in western Mexico', in *The Sage Handbook on Nature*, edited by T. Marsden, Vol. 3, 1489–1507, SAGE, London.

González de Molina, M., and V. M. Toledo. (2014), *The Social Metabolism: A Socio-ecological Theory of Historical Change*, Springer, Heidelberg, Germany.

INOSYS Réseaux d'Elevage. (2016), *Vaches, Surfaces, Charges … Tout Augmente sauf le Revenu (Quinze ans de Suivi en Bretagne, Pays de la Loire et Deux-Sévres*, L'Institut de l'Elevage, Paris.

Lenin, V. I. (1914/1972), 'The agrarian question in Russia', in *Collected Works*, Vol. 20, 375–377, Progress Publishers, Moscow.

Mariátegui, J. C. (1925), *Siete Ensayos de Interpretación de la Realidad Peruana*, Amauta, Lima, Peru.

Marsden, T. K. (2012), 'Towards a real sustainable agri-food security and food policy: Beyond the ecological fallacies?', *The Political Quarterly* **83** (1), pp. 139–145.

Meeus, J., J. D. van der Ploeg, and M. Wijermans. (1988), *Changing Agricultural Landscapes in Europe, Continuity, Deterioration or Rupture?* IFLA, Rotterdam, the Netherlands.

Moyo, S., P. Jha, and P. Yeros. (2013), 'The classical agrarian question: Myth, reality and relevance today', *Agrarian South: Journal of Political Economy* **2** (1), pp. 93–119.

Ploeg, J. D. van der. (2000), 'Revitalizing agriculture: Farming economically as a starting ground for rural development', *Sociologia Ruralis* **40** (4) pp. 497–511.

Ploeg, J. D. van der. (2003), *The Virtual Farmer: Past, Present and Future of the Dutch Peasantry*, Royal van Gorcum, Assen, the Netherlands.

Redaktiekollektief, M. (1980), 'Landbouw of natuur?', *Tijdschrift voor Landbouw en Politiek* **3**, pp. 7–48.

Rooij, S. J. G. de. (2018), 'Different farming strategies and the shaping of agricultural landscapes: The case of the Netherlands', in *The Sage Handbook of Nature*, edited by T. Marsden, Vol. 3, 1508–1531, SAGE, London.

Ruiz, J. P., and F. Gonzalez-Bernaldez. (1982), 'Landscape perception by its traditional users: The ideal landscape of Madrid livestock raisers', *Landscape Planning* **9**, pp. 279–297.

Smit, M. (2018), *De Duurzaamheid van de Nederlandse Landbouw, 1950–2015–2040*, Ph.D. thesis, Wageningen University, Wageningen, the Netherlands.

Stuiver, M., J. D. van der Ploeg, and C. Leeuwis. (2003), 'The VEL and VANLA cooperatives as field laboratories', *NJAS - Wageningen Journal of Life Sciences* **51** (1–2), pp. 27–40.

Swagemakers, P. (2008), *Ecologisch Kapitaal: Over Het Belang van Aanpassingsvermogen, Flexibiliteit en Oordeelkundigheid*, Ph.D. thesis, Wageningen University, Wageningen, the Netherlands.

Vecchio, Y., P. de Castro, M. Masi, and F. Adinolfi. (2022), 'Do rural development policies really help small farms? A reflection from Italy', *Eurochoices* **20** (3), pp. 75–80. doi:10.1111/1746-692X.12338.

Yu, T. (2019), *Energy and Labour Use on Farms: Case Studies from the Netherlands and China*, Ph.D. thesis, Wageningen University, Wageningen, the Netherlands.

7 Rural development processes

Overview: Main concepts discussed in Chapter 7

Rural development
Agency
San Nong (three ruralities)
Rural vitalization
Multi-functionality
Duality
Rural development as endogenous process
Rural development as response to the agrarian crisis
Rural development as process of transition
Rural development as re-peasantization
Rural development as the construction of new markets
Boundary shifts
Broadening
Deepening
Regrounding
The accumulation of small steps
Strategic niches
Regime shift
Partial and mutually competing transitional processes

If agricultural processes of growth chronically fail to deliver – that is, if they come with an unacceptable imbalance between positive and negative externalities (as discussed in Chapter 6) – new developmental trajectories need to be carved out. This is, indeed, happening in many parts of the world, albeit in a covert way.[1] Intriguingly, such new, and alternative, trajectories are often referred to as *rural development*. Given the many contrasting situations in which this notion is applied, it seems to be, at first sight, an umbrella concept. To many people, rural development (RD) is also a somewhat enigmatic term.[2] However, the concept of co-production helps to explain the idea of rural development.[3] Rural areas are the locus of co-production – the spaces where the interaction and mutual transformation of the social and the natural are located. RD, then, is about offering new opportunities

DOI: 10.4324/9781003313274-7

for co-production to once again meet the needs of society, nature and the actors directly involved in farming and food production (see Figure 6.11). Thus, RD tries to *redress* market failures that result in the emergence and reproduction of negative externalities. RD is not only attractive (although not always) for the actors engaged in farming – it often appeals to people other than farmers as well.

A well-balanced co-production evolves, as the Chinese say, 'in harmony with nature'. It brings sustainability. If well balanced, it is positive for the actors directly involved as well as wider society. To rephrase the same point in more technical terms: well-balanced co-production produces positive externalities and avoids and/or redresses negative externalities.

Rural development, then, is about making the rural more rural.[4] It aims at further unfolding co-production: enlarging the positive externalities and making them accessible and enjoyable to larger segments of the population and, simultaneously, reducing or completely redressing negative externalities. *RD is about better meeting the needs and expectations of society, the rural population obtaining more satisfactory (and diversified) livelihoods and establishing a better tuned interplay with nature.* The *concrete* (i.e., specific) needs and expectations will, of course, *vary* with time and space (and be different according to class, socio-economic conditions, culture and history). The same goes for the interplay with nature (which evidently will differ between the *altos* of the Andean mountains to the delta of the Danube) and for the position of the farming population. But what all of these situations have in common is that concrete improvements, especially when they are integrated and addressing wider segments of the population, might be (and effectively are) understood and represented as rural development.

Rural development includes the farming population, but not exclusively. It is both an important challenge and an opportunity for far wider segments of the population (see Text Box 7.1). But, at the same time, rural development cannot proceed without the farming population: they cannot be excluded.

Rural development is not a luxury that only rich countries can afford. Rural development also is actively aimed for – at least sometimes – in relatively poor countries (although its shape, in concrete terms, can be quite different). In this chapter, I will discuss and compare socio-geographical realities in places as different as China, Brazil and the European Union. All three have (or had, until recently, as in the case of Brazil) policies that actively and explicitly support rural development. And regardless of the many differences, there are amazing commonalities as well (just as there is, over time, a certain convergence). I will spell out some of these later on in this text.

The notion of rural development reflects the inherent link between the countryside and society at large. Urban people (and rural dwellers as well) are linked, through many different ties, to the rural and many long for the rural to be developed to better meet their expectations and needs. Text Box 7.1 illustrates some of these needs and expectations (and, indirectly, the methods that can be used to assess them).

Rural development is a dynamic process that assumes multi-level agency. It never is the simple and more or less automatic outcome of the economy. Markets do not,

Text Box 7.1   The rural as perceived by urban people (in China and Italy)

Are urban people indifferent to rural areas, or do they see them as important? What is their opinion about the quality of life in rural areas? How do they relate to rural areas? If they could move, where would they go? These and similar questions were central in a large survey (n = 2,000) held in medium and large Italian cities in 2008. The survey was part of a wider research programme, carried out by the University of Perugia, that centred on the quality of life in both urban and rural areas. The study focussed on both urban people as well as rural dwellers and farmers (for a full description of the research findings, see Ventura et al., 2007). The same questions were asked to people of different social categories.

The survey found that 78% of urban people regularly visited the countryside, with young people and the well-off (yearly income >26,000 euros) doing so more frequently than others. The quality of life of *rural* areas was considered by these *urban* respondents to be higher that of urban areas (7.26 versus 6.77 on a scale of 1 to 10).

This overall judgement was, as might be expected, a synthesis of contrasting considerations. Among the positive aspects, the relation with nature was thought to be better in rural areas than in urban ones (8.51 versus 4.06). This might be, in a way, quite obvious. This is far less the case with other findings such as 'balancing work and family' (7.03 versus 6.17), environmental quality (7.56 versus 4.41), 'the cost of living' (6.87 versus 4.76) and 'a place to raise children' (7.45 versus 5.60). Needless to say, several of these criteria relate, directly or indirectly, with rural co-production. There were also clear negative sides to the rural. Urban people perceived rural areas as worse than urban ones when it comes to employment opportunities, the presence and quality of public services (schools, health system, public transport) and communication (lack of broadband, etc.). Remarkably, the quality of social life was judged, overall, as being more or less equal (6.24 versus 6.22). One third (35%) of the urban respondents had considered moving to a rural area, although there were gender and age biases here: more men than women (42% versus 32%) have considered this, and more people younger people than those over 65 years of age. The people who had thought about moving to the countryside specifically included artists, the self-employed, people looking for work and unemployed people.

The same research programme also showed that those who had the chance to move to the countryside clearly preferred the 'new rural areas'; that is, rural areas with a strong multi-functional agriculture and actively maintained landscapes, nature and biodiversity (see Ploeg, 2008:162).

Comparable research was done in China (undertaken in 2013–2016 by the College of Humanities and Development Studies of China Agricultural University among students of four main agricultural universities). Of the surveyed students (n = 622), 36% had a rural background, 42% a strictly urban one, and 22% belonged to a kind of in-between category originating from small cities. Here, again, the survey showed frequent and many-sided contacts between the rural and the urban: 66% of these students visited the countryside several times a year (bringing back good food to town), and the majority of students with a rural background helped their parents on the farm (see Wu and de Rooij, 2022).

The respondents in the Chinese study attached great importance to agriculture as the source of food sovereignty (94%), but they were sceptical about the levels of food quality (84%). They cherished the countryside for its landscapes, nature, solitude and peace (97%) but, at the same time, were highly critical of the lack of recreational facilities (89%), good schools (85%), jobs (76%), good conditions for children to grow up in (68%) and a well-functioning internet (63%). Thus, the overall balance was negative. Asked how they would quantify the quality of life in rural areas (as compared to the towns, which received a hypothetical 6 on a 10-point scale), the respondents gave an average score of 4.98. This was a remarkable contrast with Italy (see above), where urban residents give the countryside a score of 7.20.

However, the currently negative balance between positive and negative features in China does not imply an abandonment of the rural. On the contrary: these young Chinese respondents strongly shared the opinion that rural areas should be developed (66%) and, especially, made more attractive for young people (88%). And, when reflecting about the future of their countryside, 77% of these students thought that peasant agriculture would not disappear, perceiving it as a guarantee for food security (66%) and as assuring a safer future (71%).

by themselves, bring about rural development. Instead, rural development is supplementary to, or a correction of, the 'normal' functioning of the markets. It requires explicit interventions from strong social drivers (in Brazil these are mostly social movements, in Europe it is civil society, whilst in China the state is the main, though far from only, driver). This centrality of agency (which turns rural development into a goal-oriented set of activities) is reflected in the way the concept of rural development is phrased in China. The first version was phrased as *San Nong*, literally meaning the 'three ruralities'. These are *Nong Ye, Nong Min* and *Nong Cun*. The first term regards society as a whole and implies that agriculture should result in food security (or food sovereignty).[5] *Nong Min* refers to the people involved in farming: peasants should obtain fair incomes. The last term relates to the rural as a whole and specifies that rural villages are to be nice, clean and accessible places. This *San Nong* approach[6] evolved into 'the construction of a socialist countryside' after which 'rural vitalization' became the guiding concept (see Text Box 7.2).

Europe and Brazil, in their turn, use discourses for announcing, legitimizing and outlining rural development that differ in several respects from Chinese discourse. Text Box 7.3 presents some of the major points from the *Cork Declaration* on rural development that was formulated in 1996 by Franz Fischler, at that time the European commissioner for agriculture, in consultation with a range of stakeholders. Text Box 7.4 contains fragments of a somewhat different document: a hindsight of Guilherme Cassel (2022), who previously was Brazilian minister for rural development. This hindsight not only refers to the objectives of rural development – it also synthesizes some of the major outcomes of the rural development process. In all three examples, what is declared is as telling as what they do not say. I will illustrate this in the following section (especially when it comes to the duality of rural policies).

Text Box 7.2   Rural vitalization in China

In China, the concept of vitalization is currently applied to agriculture. Nowadays there is a well-outlined policy for 'rural vitalization'. In Chinese language this concept is written as: 乡村振兴. Together, the last two characters (振兴; in Pinjin: *zhèn xīng*) make up the word *vitalization*. 振 means 'to raise, to excite, to arouse action', which comes close to energy, and 兴 means 'to thrive, to prosper, to flourish'. As a whole, the expression is about arousing action in order to make agriculture prosper. The characters and, especially, their combination, make clear that rural vitalization is assumed to be an active process that aims to bring a positive impact. Implicitly, the combination also suggests that the positive results stem from the internal energy. It is noteworthy that both characters are verbs. Thus, rural vitalization does not refer to a fixed or stable state of affairs but to a process, to a movement through time. In Chinese dictionaries the word 振兴 is explained as 'to develop vigorously'. Ironically, 乡村振兴 (rural vitalization) is often translated as 'rural *re*-vitalization'. Such a translation does not feel very comfortable because it comes with the association of a patient who is, as it were, to be 'revived'. If such a notion was to be communicated, other characters would have been used, such as those making up another Chinese word: 复兴 (复 means to repeat, to return). Be that as it may, in the authoritative opening address of Xi Jinping at the 19th National Congress of the Communist Party of China (Xi, 2017:27), the concept that is used is rural vitalization and definitely not re-vitalization.

Text Box 7.3   Excerpts from the *Cork Declaration* on rural development

Point 1: Sustainable rural development must be put at the top of the agenda of the European Union, and become the fundamental principle which underpins all rural policy. [...] This aims at reversing rural out-migration, combating poverty, stimulating employment and equality of opportunity, and responding to growing requests for more quality, health, safety, personal development and leisure, and improving rural well-being. The need to preserve and improve the quality of the rural environment must be integrated into all Community policies that relate to rural development. [...]

Point 2: Rural development policy must be multi-disciplinary in concept, and multi-sectoral in application, with a clear territorial dimension. It must apply to all rural areas in the Union, respecting the concentration principle through the differentiation of co-financing for those areas which are more in need. [...]

Point 3: Support for diversification of economic and social activity must focus on providing the framework for self-sustaining private and community-based initiatives [...]: [including] the development of viable rural communities and the renewal of villages.

Point 4: Rural development [needs to] sustain the quality and amenity of Europe's rural landscapes (natural resources, biodiversity and cultural identity), so that their use by today's generation does not prejudice the options for future generations. [...]

Point 5: Given the diversity of the Union's rural areas, rural development policy must follow the principle of subsidiarity. It must be as decentralised as possible and based on partnership and cooperation between all levels concerned. [...] The emphasis must be on participation and a 'bottom up' approach, which harnesses the creativity and solidarity of rural communities. [...]

Point 6: Rural development policy, notably in its agricultural component, needs to undergo radical simplification in legislation. [...] There must be greater coherence of what is presently done through many separate channels, [...] decentralisation of policy implementation and more flexibility overall. [...]

Text Box 7.4   Excerpts from a speech by Guilherme Cassel, former minister for rural development in Brazil (Cassel, 2022:37–38, 49)

'Family and peasant farming came to be perceived by the government of (then) President Lula as a dynamic, modern economic sector capable of better responding to the contemporary challenges of increased food production, social inclusion and environmental sustainability. This significantly changed the quality of life in rural Brazil. In this short period, the income of family farmers grew 33% in real terms, while the average income of the population increased by 13%. Direct income from agricultural work now accounts for 64% of the total income of family farmers.

Over the past eight years, the poverty rate in rural areas has fallen from 52% to 33% (5.4 million rural workers were lifted out of poverty) and the extreme poverty rate went from 25% to 14% (3.1 million people). The percentage of households considered food secure went from 56.5% to 64.9%. [...]

What actually happened to achieve this kind of results in such a short time? The answer is simple and indicates that it was necessary to simultaneously follow two pathways: 1) to face and criticize the cultural prejudices about family and peasant farming and 2) to build a set of public policies for production support in this sector. When it comes to governing, time passes quickly, and reality always tends to demoralize the technocrats on duty. At the start of the new policy it was clear that it would not be possible to proceed in a stepwise way (one step after the other). It was necessary to do everything at once. As the chorus of a Brazilian rock band says, "all at the same time now." And so it was done.

In eight years (2003–2010) the Brazilian government settled 614,088 families and created 3,544 new settlement projects. In this period, the area incorporated into the Agrarian Reform was 48.30 million hectares. To give you an idea of the magnitude, this means almost twice the area of the state of São Paulo. [...]

Brazilian rural reality changed its face with the appreciation of family farming. In addition to the increase in food production, it has become a more equal space with less poverty, more access to rights and more infrastructure. It has been proven that it is possible, with appropriate public policies and well-focused investments to change reality in a short period of time. [...]

In 2016, there was a coup in Brazil that removed an elected president from the government. The conservative governments that followed the coup have been ruthless with family farming. On the day after the coup, as their first act, the new government eliminated the Ministry of Agrarian Development, responsible for Agrarian Reform and public policies aimed at family and peasant farming. Gradually all policies were being dismantled and/or simply eliminated. Land concentration has increased again, deforestation has reached the highest level in history, the number of young people in the countryside has plummeted and intensive use of pesticides has grown. [...]

By looking back to these facts, we are facing a sad conclusion: what is difficult to accomplish is easy to destroy. What was built over ten years, with the participation of farmers and their organizations, was quickly dismantled by the government on duty to meet the interests of large landowners. What is somewhat comforting is to know that what has been done can be repeated and improved, perhaps with even more power and more features. We know what needs to be done and how to do it. But how can we prevent this kind of destruction from happening again? Obviously, there is no ready answer. Perhaps the challenge is to build up enough political strength to make public policies aimed at family and peasant farming not just government policies but state policies'.

China, Brazil and the European Union: The commonalities of rural development

The contours of rural development policies in Europe were clearly outlined for the first time at the Conference of Cork (held in 1996) and reaffirmed at the Conference of Salzburg held in 2003. At that time, it was clear to everybody that the prevailing development trajectory (and the common agricultural Policy that supported it) was producing far too many negative externalities. The main trends needed to be *reversed* (as the *Declaration* literally said; see Text Box 7.3, especially point 1). Rural economies in the peripheries of Europe (in the mountainous and hilly regions and in Eastern Europe) suffered stagnation, and employment levels had declined sharply.[7] In the typical growth centres of European agriculture (such as northwestern Europe, the fertile deltas in the Mediterranean area and the Paris basin), farming was becoming increasingly industrialized and environmental pressures were growing in uncontrollable ways. At the same time, the production of high-quality and regionally specific products that compose the heart of the different culinary traditions was increasingly under threat whilst scenic landscapes were suffering degradation. All of this triggered considerable and widespread discontent and critique – in general and towards the common agricultural policy (CAP) specifically: why subsidize, to the tune of billions of euros, a sector that does so much harm?

Thus, a policy for rural development was born, one that operated *parallel* to the CAP. It was called Pillar 2 whilst the previous CAP was continued under the name of First Pillar. The formulation of Pillar 2 policy instruments (and budgets) and, later, its implementation occurred through negotiations at many interfaces: between the European Commission and the Member States, between the latter and the many regions,[8] but also between old and new interest groups. Thus, a policy was shaped that, on the one hand, showed a limited set of clear objectives but, on the other, a wide and sometimes confusing array of modalities for implementation.[9] The objectives can be summarized in terms of *protection*: Rural development was to protect the polyvalence (see Chapter 1) of the European countryside. It was to safeguard the many positive externalities entailed in it and redress the negative ones (see Chapter 6). This was needed as the markets for food and agricultural products failed to do so. The dynamics of these markets were actually further aggravating the existing problems. Hence, policy interventions were needed. Thus, the European policy for rural development was born as a tool meant to correct *market failures* (see previous chapter). Tellingly, though, the *raison d'etre* of RD policies, to correct (or at least balance) market failures, was never made explicit.

The new RD policy not only triggered debates and socio-political struggles on what was supposed to be rural development – it also strongly interacted with debates that had already been going on for quite a while and, especially, with the many grassroots initiatives all over Europe that aimed to improve the quality of the rural. Rural development was not just 'invented' at the level of policy; the elaboration and implementation of policies interacted with all kinds of niche activities, blossoming new developmental tendencies and novel practices,[10] and this interaction again implied a series of interfaces that, in the end, turned rural development into a differentiated, decentralized and negotiated phenomenon that very strongly reflects the specificities of the local.

In the different agricultural systems of Europe, growing segments of the farming population had already started to develop their farms into multi-functional units: enterprises that, alongside their traditional agricultural produce, also produced a range of new products and services that appealed to the desires of the European citizenry (I will discuss the details of this development later under the heading 'Responding to the squeeze on agriculture'). These segments and the newly defined rural development policy started to interact (albeit often in complex, if not contradictory, ways). Schematically speaking, these new practices became symbols that justified the new policies, whereas the latter started to support the former.

Given the socio-political power balances within the European Commission and, especially, within and between the Member States, a comprehensive and abrupt reorientation of the main development tendency in European agriculture (i.e., further industrialization of agricultural production and further subordination to capital) was almost impossible. Thus, the needed correction of the many market failures took a side road: The *Second* Pillar was parallel to the first (and main) one and meant to strengthen practices and trajectories that *differ* from the dominant trajectory. In this way, a kind of *duality* entered into European agricultural policy, which reflected and further articulated the dualism existing in agriculture itself. Whether such a dual structure can really be effective remains an important question in ongoing debates and in several research programmes.

Dualism is an equally strong feature in, and of, Brazilian agriculture and Brazilian agricultural policies. Next to a segment of large-scale and basically export-oriented farm enterprises, Brazilian agriculture also includes a large and richly chequered peasant agriculture (or family farming, as most Brazilians prefer to say) that, on just 24% of the available land, generates 38% of the total value of production and 74% of all agricultural employment. Family farms supply most of the fresh food consumed in Brazil: 58% of all milk, 50% of chicken and pork, 46% of maize, 70% of beans and 87% of cassava.

The interests of the first (agro-industrial) segment are represented and served by the Ministry of Agriculture, while those of the second (family farming) segment are represented by a Ministry for Rural Development. The latter was created at the start of the Lula presidency and operated in close (albeit often conflict-ridden) cooperation with the strong rural movements (such as Movimento dos Sem Terra [MST], Movimento dos Pequenos Agricultores [MPA] and others). This ministry and its rural development policy were widely understood as a line of defence for the family farming sector. As MERCOSUR's Specialized Meeting on Family Farming (REAF) observed: 'public policies for this farming segment should be conceived as a counterweight to the unwanted effects of free trade, building mechanisms to promote a new pattern of development consistent with social inclusion, sustainability and citizenship' (REAF, 2006). REAF understands family farming as a 'bearer of diversity, expressed in their production systems, ways of life and cultural density. It is a sector with high potential and capability to balance contrasts between producing regions, to develop economic confidence and to generate political stability, which requires distinct policies comprising an integral part of economic policy' (REAF, 2006; see also Sergio Schneider, 2016).

This critical orientation, the support of social movements, the availability of considerable budgets and the design of new policy instruments made the Brazilian experience an extremely interesting case of rural development. It also brought about impressive socio-economic impacts. Text Box 7.4 includes fragments from a speech given by Guilherme Cassel, former minister of rural development, and gives some key data. These data show that rural development really does have the potential to make a difference – to achieve what definitely cannot be reached by following the path of further industrialization.

In China there is, at least at the institutional level, no duality. In a comparative perspective, the *San Nong* policy and the versions that followed later (including the current rural vitalization policy) excel as well-integrated and comprehensive policies that systematically take into account all of the dimensions synthesized in Figure 6.11. Nonetheless, at a lower level, there clearly are tensions, if not contradictions – mainly between the desire to support large-scale farming and contrasting efforts to move peasant agriculture forwards. Such tensions are omnipresent, whilst the possibility of contrasting developmental trajectories is often translated into explicit experiments that fit with the Chinese approach of 'crossing the river by feeling the stones' (as Ye et al. [2010] described the rural development process in China).

The effort to mediate, correct and/or even remedy the effects of global capitalism on the countryside and agriculture is as central to Chinese rural development as it is to European and Brazilian processes of, and policies for, rural development. It is probably even more important in China than elsewhere. This is due to the enormous size of its rural population, to the vulnerability of food provisioning and/or to the historical legacy of the Communist Party (which essentially was a peasant party, just as the struggle against the Japanese and then the Kwo Min Tang was a peasant war). There is also very much to *protect, defend* and *reverse* in China. This implies that new arenas are created that, regardless of the tensions between rural development and hegemonic trends and policies, are of considerable socio-political relevance.

In China, Brazil and the European Union, rural development is an active response to market failures. This is reflected in the *objectives* it aims for; it is also shown in the *impacts* that rural development policies are looking for (and which are, at least partly, being realized). When it comes to the main *mechanisms*, though, things are less clear and sometimes highly confusing. There are regulatory schemes and subsidies, just as there is massive spending. But underlying all of this there again is a common denominator: *the construction of new markets* comes to the fore as a central and common mechanism used in Europe, Brazil and China. Rural development tries to correct market failures and, interestingly, it tries to do so through the construction of new markets (Text Box 7.5 offers an iconic example). This apparent paradox, which currently underlies and

---

Text Box 7.5   The Brazilian programmes for school meals and the procurement of food ingredients

The provisioning of school meals for all schoolchildren throughout the country was one of the important pillars of the Brazilian Fome Zero (Zero Hunger) programme. Politicians (and involved intellectuals and social activists) understood that the acquisition of the needed ingredients could easily follow standard approaches and thus mainly strengthen large retailers and even further encourage food imports. In order to prevent this happening, the PAA (Programma de Acquisaçao Alimentar), a special programme for the acquisition of food ingredients needed for school meals, was designed. It was part of wider schemes for public procurement.

The PAA included some strategic points. One of these specified that at least 30% of the needed ingredients were to be bought locally from local family farmers. Agro-ecological food got an equal preference. This, in itself very simple, specification immediately created a range of new, localized markets throughout Brazil: a new local outlet for small family farmers and a strong stimulus for agro-ecological production methods. The PAA also led to the construction of many new local cooperatives of family farmers and smallholders whose purpose was to channel the needed food ingredients as smoothly as possible to the schools.

Both the School Meal Programme and the PAA are very well documented. An excellent overview was published by Trichet and Schneider (2010). The example later inspired several other countries in the Global South to mount similar programmes.

The FAO was very helpful in spreading the word about the Brazilian example and providing the needed administrative, technical and legal data and insights.

Theoretically, the PAA and the School Meal Programme are exemplary because they illustrate how the state can assess the required balance(s) between the social, the economic and the ecological spheres. Secondly, they also show the market to be an institution – a set of rules – that can be changed. This implies that new markets can be created by introducing new rules (which, of course, were contested at the time by vested interests, although the state stuck to its guns). Thirdly, the example illustrates that peasant struggles are partly (if not greatly) located in the sphere of circulation in which the market(s) play an important role. Fourthly, the example highlights the potential of new urban-rural coalitions (of producers and consumers of food) that are firmly grounded in common interests. And, finally, it is an example *par excellence* that shows how the concept of food sovereignty (see Text Box 9.4) can be operationalized.

characterizes rural development processes and policies, can be easily explained – at least if the *variable* nature of markets is taken into account.[11] Markets can differ in many respects: they can be governed in different ways; the distribution of value added in, and through, the market can also significantly differ; their infrastructures can be very different; and, consequently, products will 'travel' in distinguishable ways (see Table 8.1 in the next chapter) and bring contrasting consequences.

In Chapter 8 I will extensively discuss the construction of new markets. In the rest of this chapter I will present different analytical angles (different methodological approaches) to grasping and studying rural development processes.

Different approaches to rural development

Rural development processes can be studied in different ways, and each particular approach highlights specific aspects whilst leaving others hidden. Thus, combining different approaches is often recommended in order to obtain a panoramic view that encompasses the multi-dimensional, multi-level and multi-actor nature of rural development processes.

Rural development can be conceptualized and studied as an *endogenous process* that starts at the grassroots level and then (perhaps) covers increasingly extended areas. After a specific point in time, this endogenous process may start to interact with higher levels of aggregation (especially the state). Text Box 7.6 gives a Chinese example (Wu et al., 2015). Generally speaking, many different forms of interplay between the state and local initiatives are possible. These go, as explained by Colin Anderson et al. (2020:156–171), from suppression via co-option, containment and shielding to nurturing and anchoring.

It is also possible to represent and study rural development as a compound *response to the agrarian crisis* (the intensified squeeze on agriculture). When the dominant development trajectory in, and of, the agricultural sector increasingly

---

Text Box 7.6   *Nong jia le,* agro-tourism in China

Nong jia le (literally 'happy peasant home') is the branding of agro-tourism in China. Agro-tourism germinated at the end of the 1980s, more or less simultaneously in several places. One of these was Shenzhen, the first Social Economic Zone in China, where all kinds of economic experiments were possible and encouraged. One village in this area organized a lychee feast, a festival to highlight local culinary traditions and associated agricultural practices. This required having the infrastructure for visitors to have dinner and stay overnight. Thus, the first 'happy peasant homes' opened their doors. Another initiative was born in Sichuan Province (in 1986) after local peasants started to produce flowers. This resulted in blossoming fields that attracted many visitors. Xu, a local peasant, thought that he might experiment with offering accommodation, and food and after discussing this with his family he started doing so.

These first examples were quickly followed by others, and these first expressions of *nong jia le* were recognized by the state and subsequently supported through a programme that developed through three stages. The first (before 1994) centred on supporting single farms and households. The second stage (1995–2000) focussed on the village level and put architecture, folklore, culinary traditions, festivals and rituals centre stage.

After 2001 attention moved to the landscape level, embracing several villages at the same time. Scenic beauty, good roads, walking paths, panoramic viewpoints, archeological monuments and/or natural parks became the new, overarching points of reference. This stage provides infrastructure that supports the 'lower' levels, (i.e., 'happy peasant homes' and folklore villages). In this third stage, state interventions were decisive.

Together, these three stages show a progressively unfolding process of evolution for supporting and strengthening agro-tourism and help to provide an adequate response to the growing demand. In 2007, tourists spent a total of 335 million nights in the many 'happy peasant homes', and agro-tourism generated an income of 49.8 billion RMB (roughly 8 billion euros).

---

appears to be a cul de sac, rural development opens new windows – new ways forward that allow farmers to move beyond the impasse of the dominant trajectory. An interesting consequence of this approach is that it also allows for studying rural development as a kind of socio-political struggle between a dominant constellation (certain farming styles, agri-business groups, banks, state apparatuses, agricultural sciences and formal farmers' unions that are all aligned with the prospect of continuing along the lines of the dominant development trajectory) and a counter movement (that loosely groups other farming styles, new institutional patterns, critical views and new peasant-like movements together).

Thirdly, one might conceptualize rural development as a process of deliberate re-peasantization. It is a process through which farmers turn their farms back into peasant units of production – not those of the past but peasant farms of the 21st century. While there are many differences from the past, these new peasant farms also represent, from an analytical point of view, a continuity with the peasant mode of farming. Regaining autonomy and rebalancing the relations with living nature (in order to obtain new, more sustainable and more pleasant forms of co-production) are important drivers. Young people play an especially important role here.

Finally, rural development might be thought of as a transitional process – a highly complex process that embraces several decades and which will bring a shift that goes from today's agriculture towards a completely re-designed agriculture of the future whose contours are now only vaguely known but which will surely be totally different from those of today. Using transition theories (mostly developed for studying and reflecting historical processes of transition and/or ongoing transitional processes in non-agricultural sectors) can be highly useful in and for this approach.

These approaches can be combined, and doing so is one of the huge challenges for the sociology of farming (and other disciplines), for, in the end, rural development processes (at least considerable parts of them) are *simultaneously* (1) responses to the current and multiple crises (i.e., the many market failures that hamper farming, food production and the countryside) that (2) mainly start and unfold as endogenous processes, (3) proceed as re-peasantization and thus (4) constitute a major building block for large and ongoing processes of transition.

Responding to the squeeze on agriculture

Rural development processes unfold at different levels and in many different ways, just as they find their expression within different dimensions. It cannot be otherwise, because the specific agrarian questions that are to be faced differ from place to place. What all rural development processes have in common, wherever they are located, is that adaptations at the micro level (i.e., the farm enterprise level) compose the core, the engine and the power of such processes. These adaptations might emerge from one's well-understood self-interest and the (newly defined) prospects of farmers, farm women, new entrants and other actors involved in agricultural production (which is especially the case in Europe). They may also result from the actions of powerful social movements that, in the end, result in the construction of new farms that, together with existing farms, carve out new ways towards a more convincing future.[12] Even in China, where rural development processes are strongly propagated by the state and party, the micro level plays a central role as it is here that new solutions are discovered, designed and tested – after which some of these are taken up by the state and generalized over far wider areas. But, even then, these new lines of action only function if they are accepted and successfully implemented at the micro level.

The centrality of the micro level is a feature shared in common by all current rural development processes (note that I am applying here the search for commonalities discussed in Methods Box #3 in Chapter 1).[13]

At the grassroots (or micro) level, rural development occurs as a series of interconnected *boundary shifts* (this concept was developed by Flaminia Ventura and Pierluigi Milone, 2004): the original 'boundaries' of the farm are actively shifted to create a new unit of production that has increased economic activity. This is illustrated in Figure 7.1 in which the inner triangle represents the initial farm whilst the outer one

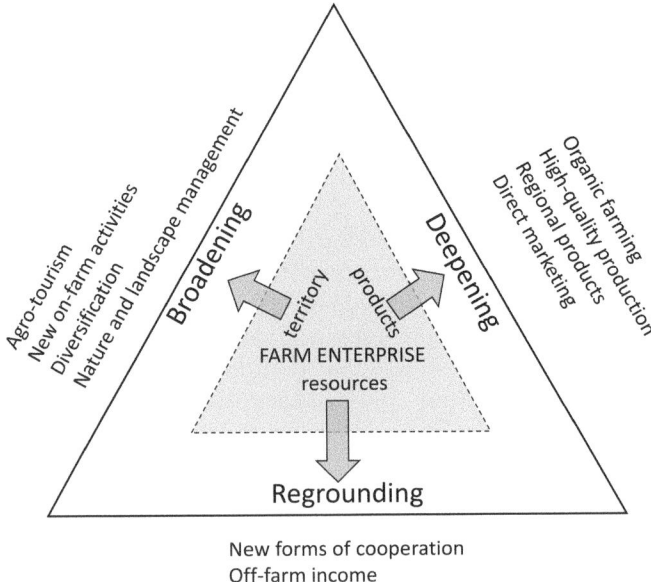

Figure 7.1 Boundary shifts (Ploeg et al., 2002).

refers to the newly created farm (that is, after the shift). Both triangles have three sides. These refer, for the horizontal side, to the mobilization of resources needed in the farm; for the right diagonal, to the agricultural process of production; and thirdly, for the left diagonal, to the farm being part of the wider economy and society.

Now, on all of these frontiers it is possible to make important shifts, which may be technical, economic, institutional or social or a combination. One set of changes can be summarized as *deepening*. This means that the agricultural process of production (and often marketing as well) is changed in a way that increases the value added (VA) produced per unit of end product (or service). This can occur through, for example, changing from conventional to organic production or to the production of high-quality products and/or regional specialties; it is also possible when direct marketing (in whatever form) is introduced in the farm. A higher VA will, of course, increase the labour income.

A second series of changes is mostly summarized as *broadening*. Basically, it is about adding more economic activities, beyond agriculture, to the farm enterprise using the same resource base that is used for farming as well as these other economic activities. These other activities are relatively close to farming (in a way they can be understood as part of co-production). Thus, synergies are created and an additional competitive advantage is created, for the cost of the used resources is now shared by two or more activities (Saccomandi, 1998). Examples of such other activities are water retention, energy production, agro-tourism, the provision of care facilities, the protection of bio-diversity, the maintenance of scenic landscapes[14] and all kinds of smaller activities.[15] What they have in common is that they link the farm in new, multi-functional ways to the rest of the (rural) economy by providing all kinds of useful and desired goods and services, whilst also providing additional income to the farming family.

Thirdly there is *regrounding*. This is a set of changes located at the bottom axis of Figure 7.1, and these changes are all related, in one way or another, with a reorganization of the farm resource base towards higher degrees of self-provisioning. Reducing the amount of external resources used and simultaneously increasing the quality and quantity of internal resources (a movement also called re-integration because it is the opposite of externalization, as discussed in Chapter 2) is one important mechanism here. Changing over to agro-ecological farming (i.e., regrounding farming on living nature) can be understood as a specific expression of such a reorganization. Other possibly relevant mechanisms are engagement in multiple job holding (or pluri-activity), which can be used to obtain the funds needed for investments and the engagement in new kinds of cooperatives (see Chapter 9). Regrounding can, if it is accompanied by, and in line with, other adaptations in the farm, result in often considerable cost reductions.

Together, deepening, broadening and regrounding activities can make for an important additional income that, combined with the income from farming as such, allows the farming family to defend and even increase their total (labour) income[16] and to thus resist the deterioration of their livelihood that would otherwise occur. In many cases, these boundary shifts introduce a new growth potential into the farm.

The building of new multi-functional farms implies an important move away from the main markets for agricultural commodities – not only because new and distinctive products and services are being produced and marketed but more because the farm is no longer solely reliant on the main agricultural markets to reproduce itself. Such farms are increasingly able to reproduce themselves through the new circuits they have engaged with (such as markets for agro-tourist facilities, regional specialties, eco-services, part-time jobs, etc.).

Building and developing a multi-functional farm is far from easy and is never a standardized routine that follows a universal blueprint. This is the case because broadening, deepening and regrounding all imply a return to the specificities of the local. What are the regional needs that can be met through broadening activities? What are the particularities of farming and eco-system that can be translated into regional specialties and/or high-quality produce? Which are the natural resources that are locally available, and how can they be (re-)integrated in farming? What are the potential and limitations entailed in the farming family? Can the new tasks be fitted into the existing schedules of the family members? Such questions (and many comparable ones) are decisive for the start, success and/or failure of the new activities and their integration into the compound web of balances entailed in the farm and farming family (Oostindie, 2015).

All of this represents a rupture with major routines and a return to the specificities of the local. Consequently, the start of new, multi-functional, activities requires *agency*. I refer here to grassroots agency: the capacity of the actors directly involved in farms *to actively make a difference* and to do so in a goal-directed and knowledgeable way (see Text Box 7.7 for an illustration). Agency is the capacity to

Text Box 7.7   Agency

The picture below shows two members of the Hoekstra family. They farm near the town of Sneek, in the midst of the Lake District of the province of Friesland. They are the mother and her elder son (the father passed away a few years ago). They are standing in front of a self-constructed small harbour for yachts. Sailing tourists can stay here overnight, and if they want they can use the farm camping site as well. In the background stands the old farm building, which now harbours a farm shop (and recreational facilities for children when it is raining). At the right-hand side is the newly build cow shed which the Hoekstra's were able to finance with earnings obtained from their agro-tourist facilities.

Photo by Adriaan van der Ploeg.

Building this small harbour was far from easy. It required considerable agency. First there was the design, which required connecting two elements that are strictly separated in conventional farming: the farm and its territory. Using the lake district as an opportunity was what *other* (non-agricultural) enterprises did and was not considered to be a farmer's business. Offering *on-farm* facilities and services to water sports tourists was a completely novel notion (and it took quite some energy to overcome the scepticism in their professional surroundings). The notion only germinated because the Hoekstras gathered the required knowledge. Then they needed the capability to deal with others (provincial and municipal authorities) and to connect with entities and persons with whom they had had no previous contact. For them the construction of this new type of agro-tourist farm represented not only a struggle (see also Chapter 9): it also went against the grain. Yet they managed to construct something anew that helped them to escape from 'the race to the bottom' in which so many others are entrapped: a new prospect. They shifted the boundaries of their farm.

Due to their agency, the Hoekstra's succeeded – regardless of all of the difficulties. They effectively repatterned their socio-material reality (see Chapter 10) and are now standing proudly in front of what they themselves created.

get things done – 'things' that would otherwise not occur (nobody *else* will build an attractive farm shop on your farm for you and actively organize a network of consumer/clients; such a shop, or whatever other expression of multi-functionality, only becomes reality when you yourself actively engage in its construction). In order to develop new functions in the farm, one has to deal with others and obtain new skills and insights. Thus, agency also requires the capacity to relate to other people, to develop knowledge and to operate on complex interfaces. Above all, agency is the capacity to create connections that previously didn't exist (and probably hadn't even been conceptualized). Following the previous example: a farm building was seen as being there to be used for farming and located in the countryside, just as a shop was located in town and meant for selling things. However, now these mutually separated entities (farm and shop) become *connected* in the concept of farm shop. Thus, rural development is, at a grassroots level, a way of repatterning socio-material realities – which critically requires agency.

When increasing numbers of farms are 'shifting their boundaries' – that is, engaging in the transition towards a multi-functional agriculture – they can together produce a considerable socio-economic impact. Figure 7.2 (which builds on Figure 6.12) summarizes the mechanics and dynamics of such an impact.

The tendencies in the main markets for agricultural commodities show (in real terms) a stagnation of price levels (or even a relative decrease) and/or low growth levels of total physical output and thus produce a (relative) stagnation of total GVP (this is sometimes also due to environmental limits and/or quota systems). At the same time, the cost levels associated with agricultural production show steady but persistent increases (partly related to increasing real prices and partly to higher levels of input use and depreciation). Together these two trends result in the squeeze on agriculture. Against this background, rural development offers a

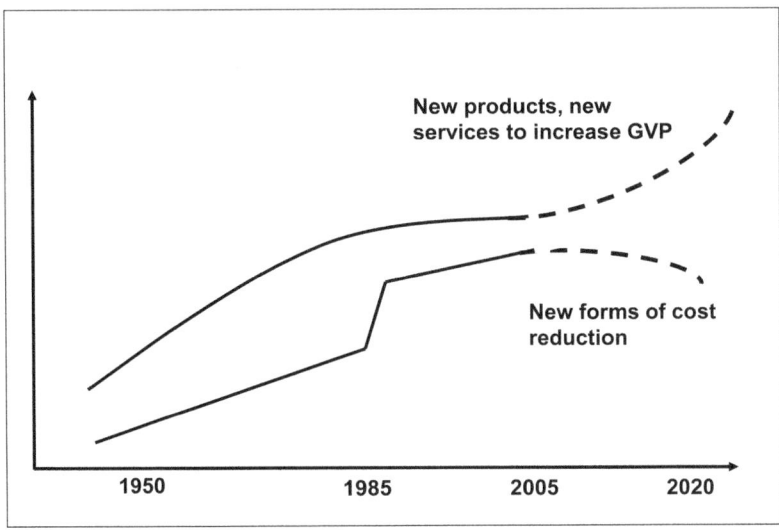

Figure 7.2 Responding to the squeeze on agriculture: The impact of rural development (author's own elaboration).

structural alternative, if not (the possibility of) a radical rupture (see the two dotted lines in Figure 7.2). Broadening and deepening bring the possibility of producing new products and services (bringing increases in GVP). Regrounding, in its turn, reduces the total (monetary) costs associated with farming. Thus, the total VA of, and in, the agricultural sector is increased (instead of being squeezed).

There are several empirical enquires into the socio-economic impact of rural development processes. Apart from showing a considerable impact (see Text Box 7.4 for Brazil and Text Box 7.8 for the EU), they also bring forward an important methodological principle. That is, that the *accumulation of small steps* (which, taken in isolation, seem to be very modest) *can make a very big difference*. This is because the agricultural sector is composed by many small units of production. When one or only a few farms develop into multi-functional units and are able to increase their income due to additional economic activities, this may just constitute interesting case studies or be seen as an interesting deviation to 'the rule' that, when related to the larger whole, seems little more than a rather irrelevant phenomenon. When, however, *increasing* numbers of farms develop such extra-economic activities, they may *together* contribute a substantial addition to the total income of the sector. Without this addition, the 'purely' agricultural income would be so low that many farms would probably be forced to close down.

Figure 7.3 offers another illustration of this methodological principle. It illustrates the situation of a sample of large, professional farms in Italy in the year 2000 (n = 795; Rooij, et al., 2013). At that time, only 27% of these farms had developed multi-functionality; the remaining 73% could be considered as conventional. When asked about the next 5 years to come, the farmers identified

Text Box 7.8 **An impact assessment of rural development processes in Europe:** **Delta net value added (in million euros) + number of involved** **farms (2000, seven countries) (Ploeg et al., 2002)**

	Netherlands	United Kingdom	Germany	Italy	Spain	Ireland	France
Organic farming	22.7	24.8	83.5	213.6	41.6	2.1	31.0
	962	1,462	9,200	43,700	7,400	900	8,140
Quality	85.0	53.6	209.0	864,7	142.0	15,8	887.0
production	3,000	3,190	40,000	143,200	223,650	160	182,500
Short chains	68.0	318.0	678.3	328.0	262.2	1.7	840.0
	6,000	14,700	35,000	800,000	90,000	786	102,000
Agro-tourism	20.0	330.8	615.0	131.5	8.8	12.7	75.5
	2,500	19,400	62,000	5,280	2,180	1,870	16,500
New on-farm	160.0	211.0	24.0	16.0	NA	18.5	3.6
activity	4,400	16,200	4,500	1,700	NA	240	1,195
Diversification	37.1	76.8	119.4	62.7	40.2	34.8	108.6
	11,800	10,740	21,000	28,650	40,600	16,600	37,000
Nature and	11.9	71.3	165.4	86.4	87.7	124.8	137.6
landscape	12,000	46,255	100,000	40,920	55,550	34,700	90,100
Total deepening	175.7	396.3	970.8	1,406.3	445.7	19.5	1,758.0
Total broadening	239.0	689.8	914.7	296.7	136.6	190.8	325.3
Deepening + broadening	404.7	1,086.2	1,885.4	1,703.0	582.4	210.4	2,083.3

The total extra value added for each new economic activity is given per country (in millions of euros), together with the number of farms involved. Thus, in the year 2000, the Netherlands had 962 farms dedicated to organic farming, which together rendered an extra net value added (that is, above the VA they would have generated by farming conventionally) of 22.7 million euros per year.

The table shows above all that *together* the many different activities rendered a considerable extra value added to the agricultural sectors of these seven countries: nearly 8 billion euros, four times as much as the total family income in Dutch agriculture in 2000. The comparison is interesting. Dutch agriculture was, at that time, considered to be an 'economic giant'. The data show that the emerging multifunctionality in European agriculture greatly outnumbered this 'giant'.

different possibilities. Some indicated that farm closure was more than probable (8% for the conventional farms; 1% for the multi-functional farms), while others indicated that they would develop new economic activities (13%). Such changes are quite small. However, Figure 7.3 also shows that, when taken together, the accumulation of small differences may make an important change. After 5 years the balance between conventional and multi-functional farms would have changed considerably, especially among younger farmers. At that time (2008),

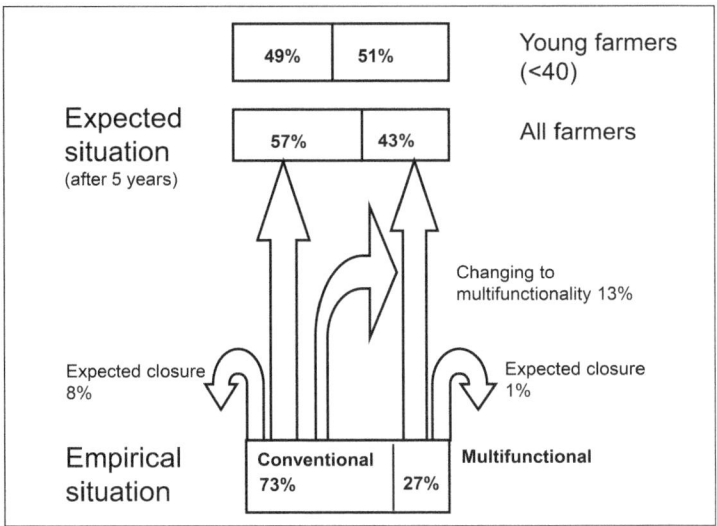

Figure 7.3 Expected changes in Italy on large, professional farms (2000–2005) (author's own elaboration).

such a 'landslide' (comparable to the one documented for France; see Figure 4.2) was hardly expected – precisely because people and institutions did not take into account that many small differences together can make for the large alteration.

Now, it can be argued that the shifts documented in Figures 4.2 and 7.3 just involve a redefinition of the notion of conventionality. The 'expected situation' in Figure 7.3 is nothing but the 'new conventional'. If conventional is understood as referring to what the majority of farmers are doing, this is undoubtedly true. However, it might also be argued that this shift from the 'old' to the 'new' conventional implies some structural, and theoretically meaningful, changes taking place. I will examine such a possibility in the next section and, especially, in Chapter 9.

Methods Box # 14   Measuring the impact of rural development processes

Although estimates about the percentage of farms engaged in different expressions of the new model of rural development differ considerably – just as data on their economic relevance are highly contested – there is no doubt that European agriculture increasingly has a dual structure. On the one hand there is a pole (see A in Figure 7.4) that groups together multi-functional farms that produce, alongside classical commodities, a range of new products and services and which try to avoid a high dependency on external inputs and credit. On the other there is a second pole (B) of highly specialized farms that are strongly integrated into input markets (including the capital market). The two poles are increasingly diverging from each other. In the second pole (B), further scale enlargement, an accelerated

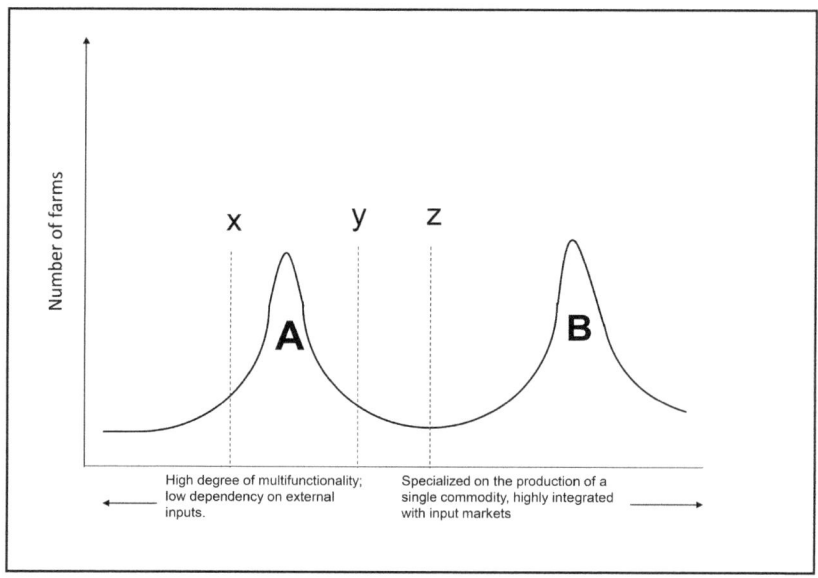

Figure 7.4 A theoretical distribution of multi-functional and conventional farms (author's own elaboration).

industrialization of production processes and integration into large 'chains' are the beacons that guide the farm development trajectory. The first pole (A) is mostly guided by quality improvements, the ongoing construction of synergy and building short agro-food chains. In terms of neo-institutional economic theory, the duality shown in Figure 7.4 can be expressed as one between 'economies of scale' (B) and 'economies of scope' (A) (Marsden, 2009). Although the two poles might be theoretically well defined as being contrasting and mutually exclusive, in practice there will be, and are, considerable overlaps and nuances (also shown in Figure 7.4). Many medium and large farms often do not have a focus on one single strategy but operate a complex 'portfolio' (Saccomandi, 1998). This means that alongside high-quality production and associated activities (on-farm processing and direct marketing) one also encounters conventional forms of production that are maintained and improved with knowledge and/or financial resources derived from the new, multi-functional activities. This phenomenon is present in all European countries and in all sectors. For the farmers involved it often represents a journey of discovery into new, as yet unknown, realities – and possibly a preparation for sudden changes in the market context.

Figure 7.4 also helps to clarify why assessments of the quantitative presence of both types of agriculture differ so much. If line x is used to demarcate the most developed expressions of rural development (i.e., those farms located on the left side of line x – the farms that are highly diversified), then their number will be relatively low. If, however, the average situation (the peak between x and y) is also taken into consideration, the number will be far higher. Taking line z (the median) as the benchmark will again give different results. Beyond that, one has to take the time dimension into consideration. Rural development and the continuing development of conventional agriculture are dynamic phenomena that are critically affected by a range of politico-economic and cultural conditions. Hence, measurements can give different results in different years.

Another major difficulty resides in the specification of the denominator (i.e., the total number of farms). The table in Text Box 7.7 shows that 43,700 farms in Italy that were involved in organic farming in the year 2000. Now, should this number be related to all statistically existing 'farms' (in Italy there are some 2.1 million), the relative presence of organic farms would be as low as 1.9%.

Should the same number be related to 'professional farms' only (in which farming produces a significant part of the income of the household), the proportion of farms involved in organic farming (in 2000) changes to 3.5%.

Conceptualizing rural development as re-peasantization

Figure 7.5 summarizes the on-farm changes that come with rural development. Most of these changes already have been discussed, but I return to them to make the point that together they imply a paradigm shift. To begin with, rural development induces deepening and broadening. Such changes diversify the output of the farm (see number 1 in Figure 7.5). Instead of being reliant on one market to deliver its products to, the farm is now linked to more markets. This increases its relative *autonomy*. Diversification offers the possibility to partly reduce supply to, and reliance on, a contracting market and increase sales to markets that offer

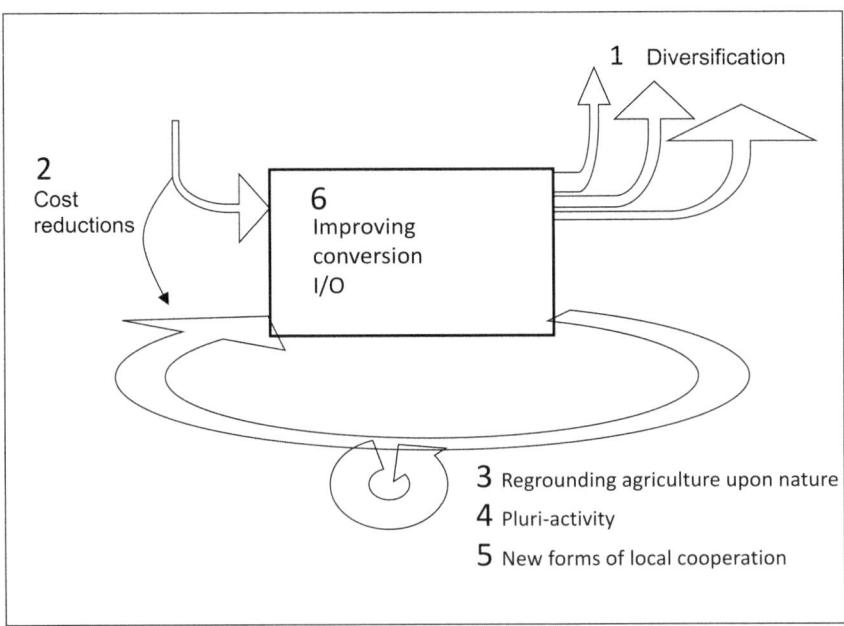

Figure 7.5 Rural development as a process of re-peasantization (Ploeg, 2008).

good conditions (*un mercato che tira* as Italian farmers say: 'a market that is pulling'), and this creates, at least potentially, the possibility to renegotiate terms. Reducing external input use (number 2) decreases dependency on agribusiness groups and thus equally contributes to an increase in autonomy. This especially applies if farming is simultaneously re-grounded on internal resources (number 3), in which case ecological exchanges replace economic exchanges (see Figure 1.1 in Chapter 1). Engaging in pluri-activity (number 4) implies new dependency relations (in the labour market) but at the same time helps to increase the autonomy in, and of, farming. So does engagement in (new) cooperatives (number 5; I will discuss some special examples of this in Chapter 9).

Finally, there is the possibility to improve the output/input ratio; that is, to produce *more* with a given set of resources (number 6). This might be the outcome of novelty production (as discussed in Chapter 2). Through such improvements in production (probably small ones but, again, by building on growing numbers of such small improvements, considerable advantages might occur), the farm also gains more autonomy and can increase the space to better face the squeeze on agriculture. At the same time, these mechanisms (notably 2, 3, 4 and 5) strengthen and enhance the autonomous and self-governed resource base of the farm.

Together, the increase in the self-governed resource base and the associated and multi-faceted increases in autonomy constitute a (hidden) process of re-peasantization. They contribute to farming becoming (once again) more peasant-like. In graphical terms they (help to) shift the farm from the situation shown in Figure 3.3 (a market- and future-dependent reproduction) towards the one shown in Figure 3.2 (with relatively autonomous and historically guaranteed reproduction). Figure 3.1 shows that this type of re-peasantization can be conceptualized as a substantial strengthening of the internal, non-commoditized flow of resources; a reduction of the inflow of commoditized resources; and a diversification of output flows.

Is this just juggling with words and images? I think not. For the preceding discussion shows, in the first place, that the current *omnipresence* and *simultaneity* of the mechanisms discussed are not random. There is an underlying logic that ties them together and which makes their presence in today's agriculture (which operates in the context of deregulated global markets) quite obvious. This underlying logic is the (re-)construction of farms that are able to resist the very harsh conditions in which they have to operate (conditions that no longer allow for reproducing the farm through the main market circuits; hence, new circuits need to be found and/or wrought). And this is what re-peasantization is supposed to be. What we are witnessing here is a kind of Polanyian counter movement: a movement away from the main and capital-controlled commodity markets because these markets place farming in a state of chronic and permanent crisis, just as they bring a dependency that will, under current conditions, annihilate growing segments of the agricultural world (Polanyi, 1957; Schneider and Niederle, 2010).

Rural development as process of transition

Finally, I want to briefly discuss rural development as a transitional process. It is a process that brings agriculture from a given situation at moment T to a new, as yet not well-known, situation at T + n. This new situation, when compared to the current one, will be structured completely different, with the many differences evident on several different dimensions. Hence, transition is a multi-dimensional process of change; it occurs at several different levels (it is multi-level); there are many different actors involved (it is multi-actor) and it mostly takes a long time (20 years or more). It is far from being a unilinear and smoothly unfolding process. Instead, it is full of contradictions, frictions, competing interests and contrasting views. Throughout history there have been many different agricultural transitions. The comparative analysis of several of these transitions can provide a series of interrelated insights (and concepts), which are summarized in Figure 7.6.

Figure 7.6 distinguishes three different levels. These are the niche level (comparable to the micro or grassroots level), the level of the reigning regime(s) and the 'landscape' level (which might also be referred to as the macro level). The niche level is that of different practices and activities and the actors involved in them. In our case it is the level of farms, farmers engaged in farming and other actors with whom farmers deal with directly. It is also the level where many new insights are born, tested in all kinds of different experiments (as discussed in Chapter 2) and new farming practices are created. These new insights and practices are called *novelties*: novel ways to think about, and to practice, farming. Novelties are modifications of, and sometimes a break with, existing routines.

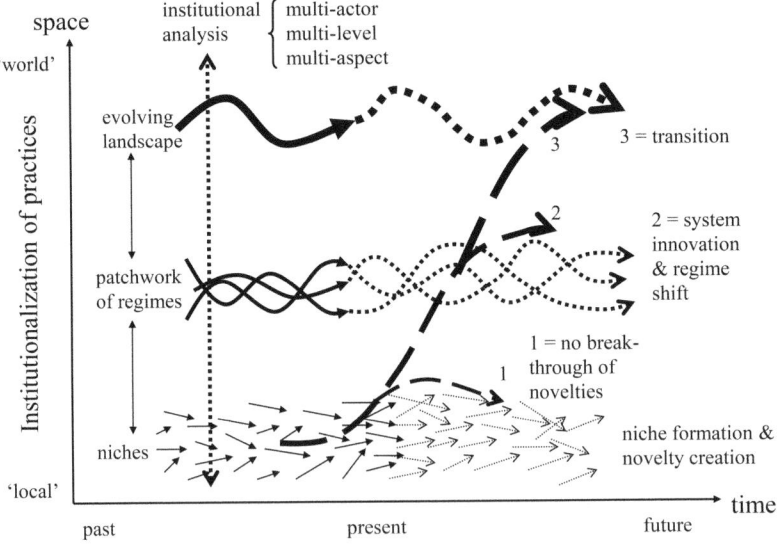

Figure 7.6 The structure and dynamics of transitional processes (Roep and Wiskerke, 2004).

They represent, in a way, ever so many deviations. Many of these novelties are hidden in the heterogeneity (the 'apparent chaos') of agriculture, and if they 'work well', why is (as yet) not well understood. Because each farm, due to its self-governed resource base, has a certain autonomy, the farm can be seen as a *niche* for the novelties it generates. The niche protects the novelties (they are the responsibility of each farmer) but at the same time limits its diffusion. Given the multitude of farms, there will be many novelties, but they mostly do not connect to each other and their reach remains limited.

The second level is the one of regimes. According to Arie Rip and Richard Kemp (1998), a socio-technical regime is 'the grammar or rule set comprised in the coherent complex of scientific knowledge, engineering practices, production process technologies, product characteristics, skills and procedures, ways of handling relevant artefacts and persons, ways of defining problems – all of them embedded in institutions and infrastructures' (331). A socio-technical regime specifies how things are to be done.[17] This regime (as it is defined in transition studies) is a constellation that is created and maintained by agro-industries, ministries of agriculture, agricultural universities, institutes for applied research, banks, etc. It acts as a technico-administrative task environment (see Chapters 2 and 3) to primary production and prescribes, regulates and sanctions the practice of farming. If we focus on the actors operating in, and representing, this regime it is a *nomenklatura*.

The third level, called 'the landscape' in transition studies, represents the values and normative orientations of society and is relatively stable.

Within the tradition of transition studies, a transition is thought of as the accumulation of many novelties that agglutinate, as it were, in new practices and new styles. If this agglutination turns out to be dynamic and self-strengthening, this can bring about a regime shift. But for this to happen, the novelties have to be aligned or clustered and integrated. Single novelties are at odds with the reigning regime; they are, as said before, deviations from the rule. But tied together in a convincing and well-functioning whole that can stand on its own (that provides the needed autonomy), they can change the rule(s). In this respect, novelties are *seeds of transition*.

The alignment of many novelties in one convincing and powerful movement of change can be supported and strengthened by 'movements' that operate at the landscape level. If the regime is increasingly at odds with the normative orientations in society at large, the flow of novelties from the niche level towards the regime level might gain additional momentum. This occurs when an alteration brought about by the flow of interconnected novelties is actively supported by civil society at large.

After having accomplished a regime shift, the upward flow in Figure 7.6 might possibly continue and bring about substantial changes at the landscape level as well. If this is the case, there is a fully fledged transition.

Further to all this, it is to be noted that the production, accumulation and agglutination of novelties can be strengthened (and accelerated) if wider and well-defended 'strategic niches' (that go beyond single farms) are constructed.

Such niches become 'field laboratories' that offer protection and thus allow for the novelties to germinate and grow. In Chapter 9 I will refer to some such strategic niches. They also play a crucial role in peasant struggles of the third kind. In transition studies, the construction of such niches is discussed as 'strategic niche management' (Moors et al., 2004). It should also be noted that changing one particular regime (for instance, the one that governs and regulates agriculture and food production) might become complicated and tenacious if such a regime is interwoven with other regimes (such as the spatial regulation of the countryside).

Conceptualizing processes of rural development as processes of transition can be very helpful. It generates a perspective that (1) indicates the strategic importance of connecting the many novelties (the small arrows in Figure 7.6) in order (2) to create a comprehensive alternative that challenges the reigning regime and (3) which aims at a clear regime shift. In the meantime, (4) exceptions to the rule(s) need to be negotiated with the representatives of the regime in order to be able to build strong and extended strategic niches (see also Methods Box #12 in Chapter 6).

The main complication that comes with understanding rural development as a transitional process is that such a transition does not stand on its own. It is far from being the only possible transition. The movement to increase incomes and reduce costs at the level of the sector as a whole through the introduction and further unfolding of multi-functionality is far from the only response to the squeeze on agriculture (see Figure 7.2). A second proposal aims at further and accelerated scale enlargement, in which the ever-diminishing VA at the level of the sector as a whole is re-distributed over a greatly reduced number of farms – thus allowing them to generate adequate incomes. This second proposal essentially aims at an accelerated industrialization of agriculture. A third proposal centres on a more or less complete outsourcing of food production to other countries (with only the production of genetic material remaining).

In recent decades what we have witnessed is these *three partial processes of transition* occurring simultaneously and engaging in complex relations of mutual competition and exclusion. There are moves towards rural development, further industrialization and outsourcing all happening at the same time: partial and mutually competing processes of transition. This rivalry is partly symbolic (Which farms produce 'safe food'? Which trajectory is 'able to feed the world'? What type of farming does society wants to see?). It is also partly material, raising questions such as where do the 'benefits' of agriculture (and the policies that underpin it) go and who pays the 'costs' that come with it (see Figure 6.13).

Weaknesses and new contradictions

During, along and as a consequence of different rural development processes, new problems, weaknesses and contradictions are emerging. One of these centres on the balance between individual and common interests. Many individual cases could develop only due to the support of others; however, when an individual case develops into a successful experience, it sometimes proves to be difficult to

---

Text Box 7.9   The unruliness of success

There is an interesting example from near Aberystwyth in Wales that informs us about the capriciousness of RD successes. A (female) farmer, running one of the longest established organic dairy farms in the UK, couldn't sell her milk for a week because heavy snowfall stopped the milk tankers getting through. She started to make yoghurt, which became a local and then national favourite (a genuine case of serendipity). She ran a family dairy firm (whose development was supported by Pillar 2 and regional development mechanisms) for many years and generated a lot (140+) of local jobs. Then she decided she didn't want to be in the world of dealing with mass retailers etc. anymore or to cash in her chips (depending whose version you like to believe) and sold the business to a US 'organic multi-national', which sold on to a French one and then to Nestle (for a fairly neutral news article, see: http://www.cambrian-news.co.uk/news/i/9341/).

---

continue providing support to other, new experiences that are in a take-off stage. Ironically, the comfort of success can tempt those involved to act greedily and selfishly. And, once this occurs, it threatens to contaminate ever-growing circles. Another, probably related, issue is the temptation to *sell* the most successful expressions – thus allowing capital to enter, once again, directly (see Text Box 7.9 for an illustration). Interestingly, the *successful* unfolding of rural development activities also generates problems and contradictions at higher levels of aggregation: how to coordinate, for instance, the growing supply of new services and goods and how to avoid major imbalances between supply and demand.[18]

There will also be other issues as what happens when state support is suddenly withdrawn or even becomes hostile (as was the case in Brazil; see the end of Text Box 7.4) and the question (especially important in Europe) of how to construct a convincing socio-political narrative that is able to attract and engage as many possible actors and which helps to tie their many experiences and practices together. Such a narrative should also specify the complicated and often contradictory relations between rural development and the other partial processes of transition that are simultaneously unfolding.

All of these are serious questions that require debate and action research. So far, experience shows that new cooperatives might provide practical solutions, at least partly, in being able to mediate between the single and the collective, as well as containing a newly argued moral economy and ethos (I will come to this issue in Chapter 9).

A societal necessity

In the decades to come, rural development will repeatedly emerge as a societal necessity. It is the counterimage of, as well as a response to, modalities of agrarian growth that are imposed by capital and which run counter to society, ecology and the interests of the farming population. The more agricultural growth follows the script of capital (i.e., moves towards food production without farmers and nature), the more rural development will come to the fore as a much-needed alternative – the more so because it can be placed on the agenda and tried out practically in a myriad of ways and by many different people in many different places.

We cannot predict *how* rural development will proceed and *what* impact and outreach it will have. This will depend on many factors. These include the degree with which future rural development processes will build on previous experiences. If these latter experiences are combined and moulded into a convincing narrative (that also identifies and anticipates the weaknesses and uncertainties), the RD experiences of the future might be considerably strengthened. In addition, the interfaces between, and possible coalitions of, urban and rural people are very likely to put an imprint on the future dynamics as well. Thirdly, the role of the state will probably be very important: is the state able and willing to condition the role of capital in the countryside and food production – or will it allow for their full conversion into money-making machines for capital? Fourthly, the power of countervailing movements will be important as well: will they be able to connect the myriad of ways the struggles for rural development will take? And will they be able to connect the many involved people?

Questions like these compose a strategic agenda for research. I believe that such research can contribute considerably to the ongoing struggles for, and processes of, rural development – just as they can contribute very much to further enriching the sociology of farming.

Notes

1. Institutional interests often strongly favour policies that emphasize continuity, with changes being presented as mere *additions* to essentially unaltered policies. This also reflects an unwillingness to address the root causes of the negative externalities.
2. Many people (and institutions) consider that 'the rural is the rural and that's how it is. It is like the sky. One cannot develop the sky, and the same seems to apply to the rural. It is simply there, static and unchangeable'. It will be evident that the reasoning I follow in this book is diametrically opposed to such a view.
3. Interestingly, the concept of rural development waxes and wanes and then reappears again. Each epoch seems to have its own 'generation' of rural development. In this chapter, I will discuss the currently used notion; that is, rural development as counterimage of, and remedy for, neo-liberal processes of agrarian growth (that have increasingly dominated the scene from, say, the 1990s onwards). The previous 'generation' of rural development (from, say, the late 1960s to the end of the 1980s) was presented as counterimage of failing processes of agrarian growth and as a remedy for these failures. It specifically emerged in the Global South where stagnating growth became a major problem. Here rural development was seen as providing the *conditions* for agricultural growth (infrastructure, credit, extension, sanitation, drinking water facilities, etc.) – conditions that were not created by, or through, the development of capitalism itself.
4 It will be evident that I do not refer here to the 'rural' as understood in geography. I regard the rural as a socio-economic reality that is different from, but at the same time part of, a wider, and increasingly urbanized, society. Paradoxically, the more society becomes urbanized, the more the need for well-developed rural counterweights.

5. I am aware that food security and food sovereignty are not synonymous. In China, however, the two largely coincide. Soy is the seeming exception. But here, the import of soy and the simultaneous export of high-value food products might be seen as an intelligent reflection of real scarcity relations.

6. It emerged after sharp critiques from the academic world (notably by Wen Tiejun, 2004) on the poverty, stagnation and misery of country life. This critique was widely echoed in society, after which Chinese authorities responded with the *san nong* policy.

7. In this respect, it is telling that the *Cork Declaration* referred to rural development policy as to be applied 'to all rural areas in the Union' (Point 2 in Text Box 7.3). This is political jargon for saying that the increased inequalities between growth poles and marginalizing areas (designating the former for 'optimal agriculture' and the latter for 'nature' as scientific jargon would have it) was not acceptable and needed to be redressed through rural development policies.

8. This followed the principle of subsidiarity, meaning that decisions should be taken at the lowest possible level.

9. Later on this was formalized as a 'menu' from which Member States and regions could make their own selection.

10. An important component of the European Rural Development Policy was LEADER: a sub-programme that aimed to strengthen grassroots initiatives. See Van Depoele (2003).

11. The perception of the socio-political relevance of rural development is strongly hindered by theoretical perspectives that strictly equate markets and capitalism to each other, with them often being regarded as synonymous. Within such a perspective, the creation of new markets (regardless of how different they are from the dominant and capital-controlled markets) cannot but further strengthen capitalism. Contrary to such perspectives is the theoretical view that markets existed before capitalism, just as there will be markets after capitalism. The key point is who or what *controls* the markets. An excellent and empirically well-grounded exposition of such a view can be found in the work of Braudel (1992).

12. That is, that they are more convincing to farmers than those that the dominant development trajectory could ever create.

13. There have also been, and are, mega-projects presented as part of 'rural development'. Especially if the latter are facilitated by the state and supported with funds, other actors and interest groups emerge that claim to be able to reach the same objectives in more efficient ways. A well-known case is farmer-managed nature conservation versus nature conservation by large nature reserves. I will not engage in the polemics around this issue. One point, though, stands out. If government opts to support large organizations instead of farmers, there will be no strengthening of the rural economy (in the form of defence of existing employment and improving the incomes of farmers). Thus, the 'heart' is taken out of rural development.

14. Some of these activities are also referred to as the production of environmental services.

15. Such as using the tractor for transport activities (as is done in China).

16. This occurs especially when the farm starts to combine an increasing number of new functions that partly build upon each other. If this is the case, considerable additional incomes can be generated.

17. The socio-technical regime in agriculture represents a shift of the ordering principle from the cultural repertoires that are central in the different farming styles to the compound whole of external institutions (that often centre on the ministry of agriculture as the main mechanism of coordination).

18. In Austria, an especially rich array of new institutional solutions has been developed, including a well-calibrated differentiation of services supplied (for instance, in the realm of agro-tourism). This may, however, also bring new forms of bureaucracy. A definitive solution has, as yet, not been found (if there is any such definitive solution at all).

Bibliography

Anderson, C. R., J. Bruil, M. Jahi Chappell, C. Kiss, and M. Pimbert. (2020), *Agroecology Now! Transformations towards More Just and Sustainable Food Systems*, Palgrave, Macmillan, London.

Braudel, F. (1992), *The Wheels of Commerce: Civilization and Capitalism, 15th–18th Century*, Vol. 2, University of California Press, Berkeley.

Cassel, G. (2022), 'Family agriculture and rural development in Brazil (2003–2010)', in *Rural Vitalization: Comparative Analysis of Rural Development Policies in Different Countries*, edited by H. Wu et al., 35–54, Social Sciences Academic Press, Beijing.

Cork Declaration. A Living Countryside. (1996), Cork, Ireland.

Marsden, T. K. (2009), 'Mobilities, vulnerabilities and sustainabilities: Exploring pathways from denial to sustainable rural development', *Sociologia Ruralis* **49** (2), pp. 113–131.

Moors, E., A. Rip, and J. S. C. Wiskerke. (2004), 'The dynamics of innovation: A multi-level co-evolutionary perspective', in *Seeds of Transition: Essays on Novelty Production, Niches and Regimes in Agriculture*, edited by J. S. C. Wiskerke and J. D. van der Ploeg, 31–56, Royal van Gorcum, Assen, the Netherlands.

Oostindie, H. (2015), *Family Farming Futures: Agrarian Pathways to Multifunctionality: Flows of Resistance, Redesign and Resilience*, Ph.D. thesis, Wageningen University, Wageningen, the Netherlands.

Ploeg, J. D. van der. (2008), *The New Peasantries: Struggles for Autonomy and Sustainability in an Era of Empire and Globalization*, Earthscan, London.

Ploeg, J. D. van der, A. Long, and J. Banks. (2002), *Living Countrysides: Rural Development Processes in Europe: The State of Art*, Elsevier, Doetinchem, the Netherlands.

Polanyi, K. (1957), *The Great Transformation: The Political and Economic Origins of Our Time*, Beacon Press, Boston.

REAF. (2006), *Bases para o Reconhecimento e Identificação da Agricultura Familiar no Mercosul*. MERCOSUR/VI REAF/DT No. 03/06. Annex IX. Reunião Especializada da Agricultura Familiar do Mercosul, Montevideo, Uruguay, http://www.mda.gov.br/reaf/.

Rip, A., and R. Kemp. (1998), 'Technological change', in *Human Choice and Climate Change*, edited by S. Rayner and E. L. Malone, 327–399, Battelle Press, Columbus, OH.

Roep, D., and J. S. C. Wiskerke. (2004), 'Reflecting on novelty production and niche management in agriculture', in *Seeds of Transition: Essays on Novelty Production, Niches and Regimes in Agriculture*, edited by J. S. C. Wiskerke and J. D. van der Ploeg, 341–356, Royal van Gorcum, Assen, the Netherlands.

Rooij, S. J. G. de, F. Ventura, P. Milone, and J. D. van der Ploeg. (2013), 'Sustaining food production through multifunctionality: The dynamics of large farms in Italy', *Sociologia Ruralis* **54** (3), pp. 303–320.

Saccomandi, V. (1998), *Agricultural Market Economics: A Neo-institutional Analysis of Exchange, Circulation and Distribution of Agricultural Products*, Royal van Gorcum, Assen, the Netherlands.

Schneider, S. (2016), *Family Farming in Latin America and the Caribbean: Looking for New Paths of Rural Development and Food Security*, International Policy Centre for Inclusive Growth (IPC-IG) Working Paper No. 137, IPC-IG, Brasilia, Brazil.

Schneider, S., and P.-A. Niederle. (2010). 'Resistance strategies and diversification of rural livelihoods: The construction of autonomy among Brazilian family farmers', *Journal of Peasant Studies* **37** (2), pp. 379–405.

Trichet, R. M., and S. Schneider. (2010), 'School feeding and family farming: Reconnecting consumption to production', *Saude e Sociedade* **19** (4), pp. 933–945. doi: 10.1590/S0104-12902010000400019.

Van Depoele, L. (2003), 'From sectorial to territorial-based policies: The case of LEADER', in *Future of Rural Policy: From Sectoral to Place-Based Policies in Rural Areas*, 79–87, Organization for Economic Cooperation and Development, Paris.

Ventura, F., and P. Milone. (2004), 'Novelty as redefinition of farm boundaries', in *Seeds of Transition: Essays on Novelty Production, Niches and Regimes in Agriculture*, edited by J. S. C. Wiskerke and J. D. van der Ploeg, 57–92, Royal van Gorcum, Assen, the Netherlands. https://edepot.wur.nl/338089.

Ventura, F., P. Milone, and J. D. van der Ploeg. (2007), *Qualitá della Vita Fuori Città*, AMP Editore, Perugia, Italy.

Wen, T. (2004), 'Jiegou xianjiang lu' [The deconstruction of modernity], *Jiegouxiandaihua* 8–22.

Wu, H., B. Ding, and J. Ye. (2015), 'The construction of new nested markets and rural development in China', in *Rural Development and the Construction of New Markets*, edited by P. Hebinck, J. D. van der Ploeg, and S. Schneider, 99–114, Routledge, London.

Wu, H., and S. de Rooij. (2022), 'Young people and the rural', in *Rural Vitalization: Comparative Analysis of Rural Development Policies in Different Countries*, edited by H. Wu, J. Ye, and J. D. van der Ploeg, 127–142, Social Sciences Academic Press, Beijing.

Xi, J. (2017), 'Secure a decisive victory in building a moderately prosperous society in all respects and strive for the great success of socialism with Chinese characteristics for a new era', Opening speech delivered at the 19th National Congress of the Communist Party of China, Beijing, 18 October.

Ye, J., J. Rao, and H. Wu. (2010), 'Crossing the river by feeling the stones: rural development in China', *Rivista di Economia Agraria* **65** (2), pp. 261–294.

8 Constructing new markets

Overview: Main concepts discussed in Chapter 8

Market as self-regulating system
Marketplaces
'Follow the eggs'
Patterns of provision
Socio-material infrastructures
Economies of opposition
Country shops
Economic relevance of new markets
Political relevance of new markets
Nested markets
Ecovida (Brazil)
Markets grounded on ecological principles
Connectedness

As argued in the previous chapter, rural development is a response to market failures. Ironically, this response often occurs through the construction of *new* markets. This apparent paradox can be explained through a critical examination of the notion of the market.

As an abstract concept, the notion of the market refers to a self-regulating *system* that creates an equilibrium between supply and demand by establishing price levels that allow for, sustain and reproduce this equilibrium. At the same time, the market is a concrete notion that refers to *specific places*: to concrete marketplaces where concrete exchanges take place between concrete people, some of whom are offering products and services and some of whom are buying them.

When it comes to understanding the market as (a set of) concrete marketplaces, the notion of multiplicity is important. Generally, in any given space and at a given time, there is a *multiplicity* of markets. This does not necessarily refer to different geographical locations (i.e., each village in a particular area having its own weekly market). The concept of multiplicity also embraces markets that are *ordered* in different ways but which simultaneously co-exist (literally side by side)

DOI: 10.4324/9781003313274-8

in a given territory. The different ordering might include the magnitude of the market, the price levels, the distribution of value added (which part is captured by the traders and which part remains for the producers?), the actors who play a role in the market, the way the market is governed, etc., To distinguish differently ordered markets it is necessary to probe into (a) the specificity of the *products*, (b) their particular *flow* through time and space, (c) the particular *actors* involved and their interrelations, (d) the *rules* that govern the making and use of the product and (e) the *position* of each market in a broader context (see also Methods Box #15).

Methods Box #15   Identifying and exploring nested markets

A first set of questions should set out to correctly assess the specificity of the product that is being marketed. Products come with a history that is sometimes synthesized in, and as, a *brand* (no matter whether this brand is made explicit or remains implicit). A brand contains messages that specify how, where, when and by whom the product has been made (or is supposed to have been made). Even the absence of such messages represents a clear signal; that is, that the origin, history and nature of the product do not matter at all. Taken together, these messages span a normative framework (condensed in, and as, a brand) that summarizes and specifies the *making* of the product.

A second set of questions regards the 'channels' through which the products flow. How does a specific product travel from the place of production towards the place of consumption? How many different stages are involved? How many different actors are involved in making the entire trajectory operative? Which, and how many, technical means are used to make the products pass through the different stages? How many places of production are delivering the product (or its ingredients) to the channel? Are these places interchangeable? Whilst travelling and traversing the different stages, how many, and what kind of, changes does the product undergo? Do the different stages and product changes have a particular 'architecture' (Fligstein, 2001; Gereffi et al., 2005) that is distinctively different from other patterns of provisioning? How is the total value added distributed along the chain, and how does this relate to the labour time invested in each step?

A third set of questions centres on the actors and their interrelations. How many people are involved in the flow (and processing) of the product? What is their specific role (e.g., wholesaler, retailer, transporter, etc.), and how do they mutually relate to each other? Do they operate on their own account or are they part of larger enterprises? Who are the ones in control? And how is that control effectuated? Is there just one centre of command or are there several? What are the interrelations (hierarchical, complementary, competitive, striving to take over the others)? Are there discernible differences in power? Are power relations associated with a specific distribution of value added? Do the (end) consumers represent a specific category with shared status, income levels and/or value orientations?

The fourth set of questions regards the rules. How are responsibilities delineated? What formal rules apply? How are they complemented (or undermined) by informal ones? Do institutions (and which ones) play a central role? Does the state have a specific role?

The fifth and final set focusses on the position of each specific market within the wider context. More specifically, how do the different channels (each making the same product flow but in contrasting ways) relate to each other? Are they complementary? Or are there relations of competition, hegemony, or subordination? What is the market share of each channel, and how do these channels relate to specific groups of producers, consumers, traders, etc.?

The answers to these questions help to assess whether we are dealing with territorial and/or peasant markets.

Territorial markets tie together local and/or regional production with local and/or regional consumption. They 'territorialize' the production and consumption of food. The responses to the second set of questions (above) are decisive here. The social definition of the relevant space (i.e., the delineation of the relevant territory) and the appreciation by consumers of, and their knowledge about, local produce are important ingredients (see the third set of questions), as is a relatively non-hierarchal structure (with no single centre of command). Such territorial markets mostly exist alongside global markets, and their operation is relatively independent from the latter.

Peasant markets are a specific form of territorial market. The food that is traded in, and through, peasant markets carries the recognizable brand of peasant agriculture: it stems from peasant farms (the first set of questions). Peasants (and notably peasant women) are important actors in these markets (the third set of questions), and they (to various degrees) position themselves as a critique, expression of resistance and alternative to the wider, conventional and/or global markets. Mutual understanding and solidarity between consumers and producers are often an important ingredient. The forms of peasant markets can differ very much (from Gruppi di Aquisto Solidiale [GAS] in Italy and community-supported agriculture [CSA] in (especially) the USA to different kinds of street markets throughout Europe and Brazil). There is, in short, a huge diversity in the architecture of peasant markets.

Eggs

In a fascinating empirical study, Marc Wegerif (2014) presented an extensive description of 'urban food provisioning' in Dar es Salaam. The description and analysis centre on eggs. The starting question regards the flow of eggs.[1] Where do they come from? How are they transported? Where and how are they sold to consumers? Who plays which role along the journey that the eggs make?

'Reading the market' in this way, Wegerif identified six contrasting patterns of provision, each with its own characteristics. Each pattern represents a distinctively different channel that makes the eggs flow in a specific way, with specific inlets and outlets (the points where the eggs enter into and leave the channel) and operated by different people who have different social positions and roles.

Through these different channels, the eggs acquire particular and distinguishable attributes (specific brands and manuals, as I will explain in Chapter 10).

The first channel is the *duka* (Swahili for 'shop'). Cyclists bring the eggs from villages located around Dar es Salaam to the many *dukas* in the city (generally within a few minutes' walk from each other). These eggs are bought in nearby villages and sold to the operators of the *dukas*. The typical *duka* offers a wide range of food and other products[2] and sells some 14 trays of eggs a week (each tray containing 30 eggs). These eggs are sold individually or in small quantities (and sometimes on short-term credit). Alongside (or in between) the *dukas* there are women selling eggs from their homes. They themselves transport the eggs from the villages to their homes by taking the bus. Third comes the chips seller, who sells cooked eggs and chips at small stalls. A fourth channel is made up of traders who sell and deliver door to door. In contrast with the others, they only sell eggs in full trays. There is another contrast in that the eggs are bought from peri-urban producers. The fifth channel consists of the supermarkets that attract different buyers and sell the eggs for the highest prices. The eggs come from large-scale producers or are imported (from as far away as France), and traders are key in connecting the producers and the supermarkets. Finally comes the people's market, where the typical stall sells 200 trays a week to small businesses and/or individuals. These eggs come from large egg farms in villages that are located further away (up to 50 km).[3]

Each channel has its own distinguishable price levels,[4] the people's market being the cheapest and the supermarkets the most expensive (233 to 250 TSh versus 325 to 560 TSh).[5] The *duka*, women selling from home and street vendors operate in the same price range (250 to 300 TSh), and the chip seller asks 400 TSh per egg.

In synthesis: even if they all circulate the same product, these channels differ considerably when it comes to flows, actors, rules and prices. They make up different markets that co-exist in the same setting. They are different in as far as they are nested[6] in different sets of social relations that bring their own geography, rules and actors.

Are these differences relevant? Wegerif argued that they are. Having (a) immediate access to a selling point (an outlet) at walking distance, (b) the flexibility to buy small quantities at a competitive price and (c) potential access to short-term interest-free credit 'are crucial for people surviving on limited and sporadic incomes' (Wegerif, 2014:3767). Wegerif added: 'in addition, these factors do away with the need for storage space, something not to be taken for granted by people who live in cramped spaces [...] and with no assets [such] as fridges' (Wegerif, 2014:3767). In short, in the case of *dukas* and the like, the eggs come with a list of requirements (a manual)[7] that is especially tailored to the poor. The eggs from supermarkets and door-to-door traders come with a very different manual.

Beyond that, the 'non-corporate provisioning patterns' (Wegerif, 2014:3747) offer more opportunities to the urban poor to generate an income than the supermarkets.[8] At the same time, they offer 'egg producers a stronger negotiation position' (Wegerif, 2014:3770), and different channels offer different levels of return to producers (3769). More generally speaking, the 'symbiotic whole' (Wegerif, 2017) of small producers in villages, bicycle drivers, *dukas*, street vendors, etc., brings an impressive gross income to the villages (see Wegerif, 2014:3761 for calculations). They help to keep value within the territory. By doing so, they operate as a countervailing power to the supermarkets that 'prioritize returns to investors and [...] debt servicing, creating a downward pressure on prices to farmers and upwards pressure on costs to consumers' (Wegerif, 2014:3773).

Given this diversity in marketing channels, can we really talk about one 'egg market' here?[9] In an abstract sense, yes. But as soon as we move the analysis to more concrete levels, the answer is evidently negative. From a substantivist point of view,[10] there are different markets here, with different actors, different rules, different routings of the eggs, different prices, contrasting distributions of the value added and different, socially constructed, meanings. They are also *different* markets in the sense that they countervail each other. The 'social justification' of the one is grounded in a negative judgement of the other and vice versa.[11] And even from a formalist point of view, they are different markets because there are different equilibria. The different channels are evidently interlinked (partly through relations of competition) but at the same time have their own specific dynamics. They also differ insofar as they are hubs that connect (albeit in an opaque way) different classes and strata: large-scale egg producers and big traders with the upper urban strata versus peasant producers and the urban poor. Finally, they differ insofar as the 'territorial markets' (i.e., the circuits closely linked to the territory) imply a more territorially bounded pattern of value distribution and retention. The value obtained through production and circulation remains in the territory (and thus contributes to regional economic growth), whilst in the case of the 'global market' (here represented by de-territorialized circuits such as the supermarkets), value is siphoned off.

Socio-material infrastructures

The different channels highlighted by Marc Wegerif are grounded in different, mutually contrasting, infrastructures. These have a socio-material nature and are defined and composed by intertwined *social actors* and *material artefacts*. One of these socio-material infrastructures is composed by small units of production in the villages, bicycles, *dukas*, single eggs, and households with low incomes and no refrigerator. Another one links large and specialized egg-producing farms, vans, supermarkets, upper-class consumers and households having a refrigerator. Whatever the specific composition, each socio-material infrastructure makes eggs flow in particular ways – in that sense, such an infrastructure is like a river bed that makes the water flow in a particular direction, with a particular velocity, Each infrastructure offers specific conditions for transport and storage and packaging.

Rural development often occurs through the construction of new markets that have socio-material infrastructures that significantly differ from those of the conventional markets. These newly constructed markets centre on the exchange of new, distinctive *products* (or products with distinctive characteristics). At the same time, these new markets make these products *flow* (from producers to consumers) in ways that differ from the ways they flow in conventional markets.

The newly constructed markets include new *actors* and also create new roles. They are governed by different rule sets. Finally, these newly constructed markets *relate* in a specific way to the dominant markets: as 'economies of opposition' (Pahnke, 2015).

Some of the new markets that emerge as part of rural development processes are constructed by the state. This applies, for example, to new markets for services as maintenance of landscapes and protection of biodiversity (in the EU also referred to, at least in political jargon, as environmental services). At one time these markets were almost monopolized by a few large nature organizations (that received massive funding from the state). Now they are demonopolized: farmers and farmers' collectives can now receive payments for the delivery of environmental services. The Second Pillar of the common agricultural policy plays a strategic role here. The bureaucracy that comes with the regulation of this specific market remains a highly contested issue.

Another interesting market is the one for care provisioning. At one time, huge and specialized institutions were paid for providing care to the elderly, handicapped people, children with specific needs, etc. In order to allow patients the freedom to choose the institute they preferred, the available budget was divided and directly channelled to the clients, who could then opt for the care institute of their own choice. This opened spaces for farms that often could offer very well-functioning and attractive care packages. Here, again, the specific regulation represents an arena where important negotiations are taking place. The same is the case with the emerging markets for green energy.

However, most new markets that are now playing a strategic role in rural development processes have been created by farmers themselves, and they centre on the delivery of specific farm products and services.

Country shops: An example

'Country shops' (known as *landwinkels* or *boerderij winkels* in Dutch) have been a familiar phenomenon in the Netherlands (and, more generally, throughout Western Europe) for many years. Forbidden during the World War II by the occupying forces and marginalized during the decades of modernization (which stressed the role of the farmer as specialized *producer* only), country shops started to re-emerge in the 1990s for a variety of reasons (some of which were discussed in Chapter 7). Figure 8.1 gives a schematic overview of such country shops scattered over the countryside, each attached to a specific farm and mainly, if not only, selling products produced on the own farm. Many of these farm shops were run by the farm women (often with assistance from the men), and they were greatly appreciated meeting points. Many urban citizens (and rural dwellers, too) love to turn the acquisition of food into an activity that includes a trip to the countryside and the opportunity to talk with others about good food. They particularly attract elderly consumers who have plenty of leisure time and often high levels of purchasing power. Together these country shops constituted a rudimentary market (grounded in a socio-material infrastructure that linked single farm shops to their clientele).

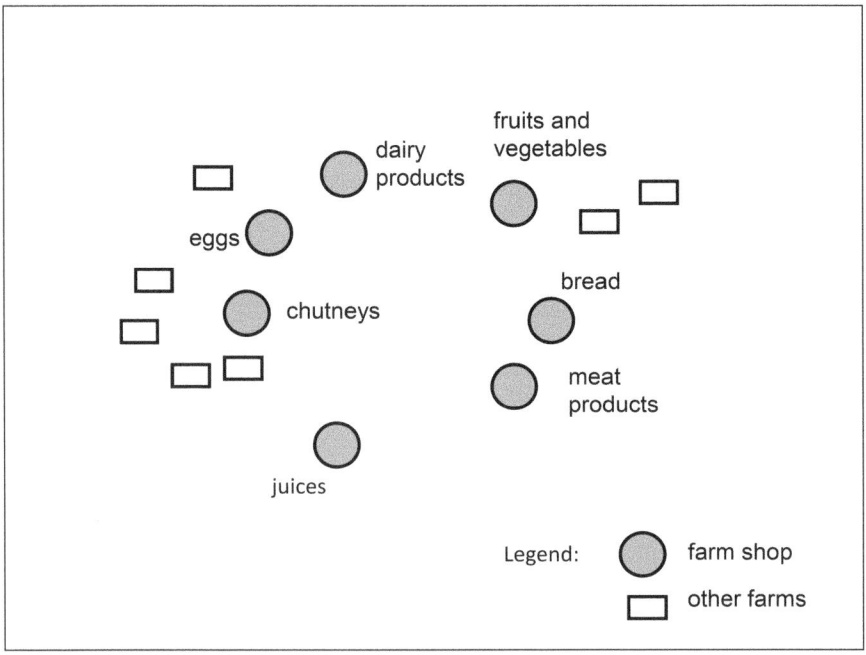

Figure 8.1 Dispersed farm shops (author's own elaboration).

This changed in the late 1990s when some 20 country shops (in the east of the country) decided to cooperate and sell each other's products alongside their own. This had immediate advantages: the sales of each farm increased significantly, and for consumers it became far easier to acquire the different products they needed or to find new products that were of interest to them. One stop sufficed. This also allowed the farms to expand the demand for their products by having more outlets. Thus, a new, more-developed farmers' market was created, built on the socio-material infrastructure shown in Figure 8.2 by the circle that links the different farms shops with each other. The material side of this infrastructure includes a refrigerated lorry that circulates selected products between the participating farms. The infrastructure also includes software (written by one of the farmer's daughters) that registers sales, channels the associated value to the relevant producer and shows when new deliveries are needed. The social side of the equation is the cooperative and jointly shared rule set that governs the transactions.

Today this specific farmers' market includes also internet ordering and home delivery (which turned out to be strategic during the COVID-19 pandemic), a box scheme, imports of farm-processed food products not produced in the Netherlands (such as wine, olives, olive oil, etc.) and delivery to urban food markets. This new market has considerably strengthened and sustained the incomes realized by

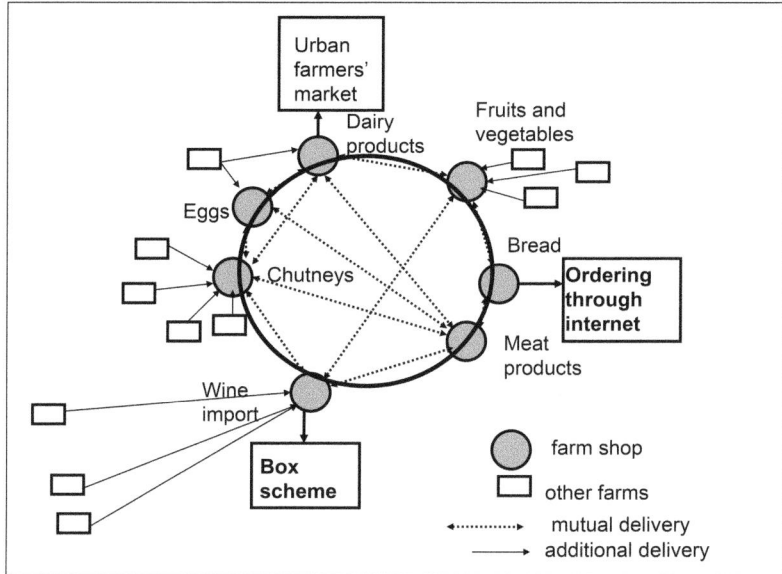

Figure 8.2 Cooperating country shops (author's own elaboration).

the participating farms and farm shops, some of which have developed into attractive meeting points (some constructed a restaurant or a pub, care facilities for the elderly people, playgrounds for children, etc.) that generate considerable additional employment and where people spend quite some time. At the same time, the cooperative structure allows each participating farming household to find the solution that best fits them (i.e., to set the balances in an appropriate way). Together the country shops have helped to increase both the accessibility and the attractiveness of the countryside in this part of the Netherlands.

All of this might appear to be, at first sight, somewhat trivial (and maybe it is). Nonetheless, there is a very remarkable feature in the socio-material infrastructure underlying this specific market: it is *circular*, and this differs very much from the socio-material infrastructure that links the large supermarkets of the country, which is essentially *radial*.

Large retail organizations critically depend on national (and regional) centres for distribution (see Figure 8.3). Here food products that come from everywhere are unloaded, repacked and loaded in lorries that subsequently deliver to the individual supermarkets. Thus, dairy products, for instance, produced and processed in the north of the country are transported to a distribution centre located in the middle of the country and some are then transported back to the north (and to other parts of the country). This radial infrastructure involves far more 'food miles' than its cyclical counterpart; it is more rigid and vulnerable and might be detrimental for food quality (because the products travel for longer time periods). It is also more costly.

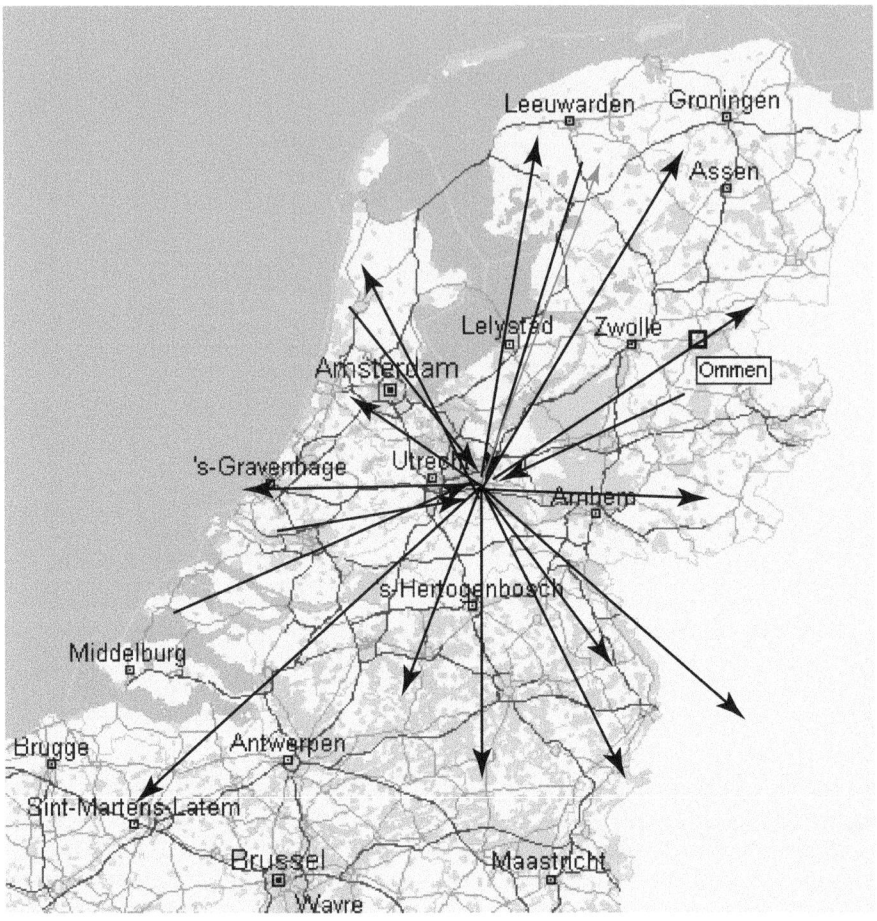

Figure 8.3 The socio-material infrastructure underlying large retail (author's own
      elaboration).

These examples make it clear that markets can be *ordered* very differently. This
applies not only to the underlying socio-material infrastructure but to several
other dimensions as well. Table 8.1 gives an overview of these.

In Text Box 6.1 I described the production of Chianina beef in Italy and how
it relates to the landscapes in which it is located. In Text Box 10.3 I will continue
the description of the production and consumption of Chianina beef. Chianinas
are at the heart of a farming style whose disappearance has been predicted several times by both scientists and politicians (notable in the heyday of the modernization epoch). Nonetheless, it is alive and kicking probably more than ever.
This is partly due the exceptional quality of the meat and the care given by the
producers to their animals but also the knowledge among regional consumers

Table 8.1 Dimensions that articulate the different structures of markets

	Dimensions	Key phrases to describe different positions
1	Socio-material infrastructure	Radial, circular, parallel; centralized, decentralized; global reach, localized; embedded in/separated from the local; poly-centric, single centre of command; etc.
2	Nature of products	Fresh, processed; local origin, anonymous; craft-made, industrial; etc.
3	Distribution of value added	Centralized in large corporations that control the processing and trade, equally distributed over many different actors; used for further accumulation, used for increase labour incomes; etc.
4	Governance	Owned and governed by corporations, owned and governed by coalitions of producers and consumers, state owned, etc.
5	Roles	Specialized roles (those who deliver raw materials, processors, traders, etc.), multiple job holding, plural roles, etc.
6	Outcomes	Contributing to or undermining sustainability, creation of employment, etc.
7	Relation to other markets	Hierarchy, subordination, autonomy, economy of opposition; extractivist relations, resistance; etc.

and *connoisseurs* elsewhere. But the endurance of Chianina production (located in Umbria and Toscana, two regions that have been, since ancient times, home to the Chianina breed) also resides in its marketing channels, which are mainly regional. The longstanding market for Chianina beef is a mosaic of highly decentralized relations that tie together local breeders with local butchers, local consumers and local restaurants. These partners directly interact and share joint definitions of quality that allow for the meat to flow smoothly from production to consumption. Together, such features make for an interesting price differential (and a distribution of value added that favours the farmers). All of this is amply documented in beautiful studies by Hielke van der Meulen (2000), Flaminia Ventura (2001) and others.[12] However, the same example also shows the vulnerability of such long-standing markets (and especially of the socio-material infrastructure underlying them). Provincial slaughterhouses had (and again have) a key function without which the market could not possibly operate.[13] In the 1980s and 1990s, when the winds of neo-liberalism blew strongly, regional authorities proposed to close down these slaughterhouses, which would have brought the end of the longstanding markets for Chianina beef. Dependency on *faraway* slaughterhouses (there would only remain three or four in the whole of Italy) is at odds with the continuation of such markets. Travelling too far inevitably causes the animals stress and, therefore, a degradation in quality. Beyond that, huge slaughterhouses operating at a national level could not possibly handle the small batches that circulate in markets such as the one for Chianina beef. Thus, local breeders defended 'their' provincial slaughterhouses. And they did so successfully.

Who, what and why?

The main questions of political economy (see Bernstein, 2010) are helpful in exposing the main (although not always explicitly stated) differences between new farmers' markets constructed in, and as a constitutive element of, rural development processes and the main commodity markets for food. *Who owns what?* is a question that reveals important differences (see 4 in Table 8.1). Farmers' markets are mostly a commons: a common-pool resource (Ostrom, 2010) that is owned and controlled by a collective of producers and (sometimes) consumers. They are not owned by individual capital groups. Nor are they for sale. *Who does what?* (see 5 in Table 8.1) refers to the roles. In the general agricultural and food markets, the farmers' role is mostly reduced to the delivery of raw materials (produced according to scripts developed by others). In the newly constructed markets, farmers figure not only as producers but also as traders in food and as researchers involved in ongoing marketing studies: they get feedback from their customers and are able to feed this into changes in the products and production methods. There is polyvalence in their performance (and this is one of the reasons young and well-educated people are attracted to this type of farming and marketing). *Who gets what?* is a question that clearly refers to the distribution of value added (3 in Table 8.1). In the case of newly constructed markets, a relatively large part of the value added is channelled back to the rural economy where it helps to sustain farms and employment levels. Finally, it can be asked *what is done with the surpluses?* (6 in Table 8.1). This is evidently related to the previous question, and in the case of the new markets discussed here, there is the prospect that the obtained surpluses might further spur the ongoing processes of rural development.

Approximating the economic relevance of farmers' markets

Although it is far from easy to quantify the economic relevance of, and the percentage of, farmers actively engaged in these new markets, there are some systematic and empirically well-grounded data available.[14] The European IMPACT programme sought to assess the overall impact of rural development practices in seven countries (Ireland, the United Kingdom, Netherlands, France, Germany, Spain and Italy). It paid specific attention to the then-existing channels for 'direct marketing' and found that they generated (in the year 2000) an extra net value added (Δ NVA) of 2,496 million euros. That, translated into micro terms, is a significant amount of money. If we take a ballpark average annual agricultural income of 20,000 euros in these countries, this would imply additional employment of nearly 125,000 people (FTE). If we further assume that these people live not only from direct marketing but also from the production of the marketed food products (say 50% from both sides), then this direct marketing could be interpreted as sustaining 250,000 units of labour force. On the other hand, though, if related to the total net value added generated in agriculture in these seven countries (92,792 million euros), direct marketing emerges as a minor element (contributing, at most, some 2% to the total net value added generated in agriculture).

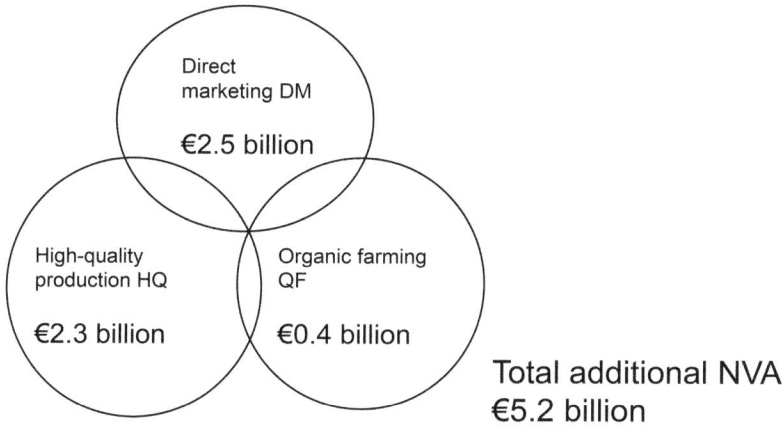

Figure 8.4 The economic impact (measured as Δ NVA) of direct marketing (DM) in relation to high-quality (HQ) and organic production (OF) (all data in millions of
euros) (Ploeg et al., 2002).

However, the contribution of direct marketing resides not solely in the value
it adds *directly and by itself* but also, if not mainly, in the *indirect* contribution it
makes to other forms of rural development and to the rural economy as a whole.
Figure 8.4 (grounded on data synthesized in Text Box 7.8) gives some overall
data regarding the contribution of the net value added of producing high-quality
(HQ) products on the farm (which often includes on-farm processing), organic
farming (OF) and direct marketing (DM). These contributions often combine
in practice (there is much overlap), although the researchers (Ploeg et al., 2002)
did everything possible to avoid double-counting. The point here, though, is a
different one: without direct marketing, much of the high-quality production and
organic production would not have been possible in the first place. And, without
direct marketing, many more of the benefits (a large part of the Δ NVA) would
have been siphoned off by large retailers. Another key point is that the *additional*
NVA thus produced and contained within the farms also sustains the standard
NVA entailed in the products. Assume that a farm produces 300,000 kg of milk
per year. If this production follows the conventional pattern (and is delivered to
the dairy industry as raw material), it might generate, at the level of the farm, an
NVA of say 12,000 euros (assuming a net value added per kilogram of milk of
4 euro cents). That would be a very meagre income. But if a part of the total production, say 100,000 kg of milk, is converted into cheese, butter and yoghurt and
then marketed directly, this might give an additional net value added (Δ NVA) of
8,000 euros (assuming a Δ NVA/kg of 8 cents per kilogram). Together with the
standard NVA (before starting HQ and DM) of 12,000 this would make a total of
20,000 euros, which would in many parts of Europe be considered a decent farm
income. Now, of course, all of this is speculative, but the point should be clear.
An *additional* income rendered through direct marketing (and associated new
activities) can turn an income that would otherwise be too low into an income

that allows for the continuation of the farm. Rural development activities sustain farms and farming practices that would otherwise probably disappear (and be taken over by industrialized farming). Such indirect effects are particularly relevant in the case of direct marketing. That is, the relevance of newly constructed markets not only resides in these markets as such – it *extends* to sustaining forms of farming that otherwise would have limited prospects.

Understanding new farmers' markets as 'nested markets'

Different terms are used to refer to the newly constructed markets we are discussing here. In some countries the expression 'peasant market' is much used. This is particularly meant to distinguish the origin of the products sold in these markets: peasant agriculture as opposed to entrepreneurial and/or industrialized agriculture. This denomination is often combined with the flag of agroecology. The international peasant movement La Via Campesina prefers to speak of 'territorial markets', thus stressing the rootedness of both production and marketing in the territory. This wording is also preferred to underline the similarities that link long-established territorial markets (especially in Africa and Asia), which still play a decisive role in the provisioning of food to large parts of the population, with the newly emerging markets. In social science debates, the concept of 'nested markets' is used frequently (albeit far from exclusively). This refers to (relatively small) markets that are *nested* within wider and far larger markets.[15] In a way, they are part of the wider market, but at the same time they are different from them. There is an additional (or alternative) set of institutions that delineates the nested market from the larger whole, which makes the nested market operate in different ways and show different tendencies than those of the wider market. Most nested markets are grounded in (literally *nested* in) a mutual understanding between producers and consumers (and sometimes traders). This mutual understanding embraces commonly shared definitions of product quality and origin, as well as mutual expectations on how to operate in the market.

I think the *nest* is a nice metaphor here. It is built in a step-by-step way and out of different materials: branches, feathers and probably even some abandoned plastic. The nest protects the eggs and the young birds. It allows for taking care of something new – something that in the end might stand on its own feet and take wing.[16]

Nested markets also represent *distinction*.[17] The products acquired in the nested markets of today (wherever located) are distinctively different from products circulating in the general food markets. They have a significantly different quality, a price that is definitely different (higher or lower), an origin that stands out and/ or a social biography that differ from those of products circulating in the general food markets. But it is not only the product that differs – the nested market itself is also distinctive. This is partly because it centres on direct transactions between producer and consumer or, at most, with one intermediary trader. The market is also distinct in that it groups together people (both producers and consumers) who share a joint, albeit mostly loosely defined, normative framework that distinguishes 'good' from 'bad', 'ugly' from 'beautiful' and, when it comes to behaving on the market, 'nice' from 'nasty'. Interestingly, the distinction entailed in the market and the products is passed on, as it were, to the consumers. By going to this particular market, they show that they are able to appreciate the good things in life. They show to themselves and to others that they are 'people who know' and/or 'people who care'. They come to the fore as knowledgeable about the value of good, or distinctive, food; the place where it is acquired; and the way it is prepared. Restaurants also like to show distinction. Preparing local food, with special local ingredients (whether Texel lamb, or Chianina beef from Umbria, etc.), attracts customers willing to pay higher prices for these special dishes. Thus, restaurants (carrying the emblem of distinctive local food) often constitute, together with their suppliers, a specific nested market – a market nested in distinction.[18]

Alongside distinction, these newly emerging markets also show connectivity, rootedness and, increasingly, relevance and transformative capacity.[19]

Distinction (or *specificity*) may reside in the product (in its taste, freshness, price, origin and/or in the way it has been produced, processed or delivered to the market), but it may also reside in the marketplace being different, attracting different suppliers and consumers who interact in ways that differ from those in the conventional markets. Distinction might also be rooted in a social definition of quality shared by the different actors involved and/or in the way the marketplace is perceived by these actors. *Connectedness* refers to the relations that link these markets, the farms that supply them, the processing units and the consumer groups connected to them into what Wegerif (2017) described as a 'symbiotic whole'. Whilst the first dimension, distinction, reflects the single differences (in products, actors, places, etc.), this second dimension, connectedness, reflects the consistency between the single differences. Thirdly, there is *rootedness*. This dimension refers to the materiality of things, such as natural resources that impose their own rhythm and calendar, products whose processing cannot be industrialized without a loss of quality, the skills and knowledge that are inherent parts of artisanal production and the socio-material infrastructures that characterize peasant markets (see also Chapter 10).

Relevance refers to the outcomes rendered by a particular market and how these differ from the ones generated by other markets. Nested markets perform in ways that differ markedly from global markets and might, therefore, produce outcomes that are not only different but also relevant to many people. The relevance might reside in improved incomes and/or increased employment levels for producers; increased accessibility of good, fresh (and sometimes high-quality) food for enlarged groups of consumers; and/or the inclusion of otherwise excluded producers that allows for production that would otherwise not occur.

Transformativity is another important dimension that helps to explain peasant markets and their significance in the wider context. This is the capacity to actively contribute to processes that transform wider society. Beyond the direct contributions of single peasant markets, the unfolding whole of interlinked peasant markets potentially carries considerable transformational power – especially when well coordinated and integrated in wider social movements.

Peasant markets are 'transformative spaces' (Pereira et al., 2018), they are able to 'catalyze change' throughout the wider politico-economic systems (Westley et al., 2013). In Chapters 9 and 10 I will come back to some of these issues.

Countering some critical arguments

Newly constructed, nested markets differ significantly from the global, empire-controlled food markets – partly because nested markets are meant and organized to *be* different. Yet the two cannot be understood in terms of a dichotomy. Nested markets may evolve into constitutive parts of global markets and the initial differences may fade away. It is also possible that the two continue to exist alongside each other, just as farmers may link to both types of market (as they often do). This latter phenomenon can be rooted in different causes. It might be a transitory phenomenon: with the growth of nested markets farmers may shift growing portions of their total production to the nested market – but in the meantime, they may depend on both.[20] Delivering to two different output markets may also be a conscious choice to reduce possible risks.

Nonetheless, over recent years there has been a clear tendency for nested markets to expand in both size and numbers. This is partly due to peasant struggles, which I will discuss in the next chapter. It is also due to the possibilities generated by the internet and new communication technologies which allow for establishing and maintaining close social relations even over large geographical distances. Quite probably the size and numbers are also growing because many social actors and movements support farmers' markets (and new ways of producing and consuming food) because they perceive them as important levers in the many-sided struggles that aim to tackle climate change, growing inequalities, poverty, unemployment and massive migratory flows.

In the meantime, there is also a growing critique of these tendencies (that stems from a wide variety of sources). One strand of critique (typically developed in Wageningen University) is that '*small* farms and markets (as opposed to large retail) never can feed *large* cites and metropoles'. This argument has rhetorical strength, especially due to the opposition of *small* and *large*. It seemingly makes the argument self-evident. The fact of the matter, though, is that in the past, *markets* (as Les Halles in Paris, Porto Palazzo in Turin, El Mercado Mayor in Buenos Aires)[21] very efficiently provided large urban populations with all the food they needed. And today the Xin Fa Di market in Beijing supplies a 26 million urban population with all of the food it needs – and this food mainly comes from smallholder producers, several of whom sell their products directly in the Xin Fa Di market.[22] A well-organized market can very well be a smoothly functioning channel linking *small* producers and the food needs of *large* cities. The opposite applies to large retailers that cannot easily deal with a multitude of small producers, because the transaction costs would be far too high.

It is also argued that farmers' markets (especially those that are recently constructed) are necessarily small and doomed to remain small. Again, this observation does not stand up to historical evidence but seems to be in line with the modest proportions of most new markets. But what applies for a single market does not necessarily hold true for a *network* of interrelated markets that covers large areas. Since time immemorial the Netherlands had had local street markets. Some of these (such as the Albert Cuyp and Noordermarkt in Amsterdam) cannot be characterized as small, but neither can they channel all of the food needed in these cities. However, 'feeding the city' (to paraphrase the often used, rhetorical expression about 'feeding the world') does not necessarily depend on one single street market.

The Netherlands has some 1,000 street markets. Some are daily, others weekly. In total they have 38,000 stalls. Sixty percent of the total turnover is in food, 8% in flowers and plants, and the rest in clothes, textiles, shoes, etc. The total turnover of these 1,000 markets is 3 billion euros per year, 2 billion of which is for food – that is, for *fresh* food, because hardly any processed foods are sold on these street markets. The visitors to the street markets spend on average 21 euros per visit.

These data might be compared with the sales of Albert Heijn, the largest supermarket chain of the Netherlands (and part of the international AHOLD conglomerate). The Albert Heijn chain has 1,000 supermarkets distributed all over the Netherlands. Together they have an annual turnover of 3.3 billion euros. A considerable part, though, of the total sales of Albert Heijn shops is made up of processed food and, especially, non-food items (tobacco, cosmetics, alcoholic drinks, cleaning materials, pet food, etc.). Thus, it is not an exaggeration to hypothesize that the fresh food sales of the Netherlands' many street markets *together* greatly exceed the fresh food sales of the largest retail chain in the Netherlands. *If well-combined, the small elements can impact very strongly – precisely due to their sheer number* (as is the case in the realm of production, as we saw in Chapter 5).

Then, there is the critique that assumes that nested markets are an expression of *campanilismo*. *Campanile* is the Italian word for 'church' or 'bell tower'. *Campanilismo* means that your attention and activities are limited to the zone within which the bell tower is visible (or audible). The expression stands for a narrow-minded type of localism, or parochialism. You only eat what is produced locally; everything that comes from beyond the reach of the *campanile* is rejected.

It is true that many new peasant markets that are flourishing in many countries are mostly supported through short supply lines. The 'local' indeed is an important, albeit somewhat flexible, benchmark. It is a logo that helps to underline the distinction vis-à-vis supermarkets and global trade. However, the new, nested markets are not destined per se to the local with its short supply lines. An excellent example is, in this respect, the Ecovida circuit[23] that links many local markets (*feiras*) and which, by doing so, makes some food products travel considerable distances – without losing their distinctiveness vis-à-vis the products being traded in the general food market.

Figure 8.5 gives a schematic overview of this Ecovida circuit. It shows some local markets and the in-between lines that allow for mutual provisioning. There are some important principles that underlie this constellation that make it operate in a fascinating way. The first principle is that every local market (or *feira*) is provided by the local producers. Only those products that are lacking (due to the specificity of the local ecosystem) might be acquired from producer groups elsewhere (whose eco-system allows for the production of such products). Here it

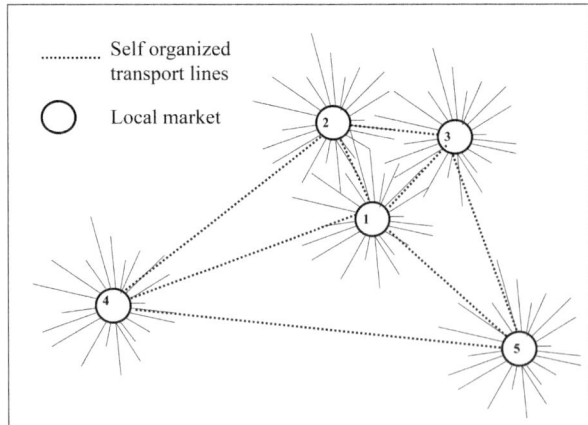

Figure 8.5 The circuit of Ecovida linking different local markets (Brazil) (author's own elaboration).

is important to know that the distance between different *feiras* might be as much as 1,000 km or more. The second principle is that only that what abounds (the amount of products that cannot be absorbed by the own local market) can be delivered to local markets elsewhere – if, and only if, the producer and consumer associations of those other places ask for it. This strongly grounds production and consumption on ecological principles.

Both local trade and trade between the different *feiras* are embedded in the eco-systems and the ecological differences associated with them. Ecology is an important ordering principle here. Associated with this comes another basic feature: trade between the different *feiras* is complementary instead of substitutive. The in-between flows complement the supply of food in the different *feiras*: by increasing the assortment they enrich the total supply (just as happens in the case of the *landwinkels*, discussed above). It is *not* possible that production in, for example, area 1 replaces (substitutes for) production in the areas surrounding other *feiras*.

There are other, highly novel, features that make the Ecovida circuit into an inspiring example. Some of these I will discuss further in Chapter 10.

Finally, there is what we can call the political relevance of newly constructed nested markets. There is a considerable polemic about this issue (see, e.g., Bonnano and Wolf, 2018). I will not enter into the associated debates but limit myself to a methodological observation: that is, that the relevance of something does not necessarily reside in the item itself – it might very well emerge elsewhere. What is decisive is the *relation* between the item itself (e.g., a newly constructed market) and the place where this item shows (probably together with other items) its relevance. Text Box 8.1 offers an example. It makes clear that the construction of new markets can induce and/or sustain important changes in the sphere of production. This applies, more generally, to agroecological transitions in farming: they go hand in hand with the building of new market channels (see Anderson et al., 2020; Gonzalez de Molina et al., 2020).

Text Box 8.1   Valdecir Vedana from the Antonio Prado Municipality, Rio Grande do Sul

The farmer shown in the picture below is a very capable vegetable grower. His tomato harvests used to be his pride and joy. However, some 20 years ago he got so annoyed by the very low prices that were paid in the market that he nearly decided to completely stop tomato production. But then there seemed to be another way out: processing the tomatoes within the farm into passata, sauces, concentrates and juices and trying to sell them directly to consumers. The first experiences of what was to become, later on, Ecovida, provided an important stimulus for reflecting on such a prospect.

As in other cases (see Text Box 7.8) there was some serendipidity involved. Or, as I would argue, the capacity to recognize a somewhat hidden opportunity and turn it into a promising and succesful novelty (more or less similar to the capacity of female breeders of potato seedlings to recognize a new variety; see the section on novelties and niches in Chapter 2). 'Having an eye' is strategic in all such cases.

Firstly, the local hospital was about to be renovated and equipped with new instruments. Valdecir became aware of this, and it occurred to him that the old equipment for sterilizing surgical instruments could be of considerable use to him. It would allow for pasteurizing the tomatoes – a crucial step in their processing – and for sterilizing the pots and bottles. He began to bargain and (following the logic of 'farming economically', see Chapter 4) was able to obtain the needed technical artefacts for a low price.

Secondly, the knowledge of his grandmother, mother and sisters (the Vedana family are descended from Italian immigrants) on how to make good sauces for accompanying the pasta turned out to be equally important. It was a capacity that had lain dormant in the family and suddenly emerged as important 'cultural capital'.

Photo by Adriaan van der Ploeg.

Thus, the family started to experiment with, and succeeded in, making very good tomato sauces and related items. The photograph shows Valdecir standing in front of the stock. One can see the pride. There is some transcultural magic in such body language. Standing with arms folded in front of the work done (see also Text Box 7.7) clearly reflects *satisfaction* (in the Chayanovian sense): it took *drudgery* to figure out the best way of on-farm processing and to produce the batch of tomato sauces shown in the background, but now there is satisfaction, and pride is a justified ingredient — simply because the obtained result is the outcome of the joint labour input of the peasant household.

If these tomato sauces were channelled through the dominant 'value-chains' (as suggested in dominant agro-political discourse), the fruits of the work done would certainly be minimal. The biggest part would go to middlemen, distributors and retailers. In such a case the on-farm processing would not have developed (or probably not even started). However, in combination with the developing Ecovida *feiras* and circuit, the on-farm processing of tomatoes (and other fruits and vegetables) was able to germinate from the seed of an idea into a coherent business strategy in the years that followed.

Thus, many different elements (tomatoes, secondhand equipment from the hospital, the culinary tradition of the family, the newly emerging Ecovida structures for marketing, the stubbornness of Valdecir, etc.) were combined: a socio-material reality was re-patterned, and this re-patterning allowed for resetting the balance of drudgery and satisfaction (see Figure 5.6): labour incomes were considerably improved and new prospects were generated for the farm.

New ways forward

In general terms, the emergence of new peasant markets (and, more specifically, the micro-level examples as entailed in Text Box 8.1) show, in the first place, that farm expansion and closure are not the only two alternatives available for farmers facing declining incomes. Accepting the 'logic of the (dominant) market' is not the only possible way forward. Through the construction of new markets (and associated alterations in the process of production), farmers can escape from the seemingly unavoidable 'dull compulsion of the market' (see also Chapter 3). They may elaborate a script that differs from the one of capital. Secondly, the example shows that engaging in the construction of a new market (and its subsequent operation) can be very interesting and empowering for those involved. It is also, I would argue, *politically relevant* because it shows them and others that resistance and building an 'economy of opposition' (Pahnke, 2015) *is possible and does matter*. Thirdly, the examples show that socio-political struggles also take place at the level of, and within, markets – they are not solely limited to the sphere of production (as, especially, old-fashioned Marxists assume). I will further elaborate on this in the next chapter.

Methods Box #16   On the exploration of possibilities for new, nested markets: Structural holes and bridging

The photograph below shows a little statue in the town of Sneek, in the northern Netherlands. The statue shows two figures. The one with the high hat is a trader – in local dialect a *lapkepoep*: a German trader selling textile. The lady standing next to him is a farm woman. The two are trading: exchanging some nice pieces of textile (probably oddments) for a duck and maybe some cheeses or butter. The scene is taken from the end of the 19th century. At that time there was a wide gap between the countryside and the (urban) circuits in which elegant clothes, nice oddments and the like were available. This gap was what would later be referred to as a 'structural hole' (Burt, 1992): Farmer's women did not go to the luxury shops where luxury clothing was sold. Going there was considered squandering money. In their turn, the owners of dress shops did not even think about going to the countryside – it was dirty over there and people would not buy their merchandise anyway.

Photo by author.

This pattern only changed when, in the second half of the 19th century, some outsiders appeared. German traders (with heavy loads on their backs) who walked from farm to farm, first to chat only and then, during next visits, to sell needles, yarn, whatever was needed. They also showed, every now and then, the oddments they carried with them. Payment with 'expensive money' (as was said at that time) was not needed: barter was sufficient.

These travelling traders bridged the structural hole: they created a new market, and by doing so they laid the foundations for what later on would become well-known and powerful economic chains such as Peek and Kloppenburg, C&A and Vroom and Dreesman.

In social sciences we think about structure as the presence of stable relations that mould social action. Through this action the structure that underlies it is reproduced. Together structure and action compose a unity that introduces regularity and stability in society. It gives the beacons that guide people and their aspirations. Because of such structure, they know more or less what can be expected.

The important point now is that social structures entail not only (stable) relations but also the opposite: the absence of relations. Where this occurs there is a *structural hole*. That is, alongside frequent communication there might be lack of communication, alongside stable patterns of exchange there might be a lack of exchange, alongside meaningful encounters, the lack of, or meaningless, encounters, etc. This is summarized in Figure 8.6.

Social actors 1, 2, 3, 4 and 5 are closely related. On the other hand, actors 7 and 10 are only indirectly (and probably in a weak way) connected, whilst 9 is connected to no one else. In these cases, there are structural holes.

Economies such as ours not only produce many relations and connections; they also produce structural holes. This occurs, for example, when high investment levels (or legal rules) bring considerable barriers for newcomers or when specific consumer groups are excluded from specific exchanges. A well-known example of the latter are the so-called green deserts: large and poor urban areas where it is very difficult, if not impossible, to acquire healthy – that is, 'green' – nutrition

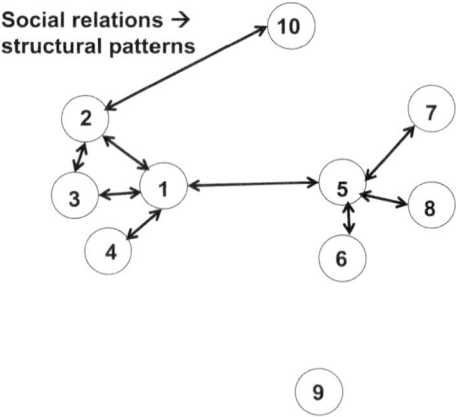

Figure 8.6 Social structure, social relations and structural holes (author's own elaboration).

(fresh vegetables, fruits, non-processed foods). Only junk food is sold. That is: the consumers in these areas are excluded: they have no (easy, comfortable and reliable) supply to healthy food.

Similar structural holes might emerge (they are, indeed, sometimes constructed on purpose) if new regulation blocks, for example, the processing of raw milk into cheeses (or the opening of a small weekly market for vegetables in the neighbourhood mentioned above). The list might be extended (structural holes abound). The point, though, is that structural holes can also be *bridged*. This has happened in the past (see, e.g., Long [1975] and the example above); it also happens today and it often occurs as, and through, the construction of a new, nested market (Ploeg, 2015).

The identification of structural holes is a major device in the construction of new (nested) markets. In terms of methods, it basically is simple. Taking the dimensions summarized in Table 8.1, the leading question is, time and again, 'what is *lacking* here?' Or, phrased differently: is it possible to organize flows and transactions in a *more satisfactory* way?

The *lack* of fresh food (as eaten in villages by farming families) in the city of Beijing, for instance, was the starting point for the construction of a new market. The lack of cheese *not* covered with a plastic film was the starting point for a new cheese brand (protected with ghee applied with laser technology to ripening cheeses) and was the beginning of a new market channel in the Netherlands. The separation of (or: structural hole between) buying food and leisure, and its consequent bridging, was the beginning of country shops (see Figure 8.2) that allowed for more satisfactory shopping.

Notes

1. Interestingly, the methodology was geared basically around the principle of 'following the eggs'. That means travelling with the different traders and transporters (cycling with them through Dar es Salaam, etc.) and talking with the different producers in the villages where the eggs come from and observing, at the final selling points, who the clients are. At the same time, a careful diagnosis was made of prices, price differentials, trade volumes, number of transactions, distances, etc.
2. The *duka* is like the proverbial *chino* encountered on nearly every street corner in most Latin American countries.
3. Such diversity in 'patterns of provisioning' (as Wegerif [2014] calls them) is not limited to the Global South. If we look at, say, the Netherlands, eggs can be obtained via (1) automated selling points at farms, (2) farm shops, (3) specialized dairy shops, (4) home delivery by milkmen, (5) supermarkets, (6) street markets and, in a way, (7) new food banks and (8) having hens in one's own garden. There is another layer of complexity because the type of product also differs: alongside 'normal' (factory-farmed) eggs, there are free-range and organic eggs. Each comes with a different normative framework that defines and supports the production method of these eggs.
4. As a matter of fact, each channel (the supermarkets apart) has a specific price *range*. The exact price depends on negotiations but also on the frequency of transactions, the number of eggs and the existing ties between seller and buyer.
5. TSh = Tanzanian shilling. At the time that the research was carried out, 250 TSh (the price of one egg at the *duka*) equalled US$0.19.

6. In the literature, the adjective *nested* is linked in specific cases to the noun *market*. The housing and arms markets are amongst these cases. The value of houses very much depends on location; that is, their position and relation vis-à-vis contextual elements. Houses in desirable locations (i.e., near a beach) have a higher value than similar houses elsewhere. If the Amsterdam subway were extended to the far side of the waterway that separates the city, the value of houses 'in the north' would suddenly go up. That is to say, the housing (or real estate) market is nested in the 'social geography' (attractions, distances, connections, etc.) of the space in which it is located. In short: it is a *nested* market (Beckman and McPherson, 1970; Esparza and Krmenec, 1996; Garcia Pozo, 2009). The same applies to sophisticated arms: they cannot be exported anywhere, nor can they be sold to anyone (and if sold or exported, they come with very strict requirements that exclude re-sale). All of this is excluded on the grounds of national security (or political interests). Hence it is, again, a nested market (Kinsella, 1995; Duch-Brown and Fonfría, 2014). In the same vein, markets can be nested as well in regulatory schemes (Van Huylenbroeck et al., 2009). These examples bring in *space* and *regulation* as key factors that make for markets being nested. The recent literature on nested food markets has extended this idea, bringing in the *actors* and representing such markets as being constructed, shaped and reproduced by, and through, the mutual understanding of the producers and consumers of food. It might be argued that such mutual understanding and agreement are decisive in these new food markets. I consider peasant markets to be a specific case of nested markets.

7. This concept is elaborated further in Chapter 10.

8. Typical here are the thousands of cyclists who supply the *dukas* and make (part of) their livelihood in this way. Supermarkets cannot be provisioned in this way.

9. The same question was asked by Hellen Kimanthi in her study on 'traditional exchange arrangements' of maize in Western Kenya. After distinguishing a range of different channels (roadside stands, home trade, open air markets, different types of barter, small shops, kiosks, *posho* mills and door-to-door trading), she argued, "th[is] multiplicity of exchange channels cannot be reduced to one abstract form of marketing [nor to] the way global markets are operated" (Kimanthi, 2019:124).

10. In formalist theories, markets are considered as self-regulating systems. They function regardless of the people who operate in them. Consequently, the operation of the market can even be described just by mathematical formulas. In substantivist theories, the market very much depends upon the folk concepts used by the involved actors, and the market(s) can be described and analyzed through the use of these folk concepts.

11. Supermarkets present themselves as a healthy, well-ordered, clean and Western-style alternative to the clumsy, crowded and, assumed, dirty street markets and small shops. There clearly is a class dimension in this presentation.

12. An overview is given in Civil Society Mechanism (2016).

13. This is also reflected in many RD experiences that are geared toward (old-fashioned) processing facilities that still are available locally. They allow for 'bypassing' (see Chapter 10) the centrality and dominant position of processing facilities controlled by food empires.

14. See also Pierluigi Milone and Flaminia Ventura (2000) on farm butcheries and Gianluca Brunori and Adanella Rossi (2000) on how to calculate the economic impact and relevance.

15. Nested markets differ from niche markets in that the former have somewhat permeable boundaries that separate them from the general market (farmers can easily move in and out), whilst niche markets are closed: only those whose products are branded can join the latter.

16. Pierluigi Milone and Flaminia Ventura (2015) developed interesting theoretical insights on how nested markets may evolve (or not) into non-distinguishable parts of the general market. See also Milone et al. (2015).

17. Elsewhere we have argued that distinction is grounded on specificity, rootedness and/or connectivity. See Ploeg et al. (2012).
18. Hielke van der Meulen (2000) gives an excellent description of the market for Chianina beef. Ploeg (2015) provides a description of the market for Texel lamb.
19. Together with Jingzhong Ye and Sergio Schneider, I have extensively outlined the first three concepts in a joint article published in 2012 in the *Journal of Peasant Studies*. The latter two concepts are developed in Ploeg et al. (2022).
20. However, not being solely dependent on the main market (i,e., partly delivering to a nested market) may already bring some space for negotiation and thus increase autonomy.
21. Descriptions can be found in Black (2005, 2012) and Viteri (2010).
22. A description is provided in Ploeg and Ye (2016, chapter 10).
23. There are 4,500 agroecological producers linked to this Ecovida network and 120 local markets that form part of it. Detailed information is given in Radomsky et al. (2015).

Bibliography

Anderson, C. R., J. Bruil, M. Jahi Chappell, C. Kiss, and M. Pimbert. (2020), *Agroecology Now! Transformations towards More Just and Sustainable Food Systems*, Palgrave, Macmillan, London.

Beckmann, M. J., and J. C. McPherson. (1970), 'City size distribution in a central place hierarchy: An alternative approach', *Journal of Regional Science* **10** (1), pp. 25–33. doi: 10.1111/j.1467-9787.1970.tb00032.x.

Bernstein, H. (2010), *Class Dynamics of Agrarian Change*, Fernwood, Halifax, NS, Canada.

Black, R. E. (2005), 'The Porta Palazzo farmers' market: Local food, regulations and changing traditions', *Anthropology of Food* **4**, 11354. doi: 10.4000/aof.157.

Black, R. E. (2012), *Porta Palazzo: The Anthropology of an Italian Market*, University of Pennsylvania Press, Philadelphia.

Bonnano, A., and S. A. Wolf. (2018), *Resistance to the Neoliberal Agri-Food Regime: A Critical Analysis*, Routledge, London.

Brunori, G., and A. Rossi. (2000), 'Synergy and coherence through collective action: Some insights from wine routes in Tuscany', *Sociologia Ruralis* **40** (4) pp. 409–423.

Burt, R. S. (1992), *Structural Holes: The Social Structure of Competition*, Harvard University Press, Cambridge, MA.

Civil Society Mechanism. (2016), *Smallholders and Markets, a Bibliography*, FAO, Rome.

Duch-Brown, N., and A. Fonfría. (2014), 'The Spanish defence industry: An introduction to the special issue', *Defence and Peace Economics* **25** (1), pp. 1–6. doi: 10.1080/10242694.2013.857462.

Esparza, A. X., and A. J. Krmenec. (1996), 'The spatial markets of cities organized in a hierarchical system', *The Professional Geographer* **48** (4), pp. 367–378. doi: 10.1111/j.0033-0124.1996.00367.x.

Fligstein, N. (2001), *The Architecture of Markets: An Economic Sociology of Twenty-First Century Capitalist Societies*, Princeton University Press, Princeton, NJ.

Garcia Pozo, A. (2009), 'A nested housing market structure: additional evidence', *Housing Studies* **24** (3), pp. 373–395. doi: 10.1080/02673030902875029.

Gereffi, G., J. Humphrey, and T. Sturgeon. (2005), 'The governance of global value chains', *Review of International Political Economy* **12** (1), pp. 78–104. doi: 10.1080/09692290500049805.

Gonzalez de Molina, M., P. F. Petersen, F. Garrido Pena, and F. Roberto Caporal. (2020), *Political Agroecology: Advancing the Transition to Sustainable Food Systems*, CRC Press, New York.

Kimanthi, H. (2019), *Peasant Maize Cultivation as an Assemblage: An Analysis of Socio-cultural Dynamics of Maize Cultivation in Western Kenya*, Ph.D. thesis, Wageningen University, Wageningen, the Netherlands.

Kinsella, D. (1995), 'Nested rivalries: Superpower competition, arms transfers, and regional conflict, 1950–1990', *International Interactions; Empirical and Theoretical Research in International Relations* 21 (2), pp. 109–125. doi: 10.1080/03050629508434862.

Long, N. (1975), *An Introduction to the Sociology of Rural Development*, Tavistock, London.

Meulen, H. van der. (2000), *Circuits in de Landbouwvoedselketen: Verscheidenheid en Samenhang in de Productie en Vermarkting van Rundvlees in Midden-Italie*, Ph.D. thesis, Wageningen University, Wageningen, the Netherlands.

Milone, P., and F. Ventura. (2000), 'Theory and practice of multi-product farms: Farm butcheries in Umbria', *Sociologia Ruralis* 40 (4), pp. 452–465.

Milone, P., and F. Ventura. (2015), 'The visible hand in building new markets for rural economies', in *Rural Development and the Construction of New Markets*, edited by P. Hebinck, J. D. van der Ploeg, and S. Schneider, 41–60, Routledge, London.

Milone, P., F. Ventura, and J. Ye. (2015), *Constructing a New Framework for Rural Development*, Emerald, Bingley, UK.

Ostrom, E. (2010), 'Beyond markets and states: Polycentric governance of complex economic systems', *American Economic Review* 100 (3), pp. 641–672. doi: 10.1257/aer.100.3.641

Pahnke, A. (2015), 'Institutionalizing economies of opposition: Explaining and evaluating the success of the MST's cooperatives and agroecological repeasantization', *The Journal of Peasant Studies* 42 (6), pp. 1087–1107. doi: 10.1080/03066150.2014.991720.

Pereira, L. M., T. Karpouzoglou, N. Frantzeskaki, and P. Olsson. (2018), 'Designing transformative spaces for sustainability in social-ecological systems', *Ecology and Society* 23 (4), pp. 1–6. doi: 10.5751/es-10607-230432.

Ploeg, J. D. van der. (2015), 'Newly emerging, nested markets: A theoretical introduction', in *Rural Development and the Construction of New Markets*, edited by P. Hebinck, J. D. van der Ploeg, and S. Schneider, 16–40, Routledge, London.

Ploeg, J. D. van der, A. Long, and J. Banks. (2002), *Living Countrysides: Rural Development Processes in Europe: The State of Art*, Elsevier, Doetinchem, the Netherlands.

Ploeg, J. D. van der, and J. Ye. (2016), *China's Peasant Agriculture and Rural Society: Changing Paradigms of Farming*, Routledge, London and New York.

Ploeg, J. D. van der, J. Ye, and S. Schneider. (2012), 'Rural development through the construction of new, nested markets: Comparative perspectives from China, Brazil and Europe', *The Journal of Peasant Studies* 39 (1), pp. 133–173. doi: 10.1080/03066150.2011.652619.

Ploeg, J. D. van der, J. Ye, and S. Schneider. (2022), 'Reading markets politically: On the transformativity and relevance of peasant markets', *Journal of Peasant Studies*. doi: 10.1080/03066150.2021.2020258.

Radomsky, G., P. Niederle, and S. Schneider. (2015), 'Participatory systems of certification and alternative marketing networks: The case of the Ecovida Agroecology Network in South Brazil', in *Rural Development and the Construction of New Markets*, edited by P. Hebinck, J. D. van der Ploeg, and S. Schneider, 79–98, Routledge, London.

Van Huylenbroeck, G., A. Vuylsteke, and W. Verbeke. (2009), 'Public good markets: The possible role of hybrid governance structures in institutions for sustainability', in *Institutions and Sustainability*, edited by V. Beckmann and M. Padmanabhan, 175–191, Springer, Dordrecht, the Netherlands.

Ventura, F. (2001), *Organizzarsi per Sopravvivere, un Analisi Neo-institutionale dello Sviluppo Endogeno nell'Agricoltura Umbra*, Ph.D. thesis, Wageningen University, Wageningen, the Netherlands.

Viteri, M. L. (2010), *Fresh Fruit and Vegetables: A World of Multiple Interactions. The Case of the Buenos Aires Central Wholesale Market*, Ph.D. thesis, Wageningen University, Wageningen, the Netherlands.

Wegerif, M. C. (2014), 'Exploring sustainable urban food provisioning: The case of eggs in Dar es Salaam', *Sustainability* **6** (6), pp. 3747–3779. doi: 10.3390/su6063747.

Wegerif, M. C. A. (2017), *Feeding Dar es Salaam: A Symbiotic Food System Perspective*, Ph.D. thesis, Wageningen University, Wageningen, the Netherlands.

Westley, F. R., O. Tjornbo, L. Schultz, P. Olsson, C. Folke, B. Crona, and Ö. Bodin. (2013), 'A theory of transformative agency in linked social-ecological systems', *Ecology and Society* **18** (3), pp. 27. doi: 10.5751/ES-05072-180327.

9 Peasant resistances and struggles

Overview: Main concepts discussed in Chapter 9

Value extraction
Subordination
Capital relation
Movimento dos Sem Terra (MST)
Micro-farm
Community-supported agriculture (CSA)
Food empires
Resistances
Peasant struggle
Targets of peasant struggle
Tomas de tierra (land occupations)
Land grabbing
Anhui rebellion (China)
Declaration of Peasant Rights
Milk strikes
Vertical cooperatives
By-passes
Cooperation-in-production
Struggles of the first, second and third kind
Alteration of processes of production (radically re-patterning processes of
    production)
Labour-driven intensification
Horizon cooperatives (production cooperatives)
Man-land ratios
Zapatista uprising (Chiapas, Mexico)
Polo Sindical do Borborema (Brazil)
Northern Frisian Woodlands (Friesland, the Netherlands)
Territorial cooperative (self-governed spaces)
Front door/back door principle
Explaining biodiversity
Farming economically (low external input agriculture)
Linking different episodes of struggle
La Via Campesina

DOI: 10.4324/9781003313274-9

Food sovereignty
Transitions (transformative flows)
Unifying and mobilizing notions
Campesino a campesino training
The (methodological) principle of symmetry
Counter power

There is a rich literature on peasant resistance and struggles (see, for example, Moore, 1967; Wolf, 1969; Huizer, 1972; Paige, 1975; Scott, 1985; Edelman and Borras, 2016). In this chapter I will build on this literature by focussing on the *roots* of both resistance and struggle and how are they anchored in, and constantly fuelled by, the everyday life experiences of peasants. I will also examine how the many-sided limitations and frustrations peasants suffer from, together with their expectations and hopes, translate into different forms of collective action, as well as how peasant movements inform the life worlds of the many men and women involved in farming.

In this chapter I will argue that the longing for, and construction of, autonomy (see Text Box 9.1 for a theoretical underpinning) is central to peasant resistance and struggles. This reflects their unique class position. As part of the third class (see Chapter 1), they control and make use of both their own labour and the means of production to which they have access. Whilst wage labourers structurally depend on capital (without capital they cannot produce), peasants have their own resource base and their own patrimony which allow them to produce independent of capital (i.e., independent of any structural relation with capital).

Text Box 9.1   The concept of autonomy

The search for autonomy can be seen as a set of practices that centre, and depend, on the social production and reproduction of resources (networks included) that allow for the pursuit of trajectories that would otherwise have been impossible. Autonomy is three-pronged:

1   it is a set of goal-oriented activities that aim to build resources;
2   the combination of these resources materially represents a distantiation from capital; thus,
3   it allows for *agency* – the capacity to define relatively autonomous courses of action.

Thus, autonomy is about constructing, defending and controlling the resources that allow for autonomous actions and practices. Having, defending and building on these resources is a social struggle (especially in times and places characterized by an imperial ordering of both the social and the natural worlds). Autonomy is also the capacity to define developmental trajectories that are in line with one's own interests, prospects, experiences and expectations – that is, in short, the capacity to deviate from the dominant script. This makes it a cornerstone of, and for, social struggle.

Autonomy is a specific junction that links the past, present and future. As argued by Karl Kosik in *The Dialectics of the Concrete* (1976), every specific reality contains a range of different possibilities for further unfolding. This implies that in every specific situation, choices (*autonomous* choices) can be made that define the trajectory that is to be built. There is no determinism, cast in stone, linking the past, present and future in a linear way. There is, instead, always the possibility to envisage a developmental trajectory (a course of action) that deviates from the interests and logics of capital and/or the dominant script (the hegemonic view – or what in farming tellingly is referred to, in the German language, as *agrardiktat*). The capacity to do so assumes autonomy and is equally rooted in culture (or 'superstructure' as a classical Marxist would possibly say). It represents counter-hegemony and is often rooted in social struggles, cultural repertoires and collective memories (and is definitely not an attribute that is held individually).

Autonomy refers to the self-organizing capacities of people, communities and movements. Such capacities assume both resources and agency. The former is not only a condition for the latter – it is also an outcome. One feeds into and strengthens the other and vice versa. Together the two generate, as argued by Joost Jongerden (2021), constructive resistance that allows for, and flows into, 'self-organized development' (11).

Of course, peasants are subject to the value of their production being extracted (through a range of mechanisms discussed in Chapter 6) and used to feed capital accumulation in other places. Equally, capital groups try to impose, directly or indirectly, their scripts in order to align the organization and development of peasant agriculture with their interests to the maximum possible extent. Yet both value extraction and subordination can be resisted, fought and altered, thus creating relations that are more favourable to those directly involved in agricultural production. In factories, labour can fight for improved labour conditions (regarding salary, the length of the working day, health and safety, etc.) but labour cannot alter the process of production itself[1]: the technologies to be used, the raw materials selected, the nature of the end products – all of these features, which are central to the process of production, are the exclusive prerogative of capital. By contrast, in farming, those who do the work also make decisions about the organization and development of production. Deciding what to produce, how to produce it, what to do with the obtained benefits, etc., is a basic right of farmers themselves. This right is rooted in their control over their resource base. The unity of (peasant) labour and a (self-controlled) resource base represents autonomy – just as it feeds the deeply rooted conviction among peasants that they can do without capital (as an organizing principle).

Autonomy defines the relations between the third class and its capitalist environment in two ways[2]: autonomy is constantly *threatened* (through, for example, the intensified squeeze on agriculture and/or rigid external prescriptions) but, at the same time, autonomy allows those involved in farming to *resist*, and actively respond to, subordination and exploitation – thus re-establishing, defending and, sometimes, increasing their autonomy.

Although there are, in practice, many obstacles (that vary with time and place), theoretically speaking, a peasant farm can be started and/or developed *without entering into a dependency relationship with capital*. This is precisely what is happening in the settlements of the Movimento dos Sem Terra (MST) in Brazil, in the *paramos* and *llanos* in Colombia where settlers (and ex-*guerrilleros*) try to open new fields and in the fields of Piura (Peru) where peasant communities and cooperatives actively tried to enlarge levels of productive employment – but also in France where youngsters are starting micro-farms and/or constructing CSA (community-supported agriculture) schemes. It is also happening in all those already-existing farms which people are trying to develop through their hard work and labour investments. Many of these people will use the expression 'building a free farm' to describe what they are striving for. A free farm is a farm that is not subordinated to heavy dependency relations – it is a farm that can be run and further developed with the means available to the farming family.

Yet, at the same time, many farms are de facto dependent on capital. But in such cases the involved farmers will try to resist capital and untie their farms from the capital relations in which they are engaged. The notion of autonomy triggers, spurs and guides such resistance and struggle. For those people, autonomy is not an abstract notion; they know from practice that it is possible and desirable.

Autonomy is the gunpowder that fuels peasant resistance and struggle. It makes the peasantry into a powerful and potentially radical adversary of food empires and other capital groups that try to control and drain farming.

The defence of, and struggle for, autonomy occurs at many different, but inter-linked, levels. It takes place at the micro level (that is, within and around the farm). It occurs at a meso level, too: in the markets and at the territorial level. The struggle for autonomy also plays out at the macro level where it is about political autonomy: the capacity and willingness to engage in the organization of power at the level of society as a whole. If this occurs, the third class is man-ifesting itself as a class *für sich*. Taken together, these different levels can make for a highly variegated and dynamic interplay: the search for autonomy at the farm level feeding the construction of new institutional patterns (in markets and territories) that allow for autonomy (at the meso level) and which simultaneously strengthen autonomy at farm level. If and when these multiple struggles find a political expression (as strong social movements), the micro-level autonomy this brings may further consolidate and strengthen autonomy at the other levels.

The micro-level struggle for autonomy that takes place in every farm – not just once but as an ongoing process – can be explained with the help of Figure 9.1.

Peasants, wherever located, always face a difficult, if not downright hostile, environment. Nobody will offer them the proverbial free lunch, and if they do not take care, even their own lunch will be grasped and consumed by others. This environment triggers, and is countered by, an ongoing struggle for auton-omy that will bring peasants the improved livelihoods that nobody else will offer them. This struggle occurs as, and through, the construction of a self-controlled resource base (land, animals, seeds, crops, housing, knowledge, networks, etc.) that allows for the production of food and other products that are (partly)

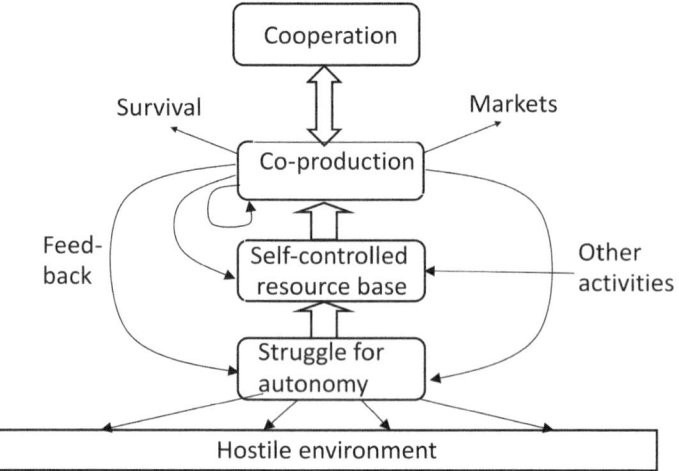

Figure 9.1 The choreography of peasant agriculture (Ploeg, 2018).

marketed and (partly) used to reproduce both the farm and the farming family. Within this scheme, farming is synonymous with struggling (*luchando la vida*,[3] as Latin American peasants will respond when you ask how they are doing). In this respect, farming is about working very hard to make sure that the family survives and the farm improves (in order to produce better in future cycles). Being engaged in agricultural production is far from a humdrum routine. It is imbued with resistance and struggle. It is identical to resisting the awkward environment (and this applies as much in the Global North as it does in the Global South) and it implies ongoing struggles. Dealing with a degraded field, bringing well-bred manure to it and/or fighting with unwilling oxen when the field needs to be ploughed – all of these are part of an uphill struggle that is physically demanding (drudgery is far from an abstract concept in this context), but in the end this bodily struggle may bring improved fields, improved potato seedlings and respect from others so that they will help one to make further progress in the cycles to come. These somewhat hidden uphill battles – and there are many of them – converge, translate into, and fuel (especially when they get blocked) struggles at other levels: building cooperatives, constructing peasant markets, organizing milk strikes and demonstrations, occupying land, blocking roads, destroying GMO experimental plots, burning establishments of McDonald's, elaborating provocative policy proposals and much more.

The peasant struggle for autonomy is many-sided and occurs at different levels and in a myriad of ways. It passes through different episodes that each have specific *targets*, which change from place to place and from time to time. They depend very much on the concrete situation. What they have in common, though, is that they all contribute to, and are inspired by, the search for, and construction of, autonomy.

First there are the specific struggles that aim to conquer, construct, maintain, defend and further develop a resource base. Access and ownership are strategic issues here. Sometimes land is constructed through enduring processes of colonization, settlement, land reclamation and/or the burdensome processes of manuring unwilling land. At other times, conquering land is central in, and to, massive peasant struggles that centre on the occupation of land (*tomas de tierra*, 'taking the land', as these occupations are termed in Latin America[4]). These struggles are often nationwide, and if land is mainly controlled by foreign states and/or capital groups, the fight for land necessarily translates into a struggle for national liberation. Under other conditions, the struggle for land can flow into massive programmes of land reform. But once land is part and parcel of one's own resource base, *keeping* it never is guaranteed. The right to access and use it according to one's own insights and interests needs to be constantly defended. Threats and menaces abound: waves of land grabbing have always existed, in many different forms.

The struggle for land also occurs in plantations, commercial *haciendas* and sharecropping systems. The forms of struggle differ (for an overview, see Paige, 1975), but equally here the search for autonomy is a shared commonality: the workers engaged in plantations, *haciendas* and the like basically understand autonomy as working the land according to one's own insights and interests. This is often (though not always) translated into a preference for an individually possessed, worked and managed plot of land. This also happened in the peoples' communes and production brigades of China. They were transformed due to, and through, the longing for autonomy. As the leaders of the Anhui rebellion (1979) explained:

> We knew it could be better [...] and when we started working as peasants again we were able to realize high yields. [...] Having the right to make [our own] decisions was very important for us. [...] When you have your own lands you care better for the plants [...] that is evident: when doing farm work the purpose is to get good results. (Ploeg, 2018:155)

Alongside land, there are other, often hotly contested, resources, such as water, genetic material, knowledge and the right to use them. Current fights that aim to safeguard the right to use and exchange seeds (through circuits not controlled by capital) reflect the multiple menaces that threaten the resource bases that are central to peasant agriculture (Kloppenburg, 2004; Badstue, 2006). Today there probably are more such threats than ever. The UN Declaration of Rights of Peasants, adopted by the United Nations (Resolution 39/12 of 28 September 2018) presents a range of concrete rights meant to address and push back these many threats (Claes and Edelman, 2020).

Secondly, next to the struggles for resources, there is the permanent struggle to keep *control* over the farm labour process. Resisting the *nomenklatura* (see Chapter 2) is as much part of this struggle as the development of one's own knowledge, skills and capabilities. This is associated with the right of the farming family

Figure 9.2 Struggling in the field (photo by Guy Ackerman).

and community to decide how to farm and how to relate to others. The image in Figure 9.2 is precisely about this particular fight for autonomy. It shows an elder peasant (probably helping his son or daughter) using a manual device (a kind of wooden pair of scissors)[5] taking out thistles from the meadow. Using this manual device (locally called *stikellûker*) is part of a struggle: it helps to avoid pesticide use and the specific dependency on agri-business that comes with it.

A third set or targets is related to farmers' share of the total value produced. This relates to the struggles situated in the markets in order to resist an excessive drainage of the value produced by farmers. Figure 9.3 looks, in first instance, as innocuous as Figure 9.2. It shows young Chinese *Nong min* trying to start direct marketing, starting with small quantities to obtain the needed experience and build, step-by-step, a network of consumers. This is, of course, just one expression of this phenomenon. In Chapter 8 I referred to the impressive Ecovida network of interrelated local markets. I could equally refer to milk strikes[6] that once were very important forms of peasant struggle. What all of these (and many other) forms have in common is that they aim to increase autonomy by fighting asphyxiating forms of value draining. In the past (that is, from the agricultural crisis of the 1880s onwards) the struggles for resisting draining flowed into the formation of many 'vertical' cooperatives (Chayanov, 1927/1991). These were meant

Figure 9.3 Struggling in the street (photo by author).

to *re-pattern* the relations between farmers and markets. Instead of having the relations with markets being structured by merchants, usurers, forgers and the like (who ordered the relations between agricultural production and markets to squeeze out the highest possible returns on invested capital), collectives of farmers started to build their own linkages with different markets (creating 'bypasses'; see Figure 10.11) – linkages that would allow for participating farmers to obtain the best remuneration for their own labour.[7]

Cooperation is a fourth target, and cooperation-in-production (a concept coined by Veronique Lucas, 2018)[8] is a key mechanism used to construct, increase and sustain autonomy. As with the other mechanisms discussed above (occupations of land, resisting experts, building cooperatives, etc.), actively building such cooperation-in-production is often a keystone in the overarching struggle for autonomy.

Figure 9.4 gives another everyday life image that, at first sight, seems to be simply one of a group of men and women planting seedlings of sweet potato in a field somewhere in China. But as is nearly always the case in the rural world, and especially in farming, immediacy is treacherous (see also Methods Box #4). The first impression probably hides far more than it reveals. Only through the sociology of farming can we unravel the seemingly non-descript and probably tedious situations that seem to make the countryside a sleepy place. The sociology of farming can do so because it provides the language (the concepts and the grammar that relates them) that allows us to enter into, and then represent, what is behind the deceptive outer appearance.

Figure 9.4 Struggling together through cooperation-in-production (photo by author).

This photo was taken in a long period of drought in this part of China (Yi County in Hebei Province). Peasants countered this drought by getting water out of the wells in the village; putting it in big plastic containers; bringing it, with their three-wheeled tractors, to the fields already prepared; and then filling the small holes with the amount of water needed for each single plant. After this, the seedlings were planted, a bit of fertilizer was added, and then the hole (with water, plant and fertilizer) was closed with soil in order to prevent evaporation of the water. All of these tasks needed to be performed very quickly and in a neatly defined sequence – if not, there would be too much loss of water and/or young seedlings. This would be impossible to do with just one or two persons. Thus, different households started to *cooperate-in-production*. This is not a simple matter of engaging in mutual help. It helped these peasant farmers avoid the need to construct an expensive borehole, or to contract wage workers, or to compensate for the missed harvest by buying food in the market. In short, Figure 9.4 shows the defence (or even enlargement) of *autonomy* in a particular place and at a specific junction in time. Doing so is not particularly soporific; it involves considerable drudgery but also renders a rich and varied amount of utility.

Fifth, the struggle for autonomy might also target broader issues such as patri-archy, environmental degradation and the quality of life in the countryside. Such issues seem to be, at least at first sight, disconnected from the specificities of farm-ing and the autonomy it requires. However, this is far from the case. Take, for example, the availability of clean drinking water within or nearby the homes of the rural. Being able to avoid waterborne diseases and not losing several hours a day fetching water might be a huge relief for peasant households, and espe-cially peasant women, and will allow them to dedicate more time and energy to the farm and the construction of autonomy. The introduction of such services reduces drudgery, just as other services (such as internet, broadband, television, swimming pools,[9] etc.) might increase people's sense of satisfaction (and help to dampen tendencies for young people to migrate). Thus, in many parts of the world, the spread of mobile telephones has helped to considerably improve the life and work of shepherds and herdsmen engaged in transhumance and given remote farmers better access to up-to-date news about market and meteorological conditions.

Finally, I have to mention here the struggles related to the inflow of non-ag-ricultural people. For many of them, starting to farm and to build, step-by-step, a self-controlled resource base are strong expressions of an emancipatory drive: going forward by building one's own farm.

So far I have addressed the *targets* of the many-faceted struggles for autonomy (see Text Box 9.2 for a synthesis). Next to these targets, the *scope and reach* of these struggles can also greatly vary. This scope and reach enter the analysis if and when we ask the questions 'what is being altered by these struggles?' and 'how are such alterations achieved?' In response to these questions, three kinds of struggle can be distinguished, which I discuss in the following section.

Text Box 9.2   The third class and the struggles in which it engages

As argued in the discussion linked to Figure 9.1, developing the peasant farm is both objectively and subjectively a struggle. Especially when this struggle takes the form of labour-driven intensification (see Chapter 5), it is very much a struggle of the third kind (see the section below): it aims to alter the process of production.

In the first chapter of this book I defined the peasantry as (part of) the third class. Peasants (just as urban craftsmen) have both their own labour force and the needed means of production. The means of production are set in motion in order to produce an income. Wherever this basic logic (this 'dance through time') is threatened or blocked, resistance and struggle will follow. Schematically speaking, this comes down to the following relations.

First, if the peasant labour force is chained (in whatever way), peasants will fight to set themselves free (or move to other locations that allow them to operate freely and independently).

Second, wherever the needed means of production are lacking or taken away (land being the main focus here for the means of production), peasants will fight for, and defend, them.

Third, whenever the autonomy to set the means of production into motion is lacking or threatened, peasants will, sooner or later, enter into struggles to create, defend and increase the autonomy they desire.

Fourth, if the income to be obtained from production is unacceptably low (due to drainage by others), peasants will engage in struggles aimed at resetting the terms of trade.

Depending on the type of society in which such struggles occur (and the responses generated in and by society), peasant struggles may translate into higher levels of aggregation.

Peasant struggles (partly) take place within, just as they (partly) occur through, the sphere of production. Peasants are constantly trying to alter the process of production (by adjusting, improving and extending the farm labour process). In a way, peasant life is part of an omnipresent and deeply rooted struggle of the third kind. Wherever this hidden struggle of the third kind – this ongoing fight for progress – is blocked, it will jump, as it were, to other levels and arenas, just as it will metamorphose into struggles of another kind.

Struggles in the markets, struggles aiming for self-governance at the territorial level, socio-political struggles at the national level, or even at the international level, etc., are not alien to the peasant farm. On the contrary, they are *rooted in*, and stem from, the threats experienced in the many single farms, just as they aim to *defend and strengthen* the many peasant farms.

Struggles of the first, second and third kind

From an analytical point of view, we may distinguish three kinds of resistance and struggle. Struggle and resistance of the *first kind* are overt, massive and, often, impressive: they include demonstrations, marches, roadblocks, milk strikes, blocking large retailers, etc. These forms of action are concentrated in time and space. They have a clear beginning and a clear end; they also include a concentration of people in a specific place – after which everybody goes his or her own way again. Struggles of the first kind (aim to) *alter* the *distribution* of the social wealth produced. They are meant to alter, as it were, the *terms of trade* between different classes. In industry, struggles of the first kind basically aim to change the exchange relations between the owners of capital and the workers. Such struggles are oriented at obtaining better wages and better working conditions. They typically occur through closing down the processes of production. In agriculture, similar struggles regard the exchange relations between capital and the third class. Milk strikes, for instance, immobilize the processes of production (located in dairy industries) and typically aim at obtaining increased off-farm prices for milk.

Struggles of the *second kind* are covert, scattered throughout time and space, and they may endure for very long periods. Examples are indifference, sabotage, foot dragging, gossip, stealing, etc. It is what James Scott (1985) masterfully described in his *Weapons of the Weak*. Modern examples of struggle of the second kind include the destruction of experimental plots for GMOs, setting establishments of McDonald's on fire, etc. An ancient form is presented in Text Box 9.3.

Text Box 9.3   Ploughing the fields from the inside to the outside

---

The *mezzadria* contract has been the main land-labour institution in large parts of the Po valley in Italy for a long period (especially in Emilia Romagna). It is a kind of shared tenancy (or *métayage*), implying that the tenants working the land had to pay to the landlord 50% of the harvest to pay for the lease. In particular circumstances – for example, very fertile lands – the *mezzadri* (leaseholders) even had to pay up to 60% or more of the harvest. The lease contracts were renewed every year in the springtime. At that time, large convoys of poor peasants crossed the countryside: evicted from the fields they previously worked and now looking for new opportunities.

To avoid the danger of the landlord asking for tightened lease conditions, the shareholders ploughed the leased lands from the middle towards the margins. Over the years this resulted in a slightly concave profile of the fields, which caused stagnation of water and thus poor germination. As a consequence, the yields remained low, so the landlords had little grounds for changing the terms of exchange.

The slightly concave fields represented a remarkable contrast with fields in, for example, the northern Netherlands. Here peasants ploughed their fields from the margins towards the midst, which resulted in slightly round fields that drained well, and this showed increasing yields. The decisive difference here was, of course, that the Dutch peasants were free farmers, who could improve their resource base and retain the fruits themselves. This markedly differed from the situation of the *mezzadri*. For them, ploughing the other way around was part of their struggle of the second kind. This only changed at the end of World War II when in the context of armed resistance (against German occupation) and land reform the *mezzadri* became free farmers as well.

Struggles of the first and second kind may *stop* the processes of production organized by capital or *slow them down* and thus reduce their profitability. They do not, however, *alter* the reigning processes of production and labour, and usually this is not the main objective. Struggles of the first kind mostly aim to improve *exchange* relations (better prices for farmers) or block further deterioration. Struggles of the second kind express anger, the desire for revenge and the search for small, individual benefits. Both of these struggles are very different from struggles of the *third kind*. Struggles of the third kind explicitly aim for, and bring, alterations in the existing processes of *production* (and commercialization) in order to better meet the interests and prospects of the involved actors. The great Italian intellectual and leader of the National Peasant Alliance Emilio Sereni wrote, in this respect, about '*le lotte che [...] i piccoli produttori agricoli hanno combattuto per il loro riscatto sociale e per il progresso della nostra agricoltura*' ('the struggles of peasants for the improvement of their livelihoods and the progressive development of our agriculture'; Sereni, 1979:439). That is, farming is not taken as it is – it is to be changed, to be developed. And the needed changes are driven by, and part of, peasant struggles. Such struggles of the third kind involve the conquest and/or construction of self-governed spaces that allow for alterations, and they have a strong power of mobilization, partly because they show that alternatives are possible. Historical examples include the construction of all sorts of cooperatives. The individual plots rural workers obtained and stubbornly defended within the former *sovchozes* and *kolchozes* of the Soviet Union also represented, in a way, a struggle of the third kind. In these plots, workers produced in ways that differed greatly from those imposed in the large fields of the state-controlled collectives. The high yields of the individual plots offered a permanent (if implicit) critique of the modus operandi of the state-controlled enterprises. In a way, the same thing occurred in the case in the *minifundia-latifundia* complex in Latin America and the Iberian Peninsula (see Figure 5.2).

In the previous chapter, I discussed the construction of new, nested markets. These might also be considered as an example of struggles of the third kind. Newly constructed nested markets go beyond simply changing the terms of exchange – they also *re-pattern* the sphere of circulation. They *alter* the commercialization and distribution of food through the introduction and consolidation of new *relations* between producers and consumers and new divisions of the produced value added. In the next three sections I will move to different examples of struggles of the third kind. The first one regards the radical reshuffling of relations in the sphere (or domain) of production during a turbulent period in Peru. The second focusses on agroecology, whilst the third example regards the construction of autonomy at the territorial level.

Radically altering the organization and development of production through collective action

As indicated before, struggles of the third kind are often located in the many single units of peasant production (see Figures 9.2 and 9.4). Peasant production and life are engrained with struggles of the third kind, albeit in dispersed and hidden ways, stubbornly resetting the different balances (see Figures 5.6 and 1.6), improving the resource base, trying out novelties, engaging in hardship (drudgery), fine-tuning the farm labour process, building new forms of cooperation-in-production, exploring new possibilities for marketing, etc. These activities are all about the *altering* of the processes of labour and production in order to better meet one's own needs and expectations. This is especially the case when the alteration occurs as, and through, labour-driven intensification, which brings together and fuses two strategic ingredients: the emancipatory aspirations of peasant families and the ongoing growth of agricultural production. The longing for a better life translates, through labour-driven intensification, into growing agricultural production, whilst the fruits of increased production help to improve the livelihood of the household. Thus, labour-driven intensification has been the engine that, throughout history, has driven agriculture towards higher levels of production and enhanced wellbeing. The prosperous development of Chinese peasant agriculture from the 1980s onwards is one of the most recent expressions of this mechanism.

If adverse politico-economic conditions block this engine, peasant struggles will – sooner or later – try to correct them. State interventions might equally try to restart or accelerate it. In China, for example, this occurred in the first decade of the 21st century when the terms of exchange between town and countryside were slowly but persistently reset in favour of agriculture. 'Peasants have given a lot; now it is their time to receive' was the leading motto.

Labour-driven intensification can be strengthened considerably through joint efforts (sometimes including state support) that aim to improve the quality of the means of production (e.g., irrigation works, breeding associations, networks for the exchange of genetic material) and through cooperation-in-production. But it can also occur through forms of collective production. When production cooperatives ('horizontal cooperatives') engage in struggles of the third kind, amazing developments are possible.

I have witnessed several episodes of peasant struggles of the third kind that occurred through the construction of such peasant-controlled production cooperatives.[10] Some of these cases I have documented elsewhere (Ploeg, 1990, chapter 5, 2006). What I will try to do here is to signal some of their commonalities and rephrase them in such a way that they may help research in similar situations elsewhere. Some suggestions are presented and illustrated in Methods Boxes #17, #18 and #19.

Methods Box #17   Walking through the fields

Labour-driven intensification can be strengthened through action research. An indispensable tool for doing so is 'walking through the fields', not alone, but together with peasants. And not once but several times (covering the whole cycle of production). The situations encountered should be carefully discussed, with the leading questions being:

- What went wrong here?
- Why did it go wrong?
- How can this kind of problem be resolved?
- Who is to resolve it?
- What has been done so far to resolve it?
- Which lessons can be learned from it?
- How can a proper solution be organized?
- What would be the benefits of a solution?

Figure 9.5 shows images from a banana-producing section in one of the cooperatives (in Peru) that was aiming for further labour-driven intensification.

In practice, many problems were often quickly identified. Following the images in the figure above, they came down to (1) insufficient care for the developing raceme (or the cluster of growing bananas) – if it is affected by leaves or branches there will be insufficient growth; (2) the development of weeds due to insufficient availability of labour, which is partly an organizational problem within the cooperative and partly a financial problem; (3) lack of support for trees so that some

Figure 9.5 Cultivating bananas (photos by author). *(Continued)*

Figure 9.5 (Continued)

fall down, plus apparent leakages in pumps that create a thin layer of oil in the canal, which makes irrigation less effective; and (4) irregular positioning of trees, which hampers several tasks. In the end, all of these problems came down to a lack of sufficient and well-skilled workers who could do the job properly. When these problems were resolved, this banana section was able to increase its production and even generated an important cash flow for the cooperative that helped to reduce dependency on the banking circuit.

The discussion about these and other problems, discussions that involved widening circles of peasants and workers, led to the elaboration of a detailed plan for reorganizing this part of the cooperative's production. The plan was discussed in the general assembly, approved and put into practice.

Methods Box #18   Reconstruct and critically reflect on one's own performance

How are we performing? This question needs to be asked about all of the dimensions that are relevant to the productive constellation being transformed. How is the tilled area developing from year to year? How are yields developing? How is productive employment developing? How is the balance of profits and losses developing? Is there progress in the consecutive years? And, if not, what is the explanation?

These questions can also be applied in a comparative sense: how are we doing in comparison to neighbouring and/or comparable situations?

These diachronic and synchronic comparisons require a rigorous and systematic approach. It is better when it is done by relative outsiders operating in close cooperation with those involved as they can bring some objectivity to the situation.

Figure 10.14 in the next chapter shows an example of 'reconstructed performance'.

Methods Box #19   Highlighting the underlying values

One of the much-used popular sayings in northern Peru during the 1970s was: '*Dile al rico que tire sus billetes al suelo – a ver si den frutas*' ('Tell the rich man to throw his money on the land and let's see whether it will give fruits'). It is a beautiful expression that, on the one hand, refers to co-production (i.e., agricultural production as resulting from the interaction of labour and living nature) and, on the other hand, implies a pointed critique. It is not capital that makes the fields blossom – only labour can do so.

These and many other commonly shared views were brought together by a small team of agronomists, leaders of the peasant community and political activists and synthesized in *principios de lucha* (values underlying the common struggles) of the peasant community of Catacaos in 1973.

The saying quoted above reappeared in the list of commonly shared values reproduced below. Together these values became a strong unifying element that helped to tie together many of the different struggles of, and within, the community.

The shared values of the peasant community of Catacaos

1 For a united, indestructible and autonomous community
2 For a community governed through the democratic intervention of all of its members
3 For a community in which all members are equal in rights and duties
4 For a community that recognizes labour as the only source of wealth
5 For a community that does not allow the exploitation of her resources and production by foreign elements.
6 For a community that struggles to secure the satisfaction of basic needs for housing, health, food, education and employment for all of its members
7 For a community that actively works for both the present and future needs of its youth
8 For a community that engages in solidarity with the entire labour class of our country in order to strive for the integral transformation of the country

Basically, struggles of the third kind aim to radically *re-pattern* the processes of production and circulation. Already available elements can be related in new, distinctively distinct ways to create new constellations that better meet the needs of society at large, the expectations of the actors directly involved and the interests of ecology. Through such means, struggles of the third kind promise to resolve the agrarian question as experienced by those on 'the front line'.

Very often these struggles centre on the creation of radically new relations between 'man and land' (that is, rural workers and living nature),[11] the two main actors that together make up co-production. This can apply in both a qualitative sense and a quantitative sense. Whereas large-scale agriculture in northern Peru had, at the beginning of the 1970s, an average man-land ratio of 1 to 5 (meaning that on average there was one worker employed for every 5 hectares of used land)[12] and the state-controlled land reform process aimed at establishing a ratio of 1 to 8 (on average), the peasant movements that raged, at that time, over this area claimed a radically different ratio: 1 to 2 and in some situations even 1 to 1.[13] In the latter relations, two emancipatory drives converged. First, the need and desire of large segments of the marginalized rural population to obtain access to the land, for employment and the possibility to earn an income, and, second, the generally felt need that farming could be organized and developed far better than was done in the pre-reform period (under the aegis of the *gamonales*, the deeply hated landowners). Both drives were intimately connected: by radically increasing the labour input in farming, the level of intensity could be greatly raised whilst the increase in production could pay for the increased levels of employment. This connection (which underlies the concept of labour-driven intensification) was far from alien to the poor of the countryside.[14] No 'rocket science' was needed to know that *trabajar mejor* (to work better) was viable. People knew it from the fields, from their own experience. They compared levels of production in the small peasant farms (*minifundia*) with those of the far larger commercial haciendas and plantations (*latifundia*), and this showed them that production could be greatly improved. At that time, this was also echoed in, and reaffirmed, by a range of socio-economic and agronomic research programmes, such as the then well-known CIDA studies on 'modalities of land use' (synthesized in Ernest Feder, 1973:162–218; see also CIDA, 1966), which greatly inspired the views of the *campesinistas* of that time.[15]

The quantitative change of the man-land ratio also implied radical changes on the qualitative dimension, for in order to materially reshuffle the man-land ratio, both 'man' and 'land' needed to be unchained: released from the capital relations that doomed 'man' to be landless (*brazos sin tierra* as the saying goes in the Peruvian countryside) and the 'land' to be badly worked (*tierra sin brazos*). The occupation of land and the subsequent introduction of workers' self-management became decisive in changing this equation.

People within the Peruvian social movements knew that a new way of farming would not be built overnight. It would take quite some time and involve overcoming many problems. Thus, right from the beginning, building a new agriculture was a multi-faceted socio-political struggle. New institutions (such as *unidades comunales de producción* and cooperatives) needed to be designed, implemented, developed and improved. People knew that, at least in the beginning of the process, cooperative solutions were needed in order to face the hostile environment. The labour process needed to be reorganized (see also Methods Box #17) and new divisions of labour introduced, the seasonality of production required new approaches, the relations between the new units of production and social movements had to be rethought, mechanization had to be redesigned, relations with the banks renegotiated, etc.[16] It was a complex and sometimes contradictory process – but it rendered strong results: an enormous growth of productive employment and an impressive intensification of production (the process was further strengthened through a critical monitoring and analysis of the results; see Methods Box #18). The process was also a source of pride and a new self-consciousness (which was strengthened by documenting the experiences and views of the involved peasants and synthesizing them into a set of shared values; see Methods Box #19). As Arturo Perez, a local leader, said at a mass rally: 'we, the peasants, are doing what the *gamonales* and the state never could; we, the peasants, are painting the countryside green'.[17]

During the 1970s and 1980s, struggles such as the one described above abounded in Peru (Diez Hurtado and Burneo, 2022). But they can be found in many other places and epochs as well. I think the *Zapatista* uprising in Mexico and the shift to agroecology that it brought (Victor Toledo [2000] gives an excellent description) is exemplary, just as the experience of the Movimento dos Sem Terra (MST) and similar rural movements in Brazil such as the Polo Sindical do Borborema (Petersen, 2015, 2017; Petersen and Silveira, 2017) and Arapongo (Berg et al., 2018). In Andalucuia, Spain, there are the longstanding and mutually contrasting experiences of Marinaleda and La Verde.[18] In Germany there is the impressive example of BESH (Baüerlichen Erzeugergemeinschaft Schwäbisch Hall), and in Novellara (in the province of Reggio Emilia, Italy) there is the CILA cooperative (initially created by *braccianti*, 'rural workers', but now turned into a vibrant multifunctional enterprise that is closely related to the territory).[19]

Radically altering production through agroecology (from movements back to the farm)

Wezel et al. (2009) conceptualized agroecology as having three prongs: embodying a scientific discipline, a social movement and a set of practices. These three aspects have different relative weights in different contexts. At the level of theory, agroecological agriculture can be clearly outlined (see, for example, Altieri, 1990, 1995; Gliessman, 2007; Sevilla Guzman, 2007; Petersen, 2017; Rosset and Altieri, 2017): it is farming that, as much as possible, is based on natural resources. Labour and knowledge are key to using these resources according to ecological principles. The centrality of natural resources (produced and reproduced in the farm and/or obtained through direct, socially regulated exchange from other farmers) implies a high degree of autonomy that in turn translates into resilience. Often, agroecological farms engage in direct selling and/or use short circuits and farmers' markets in order to relate directly to consumers.

In practice, the development of agroecology takes a variety of different, often unexpected and sometimes even contrasting trajectories (Cayre et al., 2018). These might be inspired by different motives, values and discourses, just as the particular contextual settings will have a specific imprint. The different trajectories and practices might be known under different names and the particular histories, and spatial distributions of the different experiences will vary considerably. That is, it will be hard to find agroecological practices that are completely identical. Differences abound.

In Europe, of the many practices that are close to agroecology, only a few are explicitly defined, by those involved, as being agroecological. This is due to two factors: until a few years ago, the concept of agroecology was hardly known in Europe (apart from small pockets, such as Andalucia in Spain and small groups of specialized scientists), and in discussions between farmers the term is barely used. Yet, at the same time, large segments of European farmers, all of them facing the squeeze on agriculture (see Figure 6.12), as well as the growing imperatives of sustainability, have actively developed new strategies to address these challenges.

Lucas (2018; see also Lucas et al., 2019) tellingly referred to this hidden presence as '*agroécologie silencieuse*' ('silent agroecology'). The gradual and silent process of the adoption of agroecology explains why, in practice, there are no clear delineations between agroecological and conventional farming. There are, at best, 'blurred boundaries' (Lucas, 2018:213; Wezel et al., 2009).[20] These are the result of the step-by-step conversion to a more agroecological agriculture.

Figure 9.6 refers to the farm of Taeke Hoeksma, a prominent peasant farmer (whom we met in Chapter 6). The figure reflects the amount of nitrogen, embodied in concentrates and chemical fertilizer, that is 'imported' into the farm in order to produce (a standardized production of) 100,000 kg of milk. In 1979 (the first year in the graph), in order to produce 100,000 kg of milk, Hoeksma needed 2,800 kg of pure nitrogen contained in bought feed and fodder (mainly concentrates) and 2,400 kg of pure nitrogen contained in fertilizer. Hoeksma started to document this use of external inputs because he had become very worried

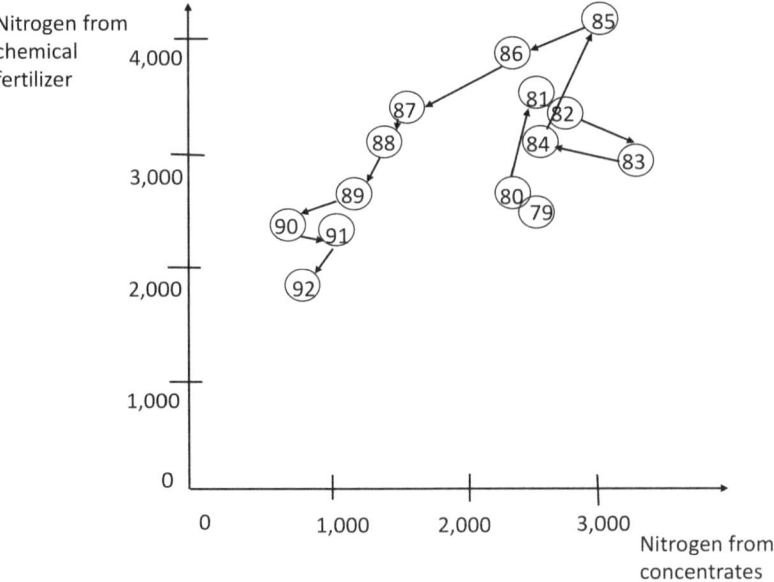

Figure 9.6 Nitrogen inputs needed to produce 100,000 kg of milk (Hoeksma's farm) (author's own elaboration).

about the degradation of the quality of his soils. The years that followed show a pattern that is characteristic for peasant agriculture. Every year Hoeksma altered the amounts of nitrogen applied through new combinations of chemical fertilizer (for the meadows) and concentrates. The changing combinations and changing amounts were in effect a series of small experiments (see Figure 2.5), which were carefully observed, compared, analyzed and then translated into new trials. Such cycles of observation, comparison, interpretation and, in the end, reorganization of farming are, in a sense, the nervous system of the farm: guiding and tuning the different activities, allowing for continual learning and for the farmer wresting back control over his production.

The early 1980s were a difficult period on the farm (partly because maize was introduced, although it was abandoned again a few years later), but from 1985 onwards there were continuous decreases in the total amount of nitrogen used in the farm (from 6,800 kg in 1985 to just 2,300 in 1992). However, this huge reduction did not result in lower grass or milk yields. On the contrary, Hoeksma is known in the area as a very good 'grassland farmer'. The explanation for the productivity of Hoeksma's farm lies in the very good (and continuously improved) quality of the manure produced on the farm, the sturdy cows and the rich soils. Thus, by steadily building up the quality of the internal resource base, Hoeksma was able to reduce his use of external inputs in a step-by-step way.

Some 25 years later, Douwe and Dictus, the sons of Taeke Hoeksma, now work together on the farm, and they continued the search initiated by their father. By 2017, they had reduced the use of externally supplied nitrogen (needed for the production of 100,000 kg of milk) by a further 1,600 kg. This brought the N surplus/ha well below 50 kg N/ha. This is how silent agroecology proceeds: a sequence of small steps that build upon each other and which radically alter the process of production.

Regaining control over the labour process through territorial cooperatives

The first nuclei of what later became the Northern Frisian Woodlands (NFW; Noardlike Fryske Wâlden in the Frisian language) were founded in the early 1990s as a reaction to a national environmental regulation introduced at the time which threatened to strongly condition the farm labour process in this area and even bring a standstill in the process of farm development. The reaction was typical for struggles of the third kind. The first groups of farmers strongly contested the new, suffocating regulatory schemes and developed a counterproposal (see Methods Box #12) outlining that they themselves would take care for the landscape, bio-diversity and environmental qualities if given the room to do so[21] – if not, they would fall back to struggles of the first and second kind. After considerable pressure, several rounds of negotiation and support from parliament, they succeeded. Thus the farmers' association that later developed into the NFW was created. It now operates as a *territorial* cooperative[22] with some 1,000 members (the majority of whom are active farmers) who manage some 50,000 hectares.

There is an interesting similarity here (see Methods Box #3) with the *sectoral* cooperatives constructed during and after the agrarian crisis of the 1880s. At that time, the main markets (notably the one for butter) were in disarray due to cheap imports, forgery, extractivist trading practices, price volatility and, more generally, a reshuffling of the international food regime (McMichael, 2009). This triggered the construction of cooperatives for processing, trading, quality control and financial services. While this first generation of cooperatives could not change the markets *as such*, they were able to effectively re-pattern the *relations* between farming and markets (i.e., by substituting the cooperatives for merchants and money lenders).

More than a century later, in the early 1990s, the relations between farming and *the state* had entered into deep disarray. The generic regulatory schemes were at odds with the heterogeneity and dynamics of farming. Moreover, the different schemes (one for environmental qualities, another for the protection of nature, etc.) did not align with each other. Thus, impossible situations were created – such as the one that the farmers of the NFW were facing in the early 1990s: operating in small fields but being obliged, at the same time, to use heavy and expensive machinery for slurry injection, which was completely at odds with both the small-scale hedgerow landscape and the biodiversity it contained. Just as the first generation of sectoral cooperatives had *mediated between farmers and markets*, the second generation started to mediate *between farmers and the state*.

The modus operandi of the territorial cooperative (that takes joint responsibility for the governance of the territory) is outlined in Figure 9.7. On the one hand,

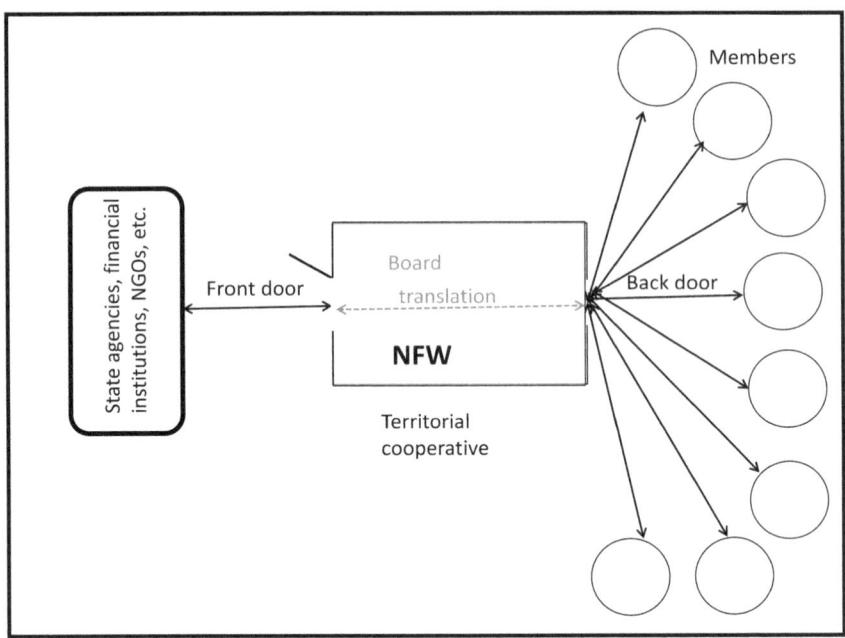

Figure 9.7 The mode of operation of the North Frisian Woodlands territorial cooperative (author's own elaboration).

the NFW negotiates with the state (and civil society organizations such as the environmental federation, the water board, nature organizations, etc.) about the objectives to be reached and the conditions permitting this. On the other hand, the NFW talks to the farmers and helps to adapt their farming practices in ways that allow the set objectives to be realized. In the NFW, this particular way of operating is referred to as the 'front door/back door' principle. At the front door the NFW is dealing with the state, at the back door with the farmers. In this way, it is made sure that the negotiated objectives are reached at the level of the territory as a whole, whilst at the back door (i.e., at farm level) the needed flexibility is maintained. With this pattern comes the needed bargaining power. To a certain degree, the territorial cooperative is able to relate to the state as countervailing power – something that individual farmers could only dream of.

Interestingly, both the negotiations with the state and the internal translation of general objectives towards specific practices and activities (one of which is summarized in Figure 6.10) are governed by the local cultural repertoire, synthesized by the NFW into ten guiding principles (compare with Methods Box #18). Figure 9.8 gives a synthesized overview of these guiding principles. The key concepts are phrased in the Frisian language with English translations added. Notions such as *lânsdouwe* and *kreas buorkje* stand out. They refer, respectively, to co-production (the unity of man and land) and to the local style of farming (farming with care).

1. *Mienskip* (community)
2. *Lânsdouwe* (the unity of man and land)
3. *Kreas buorkje* (farming gently)
4. *Eigen gerjochtigheid* (our own rights and entitlement)
5. *Wy kinne en dogge it better* (we perform better)
6. *Wissichheid* (reliability)
7. *Stadich oan foarût* (progressing slowly but steadily)
8. *Net allinnich* (not alone)
9. *Tinke oan'e takomst* (caring for the future)
10. *Mei wille en nocht* (with satisfaction and joy)

Figure 9.8 The commonly shared values of the NFW.

Currently, the NFW is engaged in a wide range of activities that range from the management of nature and landscape to the production of energy. When it comes to nature and biodiversity, the protection of meadow birds is one of the more eye-catching activities. It centres on the creation of good conditions and care for the black-tailed godwit (see Figure 9.9), a bird that is dear to Dutch society. As with many other of its activities, the NFW has also organized scientific support and research for monitoring, evaluating and improving the conservation of black-tailed godwits. Figure 9.10 offers a synthesis of this particular research (Swagemakers, 2008; Swagemakers et al., 2009). It contains a path diagram that shows that having organized a defined area, being knowledgeable about the birds (in the area referred to as 'having an eye for birds') and low external input farming ('farming economically') together explain the presence of nests (of the

Figure 9.9 The black-tailed godwit (photo by Dirk Jan van Roest).

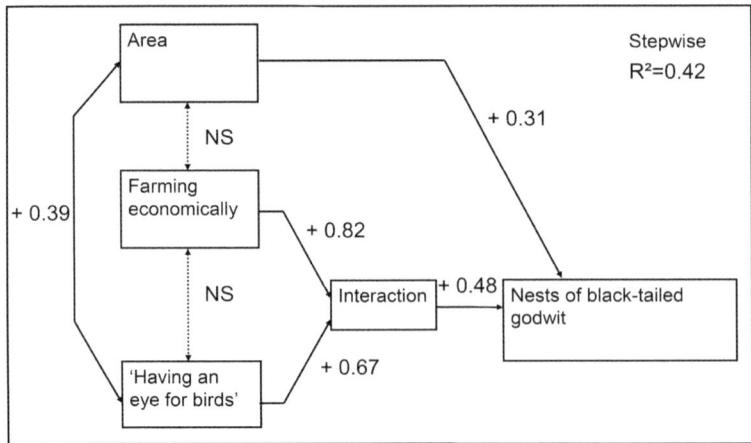

Figure 9.10 'Producing' meadow birds (such as the black-tailed godwit) (Swagemakers, 2008).

black-tailed godwit) in a statistically significant and substantial way ($r^2 = 0.42$). Having obtained such results is an important 'weapon' in negotiations with the state and in dealings with civil society organizations.

Similar approaches are used in the other domains in which the NFW operates. Together they compose an unfolding programme. Every success triggers new searches for further advances.

The struggles in which the NFW engages have had, and are having, a positive impact on the area. It has allowed the peasant farmers to defend important parts of the autonomy that underlies the structuration of their farm labour processes. This has come with considerable cost reductions: expensive but prescribed investments could be avoided, whilst the style of farming economically (with low external input levels) could be continued. Moreover, the engagement in cooperative programmes for the maintenance of nature and the landscape brings an additional income flow to the area of some 4 million euros per year.

The positive results rendered by the common struggles have made the local peasantry proud and self-assured in their dealings with the state. As two of their guiding values say (see Figure 9.8), 'we perform better [and we do so] with satisfaction and joy'. This represents a remarkable contrast with the bitterness and frustration of farmers that one notices in neighbouring areas.

Whether it is more the objective or the subjective factors that have put an imprint on the area (most probably it is the combination of the two), it is remarkable that the continuity of farms in this area is far higher than in neighbouring areas: over a 25-year period, the percentage of farm closures in the NFW area is significantly lower than in the surrounding areas. In the village where it all started (Eastermar), 34 of the original 36 farms belonging to the cooperative are, after 25 years, still fully operating (now often with the sons and/or daughters running the farm). Within the Netherlands as a whole, this is exceptional.[23] It shows that struggles can make a difference.

Tying different struggles and episodes together

Peasant life is engrained with resistance and struggle (Kerkvliet, 2009). This finds expression in episodes such as the ones noted above and in exciting new trends, such as the construction of new markets (see Chapter 8). But it also occurs, albeit in hidden ways, in everyday life. Usually, the different struggles are not articulated in politically explicit terms – especially, but not only, in the case in the many micro-level episodes of struggle. One of the consequences is that different struggles (of whatever form) remain disconnected and fail to reinforce each other.

Yet, the importance and potential of peasant movements reside in their capacity to link and tie together different episodes and levels of struggle. In this respect, it is strategic to develop a common language that synthesizes the different experiences obtained in clear key concepts, which also helps point to the way ahead. As Paul Nicholson, an early leader of the La Via Campesina movement outlined (in Text Box 9.4), key notions, such as food sovereignty, bring *coherence* to a movement that is (necessarily) decentralized. In his words: these unifying notions 'give body to the criticisms' and simultaneously provide 'an alternative to the reigning regimes'. Together these key notions outline a political project.[24] Such a project unfolds (as Paul Nicholson makes clear) on several levels and thus helps to re-pattern (or alter) significant parts of the daily realities that the different peasantries are facing – not in a distant future but now, in the many-sided practices that result from the struggles of the third kind.

Text Box 9.4   Paul Nicholson on food sovereignty and peasant struggle

The concept of food sovereignty (FS) was developed some 20 years ago. Today we can see it as a paradigm shift that has moved us beyond food security, as defined by access to the market. Food sovereignty is more than this: it is the political right to food and to produce food. This intuitive demand was initially developed by peasant organizations based on peasants' political experiences and realities. It emerged during the era of the free trade agenda, when the context was that of structural adjustment programmes, with the consequent impoverishment of rural communities.

At the beginning, FS was taken up and discussed by various peasant movements. For La Via Campesina it became the central demand. FS became the factor that allowed for, and strengthened, the coherence within a decentralized international movement. The political dimension of FS clearly resides in it being an alternative to neoliberal food and agricultural policies. It was, and still is, the basis of food policy proposals made by the global social justice movement. It is not just about eating locally and healthily. FS represents a paradigmatic change for global and local food policy. FS is the basis of the defense of peasant rights. It justifies the claim for a different trade policy, as well as the right to land in the face of corporate-driven agriculture and food policy. FS includes the right to water. It includes the defence of biodiversity and rejects monopolizing seeds. More generally, FS challenges the reigning modes of production and consumption. Thus, FS is an expression of, and response to, the Green Revolution, carbon-based agriculture and the agro-export driven model of corporate production.

The local food systems networks that are operating everywhere were very much inspired by the Nyéléni encounter in 2007 (where FS was systematically discussed) and its call for a bottom-up approach in, and for, the development of locally based FS networks.

It is astonishing to see the surge of agroecological schools and training centres run by peasants themselves. Ironically, the region that is lagging behind, in this respect, is Europe. It seems that in Europe we think that we already know all answers!

Agroecological schools based on FS paradigm are breaking with science and logic exclusively oriented at economic productivity. These schools are grounded on a new peasant-driven pedagogy that is culturally driven and unique to its own context. *Campesino a campesino*, or village to village, training in India is an exemplary expression of this new approach.

The agroecological model attracts young peasants. It offers them a future. In industrial agriculture, there is no such future for the youth.

Industrial agriculture or corporate-driven agriculture is clearly a male-orientated agriculture. It has excluded women from the food producing chain. FS has given back the roles of seed keeping, food production and processing to women. It is therefore not surprising that the participation and leadership of women in peasant movements are both clear and forceful.

At the international institutional level – for example, in FAO and IFAD (International Fund for Agricultural Development) – we see an increasing recognition of peasants and their movements. There are even timid expressions of FS. At the local level, however, the confrontation between corporate-driven agriculture and peasant agriculture is increasingly direct and violent. The imposition of monoculture crops, the continuing concentration of land and violent land grabbing are expelling millions of peasants from the land. What is more, the criminalization of resistance, the repression, imprisonment or assassination of peasant leaders and activists is becoming a reality in many places. The assassination of Berta Caceres is but one case; there are many more all over the world. In Europe right now we have the criminalization and imprisonment of activists, especially peasants, who are supporting migrant solidarity networks.

At the same time, there are the very important reactions at the local level. Solutions developed by peasant movements (solutions that mostly remain mostly outside of the focus of mainstream interests) are generating alternatives that help to defend the rural world. I will briefly mention some of the alternatives:

1   Alternative modes of production that carry different names: zero budget agriculture, organic, agroecology, permaculture, peasant agriculture, etc. What they have in common is that that they challenge Green Revolution approaches and productivist logic.
2   Diverse local food distribution systems: consumer groups, community-based agriculture, cooperatives, urban-rural distribution networks. These transform anonymous marketplaces into ones based on solidarity that are more remunerative for both producers and better value for consumers
3   *New* peasant movements: bringing together women peasants and women-run cooperatives, especially in food processing.
4   Local seed-keeping networks.
5   Locally based financial solidarity credit groups.
6   New groups recovering traditional herbal medicine.

What we are also now seeing is some local governments or institutions taking up FS policies and developing municipal food/procuration policies and schemes that promote urban and peri-urban agriculture.

However, there remain some fundamental challenges. It is still no easy task for peasants to survive as food producers for their communities. It requires:

- policies to strengthen local economy;
- the strengthening of local markets; and
- the development of social, health and educational structures.

Second, peasants need to be able to survive politically, at the local, regional, national and international levels. Via Campesina is a cohesive movement at the international level, and FS is its common vision that glues it together. We have to rekindle a class solidarity in the face of the criminalization of migration. In this context, the recognition of peasant rights by the United Nations is an important tool.

The third challenge is the defense of our commons (land, water, biodiversity) against capture by corporations.

A fourth challenge is the need to strengthen our cultural identities as peasants and people, our minority languages, our historical knowledge, our oral culture and all of those elements that strengthen our capacity to unite.

The fifth challenge is climate change. It is negatively impacting farming and needs to be resisted. But we also have to fight the false solutions that centre solely on technological fixes, for peasants, small-scale farmers, or whatever one wants to call us, are central to the challenges of sustainability in our world. The globalized food system generates hunger and warms the planet. Peasants, by contrast, feed the world and, through agroecological methods of production, cool the planet.

Paul Nicholson, Conference on the Future of Peasant Studies, Wageningen, 26 January 2017

In the discussion of peasant-driven processes of rural development (see Chapter 7) I spoke of the latter as transitions, comprehensive and structural transformations of food systems that, in an apparently paradoxical way, occur in a step-by-step way, with the accumulation of the many small steps making a big difference. Figure 7.6 graphically indicated that the many small, isolated and often hidden alterations at the micro level can be tied together into a powerful and growing flow that is able to challenge the reigning regime – especially if it is supported from the 'landscape' level as well (that is, by civil society).

It might be argued that the peasant movements of today are able to create and further unfold such a transformative flow – precisely because they are able, far more than others, to develop the unifying and mobilizing notions (such as food sovereignty) that can tie together the many experiences and episodes of struggles that would otherwise remain isolated and/or even forgotten. One reason for this is that the peasant movements of today are deeply rooted in highly diverse everyday life situations all over the countryside whilst they are simultaneously able to make the different experiences, solutions and actors involved *travel beyond the* immediacies of time and place. Another reason is that today's peasant movements have developed the capacity to engage with civil society at large by showing that peasant agriculture is able to resolve problems facing society as a whole (see Text Box 9.4). The sociology of farming can be a helpful tool in outlining and further specifying such solutions.

Peasant movements as socially constituted bodies

Social sciences set out to represent and explain the emergence and existence of social phenomena, as well as their non-emergence and non-existence. This is the (methodological) principle of symmetry. Over the ages there have been strong and progressive peasant movements that were able to effectively change parts of the world (see, e.g., Moore, 1967). In other times and places they have been weak (if they existed at all) and/or derailed into (parts of) reactionary movements centred on the interests of others and unable and/or unwilling to criticize the *organization and development* of farming (Ploeg, 2020). Such diversity still exists (and is probably larger than ever): in some places one finds well-articulated and strong peasant movements; in other places (that seem to need them as much as the others) they are absent.

The existence of such diametrically opposed phenomena excludes any possibility of socio-political or politico-economic determinism (see Methods Box #2) and critically requires a careful examination of micro-macro relations that necessarily has to take into account the views, aspirations, experiences and emotions of the actors involved (see Methods Box #1).

I cannot and will not develop here an extensive explanation of reasons underlying the emergence and non-emergence of social movements in the countryside. Nor will I seek to discuss their successes and failures. I will restrict myself to discussing some of the many and highly complicated linkages that need to be constructed by a peasant movement in order to manifest itself, endure and, maybe, become successful. I will also indicate some of the (many) internal difficulties that may arise. Such pointers might be helpful, I hope, for future research.

In order to emerge, peasant movements need, firstly, the identification of a wide (and widening) set of immediate problems and an associated set of practical solutions that are relevant at the grassroots level. Such practical solutions will probably only be partial but at least hold the promise of bringing some relief. They need to be perceived as relevant solutions that satisfy the anger and indignation of the many (O'Brien and Li, 2006). Secondly there needs to be a capable and collective leadership able to reproduce itself over time. Such leadership needs to be supported by a loosely structured team of intellectuals who contribute to the further unfolding of the different practical solutions (Desmarais, 2007). In the third place, there needs to be a programme, a set of unifying notions (and a culture), that ties the different *loci* of struggle together just as it is able to mediate the different *foci*. This programme (which operates as commonly shared language; see also Methods Box #18 and Text Box 9.4) and the collective leadership help to tie the capillary structure (that 'links the local, national and international spheres

---

Text Box 9.5   The complex dynamics of regime/movement relations

---

The joint work of Colin Anderson et al. (2020) probes into the different possible relations between reigning socio-technical regimes and social movements that aim for major alterations (illustrated here with the example of the agroecology movement). The reigning regime can affect the social movements in different ways, just as the movements may or may not alter the regime and/or exert different effects. *Containing* (keeping the movements marginal) and *shielding* (keeping the movements at a distance from the system's main dynamics) are a kind of middle ground 'that maintains the status quo and enables the dominant regime and [the movements] to co-exist' (Anderson et al., 2020:164–165). *Release and anchoring* means that obstacles are dismantled and replaced by new dynamics, values, norms and practices. *Nurturing* occurs when movements are encouraged to strengthen themselves. On the other hand, *suppressing* occurs through actively organized repression and criminalization. *Co-optation* occurs when certain symbols and practices of the movements are taken over in order to de-activate them.

---

of activism', as Edelman and Borras [2016:89] described it) of the movement into one coherent flow, potentially able to influence the reigning regime. Fourthly, the movement enters the scene as a certain kind of 'counter-power', meaning that if, for example, the state is unwilling or unable to create needed solutions, the movement itself can do so. Operating in a context of power and counterpower (see also Text Box 9.5) means that the movement needs to construct autonomy in several respects: it needs, for instance, to have its own means of communication, its own centres for the training of cadres, its own circuits for socialization, etc. Taking all of this together also implies, as point five, that the movement needs to engage simultaneously in struggles of the first, second and third kind.

If these (and probably other) ingredients are available (or capable of being developed), strong peasant movements may emerge. If one or more are lacking, such movements will most probably remain absent or, after a first trial, dissolve in failure.

However, even if all of the mentioned ingredients are available, peasant movements still have their own *scylla* and *charibdus*. There is the *scylla* of dealing with segments of the movement with entrepreneurial ambitions[25] (preferring to seek profits through the exclusion of others, thereby undermining solidarity). *Charibdus* represents another, equally permanent, danger: being captured in traditionalist and patriarchal repertoires and/or getting entrapped in right wing, populist movements (for an extended discussion, see Scoones et al., 2022).

Experience shows that there is also a range of specific problems that are nearly always very difficult to resolve within the context of peasant movements. One such problem is the issue of (internal) dissimilarities and dis-simultaneities. Certain parts of a movement might be well ahead of others, and the related differences might cause conflict in both directions. In this respect, peasant movements are 'evolving arenas of action where a movement's character and strategy may be contested and (re)negotiated over time' (Edelman and Borras, 2016:40). Issues of internal distribution are another devilish problem that farmers are not very well equipped to address. This is mostly countered by putting an emphasis on a (sometimes rigid) equality: 'everybody is dealt with in exactly the same way' (which is not very easy to realize when material differences abound). Then there is the problem of interlinking harmony and conflict. Social scientists often think that the one excludes the other (which is evidently nonsense). Instead, the two mostly go together, precisely because harmony (or 'unity', as it is called in movements) is actively established through resolving conflicts. It is in, and through, conflicts that commonalities and their limitations are assessed. Thus, the question becomes how internal conflicts can be moulded into tools for constructing unity. Something similar occurs with single-issue movements. All movements start as single-issue movements. The big challenge is how to forge them into a constitutive part of a wider and more comprehensive whole.

In short, peasant movements do not emerge or grow by chance or at random, and building a movement is far from easy. The actually existing peasant movements of this world are phenomena that should really be admired. In a world full of chaos and failure, they represent hope: they show that things can be done differently – even under highly adverse circumstances.

Notes

1. Labour can only start to define and directly control the organization and development of production if it takes power over the means of production (by, e.g., occupying factories). The theoretical implications of this were elaborated in the Italian *operaismo* tradition (see, e.g., Tronti, 1966).
2. Autonomy is not a feature of a single object or an attribute of specific persons. Autonomy is *relational*. It is also *material*, for it is based on a self-controlled resource base.
3. Which translates broadly as 'our life is a struggle'.
4. This expression has a strong connotation that 'the land is ours' and has now been taken by those to whom it rightfully belongs. This belonging might refer to having worked the land for long periods (this is as much as 'making the land' into that what it is) and/or to common ancestral ownership (embodied, e.g., in indigenous communities).
5. The image shows a skill-oriented technology. One needs to take out the thistles very carefully in order to avoid parts of the roots staying in the subsoil. This needs to be done at the right time (before flowering when there are plenty of seeds) and in the right way (to avoid pain in the back).
6. See, for example, Lambert (1970) and Scheringer and Springer (1970).
7. Over time, many of these cooperatives have become mainstream business, sometimes due to fusions and peasant control over these cooperatives becoming eroded.
8. Cooperation-in-production is neither horizontal cooperation (that is, joint control over land and the other means of production combined with collectively working it) nor vertical cooperation (individually owned and controlled farms that together run joint enterprises for processing farm products or obtaining needed inputs). Cooperation-in-production is about individual farmers (each with their own farm) belonging to a machine pool and/or assisting each other in the performance of certain tasks in the farm labour process (Lucas, 2018).

9. Sometimes, irrigation schemes are constructed in such a way that they include swimming pools.
10. There is always a balance in peasant life and production between the individual and the collective. Producing without some form of cooperation is fairly impossible. Those who are not convinced of this should try to catch an escaped bull on their own. The balance is mostly set in such a way that the responsibility for organizing production belongs to the single household. There are, however, many situations in which there are extreme threats (from the state, or from paramilitary gangs, or as a consequence of extreme weather conditions) that bring the possibility of devastation. In such circumstances, peasants often opt for a collective organization of production as the best possible line of defence. Thus, individual versus joint production is not solely a matter of principle but also of circumstance.
11. Once again: 'man and land' can easily be now taken as a somewhat myopic, if not ugly, expression that is seemingly gender biased and apparently reduces living nature to just land. Nonetheless, I use this classical expression in order to not lose sight of the historical continuities.
12. This varied greatly according to the main cropping patterns, local history, etc.
13. As a matter of fact, they succeeded in doing so – thus materially altering both the processes of agricultural production and the regional economies.
14. In peasant communities, people referred to themselves as *nosotros los pobres del campo* (we, the poor people of the countryside). This notion cross-cut the many differences between fully employed rural workers (a minority), part-time workers, migrant workers, peasants and landless people – it offered a common denominator and functioned as a unifying clarion call.
15. In the heated polemics of the time about land reform and peasant struggles, there were basically two orientations: that of the *campesinistas*, who considered the peasantry as a central progressive force that could transform the rural economy, and that of the *descampesinistas*, who were of the opinion that the peasantry was the main obstacle to transformation. These positions coincided with the classical views of Chayanov and Lenin, respectively.
16. The many, often exciting, details are described in Ploeg (2006, especially chapter 9) and Diez Hurtado and Burneo (2022).
17. The origin of this expression comes from Piura, a department in northern Peru, an extremely arid area that without farming (and the required organization of irrigation) looks yellow and brown, like a desert.
18. See https://en.wikipedia.org/wiki/Marinaleda and https://www.accesstoland.eu/IMG/pdf/cs_andalusia_final.pdf, respectively.
19. See https://www.besh.de/ and http://www.coopcila.it/en/homepage-en/, respectively.
20. Consequently, it is impossible to delineate an 'agroecological segment' or to point to single farms that are 'completely agroecological'. Agroecology is always 'under construction'. There are, in practice, no dichotomies that separate the agroecological from the conventional or the organic. In practice, there only are degrees of being more (or less) agroecological. It is the degree of agroecology that matters, not any assumed, essentialist 'purity'.
21. Thus, they proposed, as it were, a kind of exchange – what they asked for was autonomy and local self-regulation on the condition that they would effectively, and in a measurable way, maintain and develop environmental qualities (notably low N emissions), biodiversity and landscape.
22. A territorial cooperative is not only rooted in, and defending interests and prospects linked to, the territory; it also aims at the defence and valorization of services produced by the territory (such as landscape management, the protection of biodiversity, employment levels, etc.). In order to do so, these cooperatives claim (legally conditioned) self-government and the required degree of autonomy.

23. In the realm of organic farms, similar deviations from the rule can be found. The percentage of children taking over the farm is far higher than that in conventional farming (some 75% versus 50%), whilst many organic farms are able to generate even two or three farms for the next generation. This is due partly to subjective factors (the children see their parents work with far more joy) and partly to objective ones. Apart from price differentials, the big difference resides in indebtedness (which is far lower, on average, for organic farms because they experience less pressure to continuously expand the farm). Similar exceptions can be found on the Dutch isles where farming became multifunctional far earlier than in the rest of the country.

24. Interestingly, the very name Via Campesina implicitly refers to such a political project: agrarian and rural development are to be unfolded along, and in, a peasant way. This means that they are to be inclusive, to unfold as labour-driven intensification, to enlarge autonomy at different levels, etc. In the 7th International Conference of La Via Campesina held in 2017 in Derio, this was made explicit as: 'we feed our peoples and build the movement to change the world'.

25. According to Shanin and Galeski (personal communication), the pre–World War II agrarian movements in Central and Eastern Europe faced this problem. Later on, taking a distance from entrepreneurial positions was a main driver for the construction of La Via Campesina (personal communications from Paul Nicholson, Nico Verhagen and en Jun Borras; see also Edelman and Borras, 2016).

Bibliography

Altieri, M. A. (1990), *Agroecology and Small Farm Development*, CRC Press, Ann Arbor, MI.

Altieri, M. A. (1995), *Agroecology: The Science of Sustainable Agriculture*, Westview Press, Boulder, CO.

Anderson, C. R., J. Bruil, M. Jahi Chappell, C. Kiss, and M. Pimbert. (2020), *Agroecology Now! Transformations towards More Just and Sustainable Food Systems*, Palgrave, Macmillan, London.

Badstue, L. (2006), *Smallholder Seed Practices: Maize Seed Management in the Central Valleys of Oaxaca, Mexico*, Ph.D. thesis, Wageningen University, Wageningen, the Netherlands.

Berg, L. van den, P. Hebinck, and D. Roep. (2018), '"We go back to the land": Processes of re-peasantisation in Araponga, Brazil', *The Journal of Peasant Studies* **45** (3), pp. 653–675. doi: 10.1080/03066150.2016.1250746.

Cayre, P., A. Michaud, J.-P. Theau, and C. Rigolot. (2018), 'The coexistence of multiple worldviews in livestock farming drives agroecological transition: A case study in French protected designation of origin (DOP) cheese mountain areas', *Sustainability*, **10** (4), pp. 1097. http://www.mdpi.com/2071-1050/10/4/1097.

Chayanov, A. (1927/1991), *The Theory of Peasant Co-operatives*, Ohio State University Press, Columbus.

CIDA (Comite Interamericano de Desarrollo Agricola). (1966), *Tenencia de la Tierra y Desarrollo Socio-Eeconomico del Sector Agricola, Peru*, CIDA, Washington, DC.

Claes, P., and M. Edelman. (2020), 'The United Nations Declaration on the Rights of Peasants and Other People Working in Rural Areas', *The Journal of Peasant Studies* **47** (1), pp. 1–68. https://doi.org/10.1080/03066150.2019.1672665.

Desmarais, A. (2007), *La Via Campesina: Globalization and the Power of Peasants*, Fernwood, Halifax, NS, Canada.

Diez Hurtado, A., and M. L. Burneo, eds. (2022), *Nuevas Miradas sobre la Reforma Agraria Peruana*, Cisepa-PUCP, Lima, Peru.

Edelman, M., and S. M. Borras Jr. (2016), *Political Dynamics of Transnational Agrarian Movements (Agrarian Change and Peasant Studies 5)*, Fernwood Publishing, Vancouver.

Feder, E. (1973), *Gewalt und Ausbeutung, Lateinamerikas Landwirtschaft*, Hoffmann und Campe, Hamburg, Germany.

Gliessman, S. R. (1997), *Agroecology. Ecological Processes in Sustainable Agriculture*, Ann Arbor Press, Chelsea, MI.

Huizer, G. (1972), *The Revolutionary Potential of Peasants in Latin America*, Lexington Books, Lexington, MA.

Jongerden, J. (2021), 'Autonomy as a third mode of ordering: Agriculture and the Kurdish movement in Rojava and North and East Syria', *Journal of Agrarian Change*. https://doi.org/10.1111/joac.12449.

Kerkvliet, B. J. T. (2009), 'Everyday politics in peasant societies (and ours)', *The Journal of Peasant Studies* **36** (1), pp. 227–243. doi: 10.1080/0306615090282048.

Kloppenburg, J. (2004), *First the Seed: The Political Economy of Plant Biotechnology*, 2nd ed., University of Wisconsin Press, Madison.

Kosik, K. (1976), *Dialectics of the Concrete: A Study on the Problems of Man and the World (Boston Studies in the Philosophy of Science*, Vol. 52), Reidel, Dordtrecht, the Netherlands.

Lambert, B. (1970), *Les Paysans dans la Lutte des Classes*, Seuil, Paris.

Lucas, V. (2018), *L'Agriculture en Commun: Gagner en autonomie grâce à la coopération de proximité. Expériences d'agriculteurs en CUMA à l'ère de l'agroécologie*, Ph.D. thesis, Université d'Angers, Angers, France.

Lucas, V., P. Gasselin, and J. D. van der Ploeg. (2019), 'Local inter-farm cooperation: A hidden potential for the agroecological transition in northern agricultures', *Agroecology and Sustainable Food Systems* **43** (2), pp. 145–179. doi: 10.1080/21683565.2018.1509168

McMichael, P. (2009), 'A food regime genealogy', *The Journal of Peasant Studies* **36** (1), pp. 139–169. https://doi.org/10.1080/03066150902820354.

Moore, B., Jr. (1967), *Social Origins of Dictatorship and Democracy: Lord and Peasant in the Making of the Modern World*, Penguin University Books, London.

O'Brien, K. J., and L. J. Li. (2006), *Rightful Resistance in Rural China*, Cambridge University Press, New York.

Paige, J. (1975), *Agrarian Revolution: Social Movements and Export Agriculture in the Underdeveloped World*, Free Press, New York.

Petersen, P. (2015), 'Hidden treasures: Reconnecting culture and nature in rural development dynamics' in *Constructing a New Framework for Rural Development (Research in Rural Sociology and Development*, Vol. 22), edited by P. Milone, F. Ventura, and J. Ye, 157–194, Emerald, Bingley, UK.

Petersen, P. (2017), *Arreglos Institucionales para la Intensificación Agroecológica; Una Mirada al Caso Brasileño desde la Agroecología Política*, Ph.D. thesis, Universidad Pablo de Olavide, Sevilla, Spain.

Petersen, P., and L. M. Silveira. (2017), 'Agroecology, public policies and labor-driven intensification: Alternative development trajectories in the Brazilian semi-arid region', *Sustainability* **9** (4), pp. 543–569. doi: 10.3390/su9040535.

Ploeg, J. D. van der. (1990), *Labour, Markets and Agricultural Production*, Westview Press, Boulder, CO.

Ploeg, J. D. van der. (2006), *El Futuro Robado: Tierra, Agua y Lucha Campesina*, Instituto de Estudios Peruanos, Lima, Peru.

Ploeg, J. D. van der. (2018), *The New Peasantries: Rural Development in Times of Globalization*, 2nd ed., Routledge, London and New York.

Ploeg, J. D. van der. (2020), 'Farmers' upheaval, climate crisis and populism', *The Journal of Peasant Studies* **47** (3), pp. 589–605. https://doi.org/10.1080/03066150.2020.1725490.

Rosset, P. M., and M. Altieri. (2017), *Agroecology: Science and politics*, ICAS small book series, Fernwood, Vancouver. http://practical-action.org/10U-588QE-B869K6V111/cr.aspx (international edition).

Scheringer, R., and W. Springer. (1970), *Arbeiter und Bauern Gegen Bosse und Banken. Bauernproblemen der Bündesrepublik*, Marxistischer Blätter, Frankfurt, Germany.

Scoones, I., M. Edelman, S. M. Borras Jr., L. Fernanda Forero, R. Hall, W. Wolford, and B. White, eds. (2022), *Authoritarian Populism and the Rural World*, Critical Agrarian Studies, Routledge, Oxon, UK.

Scott, J. C. (1985), *Weapons of the Weak: Everyday Forms of Peasant Resistance*, Yale University Press, New Haven and London.

Sereni, E. (1979), *Storia del Paesaggio Agrario Italiano*, Universale Laterza, Rome.

Sevilla Guzman, E. (2007), *De la Sociologia Rural a la Agroecologia: Perspectivas agroecologicas*, ICARIA Editorial, Barcelona, Spain.

Swagemakers, P. (2008), *Ecologisch Kapitaal: Over Het Belang van Aanpassingsvermogen, Flexibiliteit en Oordeelkundigheid*, Ph.D. thesis, Wageningen University, Wageningen, the Netherlands.

Swagemakers, P., J. S. C. Wiskerke, and J. D. van der Ploeg. (2009), 'Linking birds, fields and farmers', *Journal of Environmental Management* **90** (Suppl. 2), pp. S185–S192.

Toledo, V. (2000), *La Paz en Chiapas: Ecologia, Luchas Indigenas y Modernidad Alternativa*, Ediciones Quinto Sol, Mexico City.

Tronti, M. (1966), *Operai e Capitale*, Einaudi, Torino, Italy.

Wezel, A., S. Bellon, T. Doré, C. Francis, D. Vallod, and C. David. (2009), 'Agroecology as a science, a movement and a practice. A review', *Agronomy for Sustainable Development* **29** (4), pp 503–515.

Wolf, E. R. (1969), *Peasant Wars of the Twentieth Century*, Harper & Row, New York.

10 Dealing with socio-material practices

Overview: Main concepts discussed in Chapter 10

Socio-material practices
Artefacts
The biography of things
The manual of things
Socio-material constructs
Connectedness
Diffusion on innovations
Hierarchy of breeding criteria
Farming style as relational pattern
Brand (of food products)
Web of dependency relations
Altering patterns of consumption and production
Food products as actants
Transformativity (as feature of food products)
Uncaptured food
Extended co-production
Intertwining different forms of co-production
Centre of command
Control at a distance
Food empires
Structural holes
Obligatory passage points
By-pass
Collectives functioning as brand
Floating narratives
Differential processes
Identifying grounded alternatives

Farming is about producing *things* (food and other useful products). It does so by using other *things*: resources that are converted into food, fibre, fruits, fragrances, etc. Within this conversion yet another category of *things* is of strategic importance: the technical instruments that facilitate this conversion of resources.

DOI: 10.4324/9781003313274-10

As part of the production process, a variable, but often significant, portion of the things involved (seeds, sires, soil, etc.) is reproduced for the next cycle(s) of production. The centrality of *things* makes farming a very *material* activity. Farming is not about dreaming or making nice discourses. Neither dreams nor discourses can satisfy the stomach or eradicate hunger. If farming did not result in material end-products (things you can touch, see, smell, eat, drink, store, process or even spill), it would have no meaning. And if those involved in farming do not have access to the things they need (land, water, cattle, seeds, machines – again, things that you can see, touch, caress, whip, start, operate, etc.), they would be unable to farm. Grain does not grow in the clouds; it needs earth (and water and sunshine).

Yet, as hinted above, the many things that are necessary to the farm labour process do *not* produce themselves and do not, in or by themselves, constitute a well-working and well-balanced farm. They need to be combined and set in motion according to a well-thought-out plan. While farming requires different material objects, it also requires social actors able to operate and alter the material objects.

In short: farming is a *social-material* practice. It involves both social and material elements,[1] and the two interact and mutually shape each other. The material elements are *artefacts*. They are made ('constructed') in a particular way, with particular components, in a specific place, by specific actors, who have specific objectives. They have, in short, a biography. Artefacts also carry a mostly hidden, but sometimes explicit, code or manual (see also Chapter 3) that specifies how the artefact should be used, with which objectives, by whom and under what conditions. Often this manual is literally 'built into' the objects. If specific things are not used according to their in-built manual, there will be frictions, accidents or failures. When the biography is neglected (or unknown), all kind of things can go wrong. To work a field properly, one has to know its history: the quantities and qualities of fertilizers (manure included) applied and the rotational patterns followed in the past; the depth and sequence of the different soil types (resulting from historical geological processes); the way the field has reacted to different types of tillage; the previous owners and how, and under what, conditions the field was passed on to those currently working it. Thus, the biography of the field is part of its manual.

The material objects used in, and produced by, farming are *socio-material constructs*. Their materiality is shaped and reshaped by social actors (often over long periods and by many different actors whose innumerable decisions flow together into the particular design that characterizes each of the many objects). The same applies to the identity of the actors, which is shaped and reshaped in, and through, the ongoing and long-lasting encounter with the material objects. The essence of a farmer is that he or she knows how (and when, where, with which objectives, in which sequence, etc.) to deal with the many material objects involved in the farm labour process. Without such knowledge and the associated skills,[2] a farmer is not a farmer: he or she is not yet 'moulded' by material experiences (working the land, caring for animals, etc.) into a farmer and is at best a 'would-be farmer'.

Social actors turn material objects into artefacts that are indispensable parts of the farm labour process. At the same time, it is only through operating these specific material objects that these actors turn into farmers. The farm labour process not only ties the social actors and material objects together but, by so doing, also produces farms, farmers, farm resources, farm products and styles of farming. Resources, farmers, food products, etc., are tightly related in, and through, the farm labour process. It is this *connectedness*, these specific sets of relations, that define the different components, that make the involved actors into farmers, resources into agricultural resources and outcomes of the process of production into food. 'I am farmer because I work the land and produce food'. The mutual interrelations define the different components. The relations with the land and the food produced define the farmer – just as the farmer defines the land and the food.

This much may be completely obvious, and it may seem as if I am labouring the point, yet its centrality is often lost from sight (just as fish are not aware of the relevance of water). This has practical and theoretical implications because the socio-material nature of things and the interconnected and mutual shaping of men and things opens the door for important methodological approaches. Unravelling materiality – that is, understanding it as a compound conglomerate of socio-material constructs – can help very much to grasp society, its dynamics and its problems. This applies, of course, to society as a whole – but also, and perhaps especially, to farming. This is partly because agricultural production is decentralized, with millions of productive units and even more actors involved in them. This results in an abundant heterogeneity of the many 'things' that are part and parcel of farming. A thorough and rigorous exploration of these things and the many shapes they take can help us better understand agriculture and farming.

The previous chapters have already showed some examples of such an approach. When confronted with specific tools, machines and other technical artefacts (see Chapter 3), we may try to unravel the *code*[3] they entail (the code that 'has been built into it'), and this provides highly relevant information about who is supposed to work with a particular technique, what kind of knowledge is needed to operate it, the reasons why this particular technique is used (and not another one), the results it is expected to render and the conditions under which it becomes useful (or not). Just as techniques and technologies have their storytellers (Staudenmaier, 1985), they themselves also tell their own stories. This is especially the case when it comes to the techniques and technologies meant for and/or used in farming: they render impressive amounts of information, and it is the task of those studying farming to collect and analyze this information. Indeed, 'technology is a language' (as Bruno Benvenuti [1982] argued). It tells you what to do, how to do it, when and where and which conditions need to exist or are to be created for the proper functioning of the system. This is also the reason why specific technologies are often so bitterly contested – why farmers feel uncomfortable with some technologies and more comfortable with others. It also explains why simple, 'innocent-looking' innovations are, at least sometimes, only very slowly adapted, or not at all (see Text Box 10.1).

Text Box 10.1   Reconsidering theories on the diffusion of innovations

Discussions about the 'diffusion of innovations' have played an important, albeit highly problematic, role in the history of rural sociology. The concept was much used in the modernization and Green Revolution programmes of the past and still underlies the construction of the second generation of Green Revolution programmes, especially in Africa. The theory centres on the willingness to adopt innovations as an attribute of individual farmers. It perceives willingness to adopt specific technologies as conditioned by a range of social and cultural particularities but typically leaves the features of the innovation itself outside of the analysis. Thus, the central categories classify and distinguish different types of farmers into 'early adopters', 'followers' and 'laggards' (just as agricultural policies generally distinguish between 'stayers', 'leavers' and 'in-between farmers'; i.e., those 'sitting on the fence'). This classification coincides with the typical S-shaped adoption curve that synthesizes the adoption of innovations over time: initially there is a small group of daring and entrepreneurial farmers who supposedly understand the benefits of an innovation and quickly apply it. These farmers are 'open for change' and reap the benefits of the innovation (e.g., higher productivity). Seeing the success of these early adopters, a next category (the followers) starts to imitate the pioneers and, finally, even those who were initially unwilling to even consider any change (the assumedly traditional farmers) are more or less forced to follow suit and apply the innovation as well (they are forced to do so, it is thought, because the production increase by the early adopters leads to lower prices).

In parts of the literature this process as a whole has been characterized as a 'treadmill'. In yet other blocks of literature these different attitudes have been related to, and supposedly explained by, contrasting cultural repertoires: a modern-dynamic culture versus traditional cultural patterns.

These explanations have been subject to much criticism. Out of all of these, mostly quite convincing, critiques I want to take two elements. A first critical finding (obtained through, and consolidated by, many empirical studies) is that individual farmers do not occupy a fixed position with regard to their propensity to adopt available innovations. A farmer can be an early adopter when it comes to one particular innovation but a complete laggard when it comes to another. A second finding is that many farmers may try to adopt particular innovations but that only a minority are able to meet all of the requirements or conditions to make the innovation a success. Green Revolution programmes typically introduced *packages*: an assemblage of improved varieties, chemical fertilizers, pesticides, tractors, new ploughs (allowing for more depth), credit facilities, technical assistance and the like. Mostly, however, only a minority of peasant farmers are able to apply all of the required changes (that come with a new variety) at the same time (see also Text Box 3.5). Consequently, the uptake of the package is disappointing and considerable numbers of farmers refrain from further following the recommended trajectory.

The point here is that uneven rates of adoption can be as much, if not more, due to the materiality of things as to the (assumed) psychology of farmers. For innovations contain a code: they assume specific conditions (specific growth factors) that may or may not be available in the reality of farms and farmers. If not, the mismatch between the material conditions assumed by the innovation and those existing within the fields and farms of the 'target group' will make many farmers wary of adopting an innovation (i.e., be 'laggards'). Only if there is a good match will a group of early adopters emerge. So, theories about the 'diffusion of innovations' put the cart before the horse.

In the *socio-material perspective* suggested here, tackling the issue of innovations and their uneven distribution raises the question of the assumed (or preferred) beneficiaries targeted in, and during, the process of designing different innovations. Whom did the engineers, agronomists and others have in mind when they started to draw the outline of an innovation? What was the horizon of relevance they had in mind (see Figure 1.7)? Would it be worthwhile to redesign the innovation by taking other, distinctively different, groups of beneficiaries as the target?

Social scientists, agronomists and others studying agriculture (regardless of whether they do so for practical or more theoretical purposes) are to be able to understand the artefacts used in farming as a 'language' and, if needed, they are to be able to translate it.

Large parts of social life are *shaped by artefacts*: these things not only (co-)shape the many different practices in which social actors are involved – they also *tell* other social actors what to do, when, where and how (just as they inhibit others from doing what they want to do). This means that studying the *materialities* of soils, crops, techniques, cows, food, etc., is a crucial methodological device for delving into the realities of farming and food.[4]

Understanding artefacts as a language often implies an inquiry into the *authors of the script*. *Who* has written the script that is built into a particular artefact, and whose interests and prospects does it promote? Has there been any anticipation of the situations in which the particular artefact will be used? Has there been any consultation with the potential users? Have they had any chance to discuss any possible shortcomings in the design?

Applying such a socio-material perspective and asking questions such as those outlined above can help to avoid major mistakes and mismatches. One of the awful, but nonetheless persistent, mistakes made in programmes for agricultural development and associated technological R&D is the construction and subsequent distribution of new plant varieties that do not match the taste or culinary preferences of local populations. This, at first sight, nearly unbelievable mistake (how stupid can you be?) has nonetheless proven to be a chronic feature of 'planned development'. It was first noted in the 1960s (Haverkort, 2021) but has continued to show up, in all continents, throughout the first 2 decades of the 21st century (Kimanhti, 2019).

There are different aspects to this issue. First, it has been repeatedly shown that introducing new cultivars that do not fit with local tastes and habits seriously undermines well-intended development efforts. Adoption will only be partial and often temporary, to be abandoned when 'the jeeps have rolled away'. In the meantime, the innovation does not help, in any possible respect, to improve the situation of local populations and/or reduce poverty. The efforts are just in vain. Secondly, one needs to ask why this 'mistake' is so persistently made. It cannot be maintained that all agronomists are stupid or fail to learn from their mistakes.

On the contrary – they are mostly very motivated and capable people (but then, the problem is not an individual problem but an institutional one). At the institutional level, it is known that matching local taste and habit is a positive criterion, but it is regarded as secondary to a range of other criteria. These are the yields (the more so because low productivity is considered as *the* main agricultural problem), the possibility to conserve the harvested output for longer periods (thus allowing for longer distance trading), the uniformity of the product, etc. Currently, agricultural sciences establish a clear and fixed hierarchy of breeding criteria, and this hierarchy helps to determine the horizon of relevance (see Figure 1.7). Underlying these criteria are more general assumptions (that, together, have an inbuilt bias towards the desirability or inevitability of the entrepreneurial model of farming). It is assumed, for example, that production for self-consumption will inevitably come to an end (implying that local taste preferences will become less relevant). Secondly, it is assumed that all production will (and should) be completely market oriented and that R&D has to produce the genetic materials needed within the framework of main export markets. Thirdly, there is the conviction that progress implies a reduction of diversity: that there can only be one variety (or just a few) that is (are) the best. The notion that a potato farm producing 10 to 12 varieties with fields with up to 4 to 5 different varieties and a small plot near the house that harbours up to 50 different tubers might be the best possible solution in the given circumstances is difficult to digest in the world of agricultural R&D.

I think a socio-material perspective – that understands material things as social artefacts and which consequently probes into the question of how such things (e.g., food products) and their characteristics (e.g., their taste) fit into, and shape, society – is extremely useful in helping to avoid the chronic failures discussed above.

In more general terms, the systematic and in-built bias of much agricultural R&D goes back to two major (albeit hidden) ordering principles. These are *centralization* and *control*. Over the years, agricultural R&D has become increasingly centralized (e.g., in the CGIAR [Consultative Group on International Agricultural Research] framework) and it is nigh-on impossible for a few, centralized, R&D institutions to relate to the impressive variety of different ecological micro situations and the many different farming styles. They are not equipped to produce the range of place-specific solutions required. This can only be done by a highly de-centralized system of local breeding and selection, which could produce many small steps forward but never will come up with the 'silver bullet' that the centralized institutions are looking for. The search for control strengthens this tendency towards centralization (it is easier to control and align one central institute than a myriad of local systems) but also has its own momentum and agenda. If there is one 'silver bullet' solution, it is far easier to control its application, especially if this solution is constructed as a series of specific *objects* (an improved variety, inputs, machine services, credit) to be applied – instead of a range of recommended *practices* (local selection, better preparation and use of manure, etc.). The application of objects can be controlled – not so the application of site-specific practices.

Control coincides with the interests of capital (the objects to be applied are commodities that can be marketed; farmers' practices, by contrast, are non-commodities and not marketable), the state (which wishes to appear to be firmly in charge), political elites (who seek to bring benefits to their following/clan) and institutional interests (R&D as a driver of agricultural growth and thus deserving of more funding). Thus, a jumble of interests coincide and flow together in a powerful structuring principle that unavoidably generates and reproduces what I have discussed earlier as the vanguard approach (see Chapter 5), which is so much at odds with inclusive development.

In the remainder of this chapter, I will indicate how different material objects and constellations can be 'read' in order to better understand the social contexts in which they form an essential and constitutive element. I will discuss farming styles, techniques, food products, food markets (and especially their socio-material infrastructure) and, finally, development as the emergence, accumulation, evolution and/or disappearance of material artefacts.

Reading farming styles

In Chapter 4 I discussed farming styles by going from the social to the material: from farmers' strategies towards the organization and development of farms and farming practices. Here I follow the opposite direction and try to show that reading and de-ciphering the farm as a complex constellation of material elements (cows, fields, concentrates, fertilizers, income flows, etc.) can also help to inform us about the strategy that has helped to (co-)construct this specific (material) reality. The starting point for doing so is understanding farming styles as *relational* patterns. In each style of farming, the composite elements are *related* to each other in different ways. Animals are *related* in different ways to the available land (thus resulting in different cattle densities). This *affects* both the land (because of the grazing pressure and the amount of manure that is produced per hectare) and the animals (partly because the amount of roughage per animal will differ). Such effects are not only short term – in the longer run they materially reshape the land and the animals (for the better or the worse).

The material objects (fields, animals, roughage, manure) differ because they are shaped and related by, and according to, the relational patterns through which they function and are developed. This implies that we might get to (better) know the socio-material objects *and* social actors through these relations, just as we might explore the relations through the objects, artefacts and actors that compose them (see Methods Box #20).

Figure 10.1 gives an (admittedly rough) model of a dairy farm: it centres on the interlinkages between the level of N fertilization; the amount of self-produced fodder, bought feed and fodder; cattle density; production per milking cow; and labour income per milking cow.[5] For each of these variables the average value and the standard deviation (in brackets) are given. Each variable shows a certain variation that is (partly) associated with style differences. Thus, each style shows a specific level for each variable shown in Figure 10.1. These levels are often telling

Methods Box #20   Using objects as entrance for studying styles and strategies

Careful observation of different farms (representing different styles) shows an amazing variety in the way fields are worked, how the livestock is cared for or the machines and machine operations combined, as well as the architecture of the buildings, the type of inputs acquired, etc. A basic approach to understanding these differences is to take, say, a specific machine that is used on, say, farm A but not on farm B (probably another type is used there) and then ask *why* the farmer opted for it; *what* would happen if he or she had to do without it (or had to use another type); what the use of the *alternative* tells about the farmers using it; etc. This can be done equally well for cattle breeds, operations in the field, etc.

In short: by unravelling the code of techniques that fit (or do not fit) into the farm, you can learn an awful lot about the farm and the way it has been developed by the farmer.

Farmers rely on a wide range of different techniques for getting roughage into the barns (to feed the cattle during wintertime), each implying specific artefacts. The mown grass can be (1) converted into hay by drying the grass in the field with the help of the sun and by regularly tedding it (hay tedding always is a much-debated sub-task); (2) it is, however, also possible to blow the semi-dried grass into the barn and then have it dried artificially. A completely different approach is to make silage: gathering green, semi-dried grass (3) in large, flat silos (that can be filled in different ways) or (4) put into small, round plastic bales. Drying and silage making can also (5) be externalized. Beyond that, (6) it is possible to combine these different techniques.

A different, but equally rewarding, inroad is to accompany farmers to agricultural exhibitions, demonstration farms, open farm days, etc., and ask, time and again, whether a specific machine, or approach, etc., would fit in their own farm and then delve into the reasons for it.

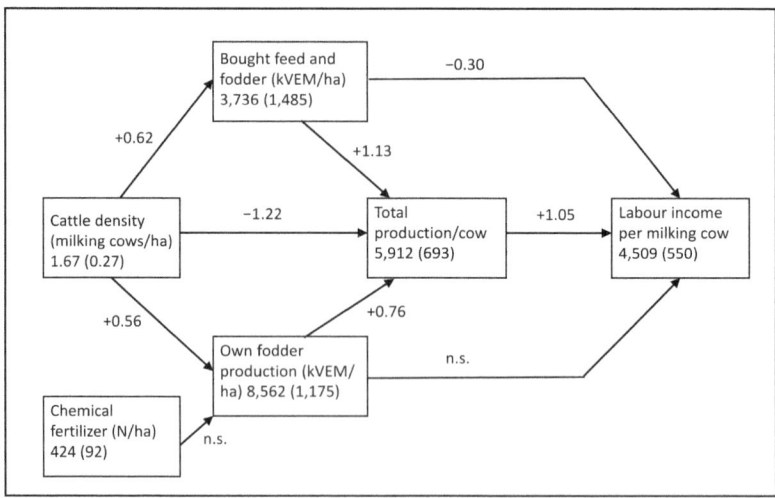

Figure 10.1 Interrelated socio-material entities (dairy farms, Friesland, n = 300, 1990) (Ploeg et al., 1994).

(especially in combination with each other). In terms of Figure 10.1: a high level of production per cow, in combination with an average level of concentrate use, most probably points to an underlying strategy that centres on fine-tuning (i.e., to 'cow men'). Low levels of bought-in feed and fodder, together with low levels of fertilizer use, suggest the strategy of farming economically. Beyond that, the *relations* may also vary significantly, which is also indicative of the style of farming.

The model in Figure 10.1 shows the main interrelations that were calculated here as β (beta).[6] These are the average betas. The style-specific betas were calculated as well. I will limit myself here to discussing just a few telling points. The relation between one's own fodder production per hectare and total production per milking cow is +0.40 for large farmers and +0.95 for cow men (those who aim at high intensity levels based on the quantity and quality of labour). In other words, cow men *make* the production per cow far more dependent on their own fodder production than large farmers. The improvements that cow men make in their own fodder production (improved grassland yields, pasturing and silage making) *translate* far more strongly in improvements in the production per milking cow (and hence in the labour income per milking cow) than is the case in the style of large farmers. Another telling example regards the relation between cattle density and the amount of bought-in feed and fodder: +0.72 in the high-input/high-output style focussed on a high level of cattle density and only +0.48 for economical farmers. The economical farmers are cautious about spending money on external inputs, so they will acquire less feed and fodder than the former with the same density of grazing cattle.

Farming styles are relatively stable over time as they are rooted in (and dependent on) *socio-material* entities: the specific layout of the fields, the specificity of the selected animals, the type of technologies chosen and the specific skills obtained by working the fields, caring for the animals and operating specific technologies. At the same time, these 'things' (together with the associated skills) create a kind of path dependency that enables the development of specific trajectories (see Chapter 5) and makes other trajectories less probable, or even unthinkable. The introduction of an 'alien' element (e.g., a new technology) can also introduce a persistent disequilibrium into the productive constellation as a whole which might, as many farmers can tell you, take years to rebalance.

Reading techniques and technologies

In Chapter 3 I discussed the concepts of technique and technology. A technology needs to fit into the farm where it is used. It also needs to correspond with the landscape and ecosystem in which the farm is embedded. Unexpected problems can emerge if its in-built code is at odds with farm, landscape, ecosystem and/or the surrounding society. It sometimes occurs that technologies that initially look highly efficient later turn out to be highly inefficient. Reading the code that is intrinsic to technical artefacts and solutions can help to avoid such problems (see Text Box 10.2).

Text Box 10.2   Materiality striking back

In previous chapters, I have described potato production in the Andean mountains (see Methods Box #5, Text Box 3.5 and Methods Box #7) and indicated how the introduction of newly designed potato seedlings comes with a package: a series of additional elements (inputs, practices, conditions, etc.) needed to meet the assumptions (the code) built into the new seedlings. Ploughing more deeply than is possible with a pair of oxen is one such commonly needed change. However, operating a tractor on the steep hills is no picnic: the danger of turning over is considerable (and often fatal) and is most likely to occur when the contour lines are followed. Thus, peasants (and especially contractors, who want to work as fast as possible) choose to plough at right angles to the contour lines – which evidently accelerates erosion and the loss of topsoils.

The irony of all of this is evident. Oxen are, for sure, dull and probably stupid animals, but they are able to work along contour lines. However, the more powerful tractors, which come as part of technological development, are unable to do so and thus often produce a disaster. Materiality, here present as the ecology of the Andean mountains, strikes back and turns this specific case of 'development' into a farce, if not a tragedy.

I also referred earlier to the *radu*, a 'traditional' and seemingly simple technical artefact used in the tropical rice polders (*bolanhas*) of West Africa (see Text Box 3.4). Like the *yunta* (the pair of oxen) used in the Andean mountains, the *radu* apparently resembles the opposite of development. Replacing it would seem to be a great benefit to those who work with it. Thus, when broken dikes needed to be repaired, it appeared evident that bulldozers would do a better job. However, here, again, the materiality of local ecology and the *bolanha* struck back – in a way that, again, turned out as farce if not a tragedy.

Figure 10.2 is a (simplified) map of a *bolanha*. The dikes that are there to protect the rice fields have suffered several holes (which might be due to damage by

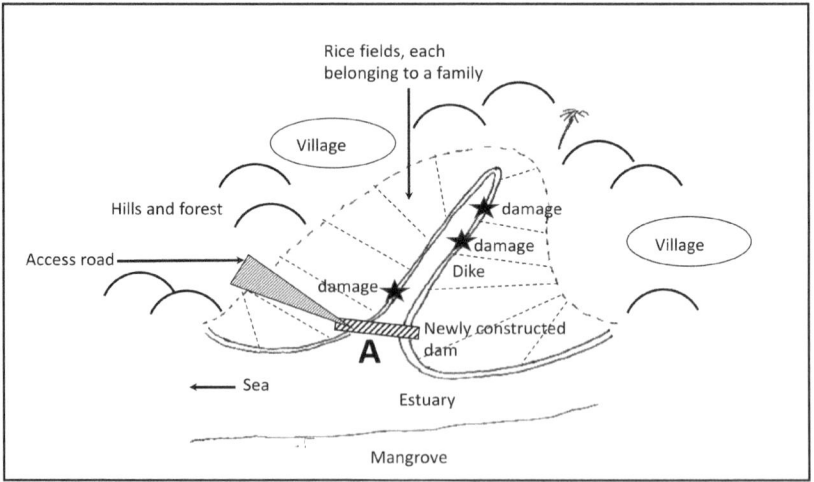

Figure 10.2 Bolanha (tropical rice polder) (author's own elaboration).

livestock/wild animals, erosion, extreme weather or warfare). Whilst the peasants working in these *bolanhas* can easily walk towards these holes in order to fix them, a bulldozer cannot (it would get stuck in the mud). Thus, it needs an access road, and this has to be made by the bulldozer itself and is a costly and time-consuming business that can only be done once. Hence, opting for one new dam at point A seems to be the logical solution (it is indeed logical in terms of the in-built code of the bulldozer). However, having a central dam at point A will stop the tidal movement in the part behind the dam and thus bring the need to standardize irrigation practices in all of the rice fields in the *bolanha*: all families have to do the same thing at the same moment, and this blocks the possibility of fine-tuning production and assessing the right balance between fields and family (between the demand and supply of labour). A French study tellingly carried the title 'Barrages contre le développement' and recounted how this practice eventually led to many dams being destroyed ('deactivated') by local producers (Reboul, 1984). The chosen solution (more specifically: the code of the used technology) was at odds with the local socio-material reality, which, in the end, struck back.

Reading nature

One of the storylines in this book concerns the peasants of the Northern Frisian Woodlands (see Methods Box #12, Figure 6.10 and the discussion that accompanies Figures 9.6 to 9.10). They attentively observed different meadows (some producing far more biomass than others, even with far lower levels of fertilization) and different samples of manure (the structure, colour, smell, composition, solubility, etc., of which varied considerably) and it occurred to them (see Figure 2.2) that there probably was a connection. This was tested in what was called 'peasant-managed action research', which rendered many new insights about the animal-manure-soil-fodder cycle (see Methods Box #12). This, in turn, helped to considerably increase the sustainability (see Figure 9.6) and resilience of farming in the area. Their careful reading of nature thus helped to change farming – and helped reveal a new way forward.

But the story did not stop there. The changes in farming practices, in turn, affected nature. They considerably ameliorated the quality of soil life (and thus the capacity of the soil to deliver nitrogen). Soil biology involves a complex food chain, or pyramid, with each level characterized by specific micro-organisms – fungi, bacteria, nematodes, etc. – and each level 'eating' the lower levels and being consumed by the upper levels. This complex system effectively converts dead roots, leaves and all kinds of detritus into nitrogen, which feeds plant growth. The farmers' action research identified that the soil biology was positively affected when they used improved manure (see Methods Box #5). N-poor manure increased not only the autonomous flow of nitrogen (produced by soil biology) but also, as a consequence, the amount of feed and fodder produced per hectare (expressed as dry matter production/ha: DM/ha), as summarized in the three-quadrant diagram in Figure 10.3 (Groot et al., 2003, 2006, 2007).

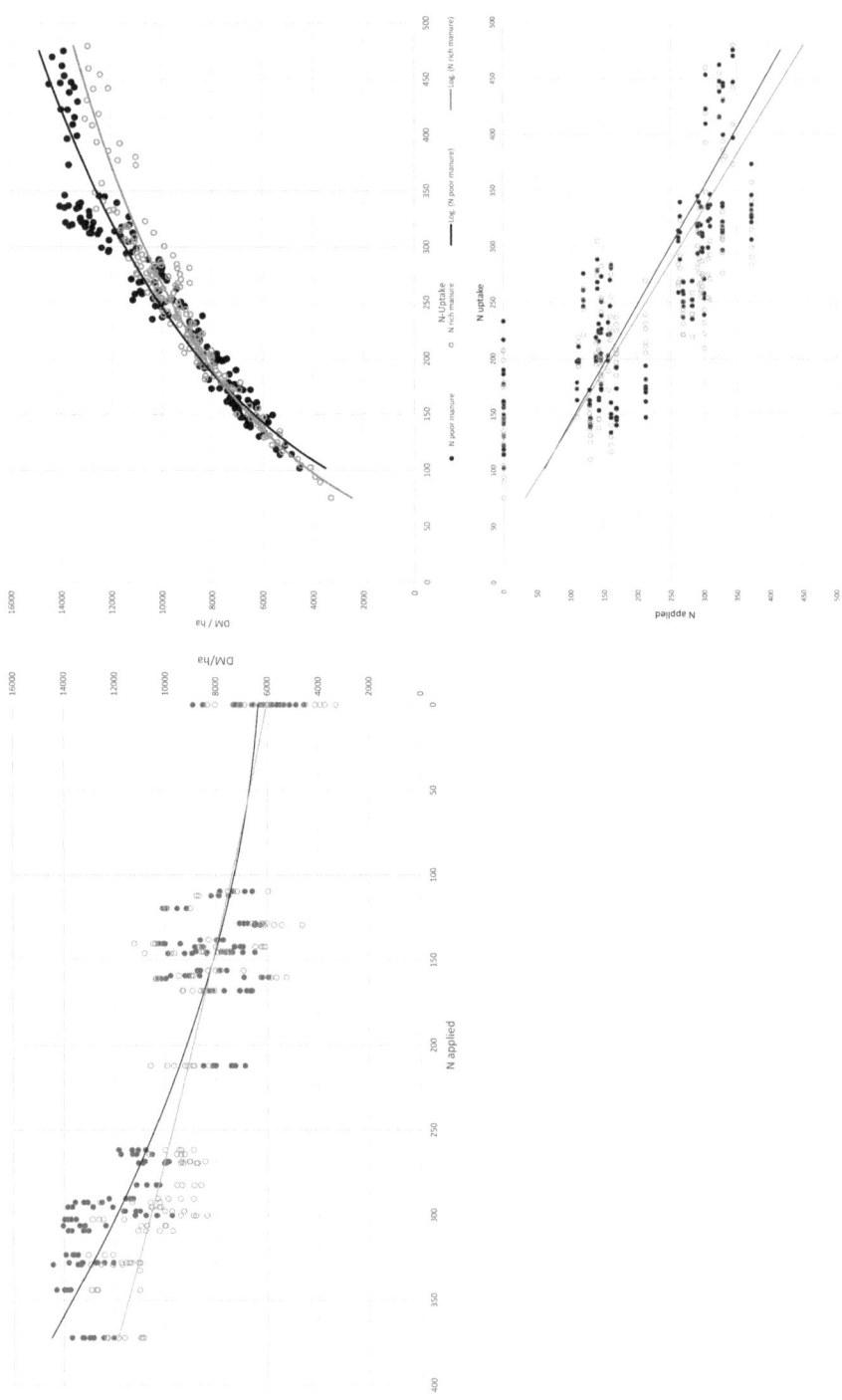

Figure 10.3 The uptake of nitrogen by grass, its conversion to DM production and the input/output ratio (author's own elaboration).

Figure 10.3 follows the standard representation of agronomic research. The lower-right quadrant shows that using improved, N-poor manure translates into a higher uptake (by the growing grass) of the nitrogen applied. The upper-right diagram illustrates that this nitrogen, once taken up by the growing crop, results in a higher DM production/ha. Finally, the upper-left quadrant shows the overall efficiency: the DM produced per kilogram of applied nitrogen: with the improved manure this was 51.2 kg DM/kg of applied nitrogen compared to 42.9 for conventional slurry (p = 0.002), an increase of more than 20%.

Changed nature (i.e., soil biology) also contributed to increasing biodiversity, with the enhanced subterranean soil life producing more feed for the visible side of biodiversity, which includes insects, slugs, beetles, butterflies, birds, deer and top predators such as owls and buzzards.

In the same vein, changed nature equally improved the quality of resources (notably water and air), the strength of the local economy (by reducing costs and thus increasing incomes), the welfare of animals (less stress as a consequence of the reorganized feeding) and the quality of the milk produced. All of this is synthesized in Figure 10.4 (which builds on Figure 6.1) and shows how a reorganization of farming can positively induce a range of alterations at higher levels of aggregation. The compound and socio-material nature of farming is key here.

Probably the most interesting (but also quite enigmatic) effect of the material on the social is that peasant agriculture, in this area, is increasingly *protected* by the hedgerow landscape that has been co-constructed over the course of centuries. The hedgerows have helped to (1) improve the quality of resources, (2) turn landscape maintenance into an additional source of income and, above all, (3) slow down scale enlargement. Consequently, variable costs and debt levels are

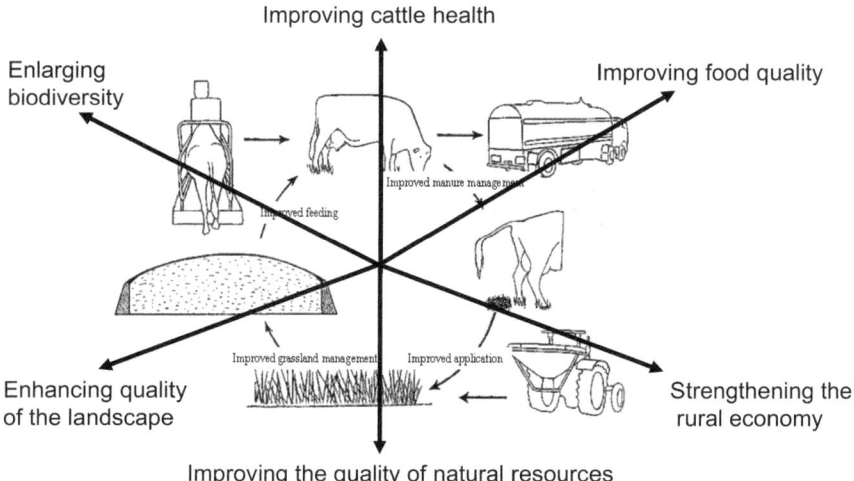

Figure 10.4 Enhancing the values generated in the area (author's own elaboration).

far lower in this area than in surrounding areas – which, in turn, explains the resilience of farms in this area, evidenced by the substantially lower levels of farm closure here than in the surrounding areas. Thus, in this instance at least, the landscape forms an effective line of defence for the farmers (or whose forebears) who co-produce(d) it.

Reading food products

Farming converts natural resources into food. Like all other products, food products come with a *brand* (no matter whether this brand is made explicit or remains implicit), and each brand represents a normative framework. A brand contains messages that specify how, where, when and by whom the product has been made (or is supposed to have been made). The absence of such messages equally represents a clear signal: that the origin, history and nature of the product do not matter. Taken together, these messages generate a normative framework (condensed in, and as, a 'brand') that synthesizes the *making* of the product. They also regulate inclusion and exclusion (they specify who is allowed to produce it and who cannot), just as they carry (or at least promise to bring) trust – that the product is indeed as good, cheap or sustainable (or whatever) as suggested.

As argued before, products not only carry a brand but also come with a *manual* (which, again, is sometimes made explicit, whilst it may also remain more or less implicit). Each product contains a series of requirements regarding its *use*. The manual specifies these requirements: it prescribes how to use the acquired product in a correct way.

Products make up a concrete nexus between production and consumption, and these are ordered through the brand and manual that come with the product. As such, a market transaction is far more than the simple exchange of a commodity for money. The transaction assumes and activates a normative framework that describes how the product was made, just as it communicates a series of requirements to be followed when the product is used. Thus, the products that circulate in markets are socio-material constructs: besides their physical, chemical and/or biological properties, *they also contain a range of messages* that reproduce, and often modify, large parts of the social and natural worlds (see Figure 10.5 for a graphic summary). In this respect, products are *actants*: non-human actors (Latour, 2005; Delanda, 2006; Long, 2016) that interact with other actors and other actants to co-order significant parts of the world. They are actants because each artefact is wrapped in a double set of messages (a brand and a manual) that concern the production and consumption of each particular product. Such messages travel in, and through, the market. They travel to, from and between the different actors involved, just as they may pass from one actant to the other (through processes of assemblage). Potentially, new products (that carry disruptive manuals[7]) can contribute to changes in consumption patterns, just as new brands may induce considerable changes in the sphere of production, provided that they are widely accepted.

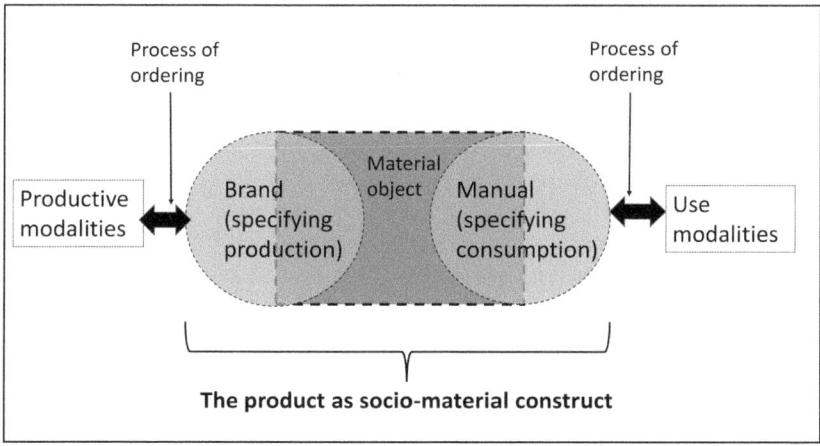

Figure 10.5 Understanding marketed products as actants (author's own elaboration).

In a way, Figure 10.5 refers to the ordering of social life through things – things that are exchanged in marketplaces. Social life is not outside the world of things, nor is it separated from the movement of things. On the contrary: a considerable part of social life is ordered through and around the movement of things. That is precisely why 'reading' products is important.

Alongside material negotiations (which centre on *exchange value*), market transactions also include symbolic negotiations: a brand reflects and embodies certain productive practices, just as the manual specifies the consumptive practices. But the reverse is also true: the brand also seeks to invoke a desired form of consumption, whilst the manual specifies (albeit sometimes implicitly) the characteristics built into the product by the producer. Through changes in the manual (even minor ones), producers are able to modify the behaviour, expectations and practices of consumers – and changing consumer preferences (i.e., vis-à-vis different brands) will impact production practices. These complex and two-way patterns of communication are all related to the *use value* of the negotiated goods and services. In our societies, they mostly are non-verbal and indisputable. The exception to this is encountered in marketplaces grounded on short circuits that involve direct, face-to-face meetings of buyers and sellers.[8]

Theoretically, market transactions ensure coherence between productive and consumptive practices. This occurs not only through assessing the right price but equally by aligning the brand and the manual (the latter is usually a precondition for the former). However, such coherence is not always effectively established in a sustainable and mutually acceptable way (Büntzel and Marí, 2016), and there are no a priori answers as to what contributes to, or detracts from, such coherence.

Probing into the brands and manuals of the things that circulate in the market is an important methodological device within the sociology of farming. Do these brands and manuals coincide with the interests and expectations of producers and consumers? Or do they run counter to their interests and expectations? Who (or what) formulated the brand and manual and influenced its development and wording? Interfering with the genesis or evolution of brands and manuals influences the circulation of the associated commodities and the control over large parts of the social and natural worlds. Through acquiring specific commodities (and neglecting others), consumers participate (willingly or unwillingly, consciously or unconsciously) in the design and reproduction of particular modes of production. And, by selling particular commodities, producers strongly influence the sphere of consumption. Control over the market thus gives enormous power. The market needs to be understood as an important arena where conflicting social interests interact, struggle, negotiate and fight for hegemony. Who and/or what controls the market becomes a strategic question. In a beautiful and convincing Ph.D. thesis, Louis Thiemann (2022) wrote, 'Capital[9] functions – and therefore *is* – only when and where it succeeds in making labour depend on it. It must create [...] a web of dependency relations that makes workers and consumers participate in its designs' (1). The construction of brands and manuals plays an influential role in determining how such webs of dependency are constructed. It is also through the construction of new products (containing a different or altered brand and manual) that alternatives can be created that bypass the dependency relations that are central to capital. In this respect, new products (with radically different brands and manuals) can become catalysts for change in disguise: 'undercover agents' that start to alter patterns of consumption and production. The difficulty, of course, is to materially align the brand and the manual and to materially change (parts of) the spheres of production and consumption.

Is it utopian to assume that resistance and transformative capacity can be built into particular products, thus transforming them into actants that struggle for change?

This question still is far from satisfactorily resolved, but here I suggest some conceptual and methodological guidelines that may be helpful in tackling this issue. First, the comparative approach (see Methods Box #1) might be helpful here. There are (still) many food products that have not (yet) been industrialized – they have *resisted* direct subordination to the logic of capital. Some of these products have been widely documented (for some examples, see Text Box 10.3). Comparing these products (which share such in-built resistance)[10] to fully industrialized food might help to reveal at least some of the answers.

A first common element is *craft* (Sennett, 2008), which is a decisive feature in the production of these types of food. Their production is organized in an artisanal way, which critically requires craftmanship: the capacity to care for, and deal with, co-production. Living nature, as embodied in raw milk, Chianina cows or the *terroir* for Parmesan cheese, is decisive for the organoleptic and nutritional qualities of these food products, whilst craft-like labour (the opposite of industrialized labour) is strategic in fully unfolding these specific expressions of living nature. Their brand fully depends on, and faithfully reflects, the craft-like fine-tuning of the labour process and the care that comes with it. It is impossible to standardize the production of these types of food products to (partly)

Text Box 10.3 Uncaptured food

Chianina cattle (see Text Box 6.1 and Figure 10.6) is a precious breed with ancient roots. It was used in Roman times to transport marble from the Carrara caves to the nearby harbour from where it was sailed to Civitavecchia in order to be brought (again by Chianinas) to Rome to be used in the many monuments, some of which are still standing today. In later periods, the Chianina was also used for traction power (especially ploughing heavy clay grounds). It was bred and selected for its traction power, and this is still reflected in the height and long legs of the animal. Alongside its power, the Chianina has an exceptional meat (with an intramuscular dispersion of fat – what connoisseurs call *marbled meat*). Raising and caring for the Chianina is far from easy. It requires the on-farm production of a wide array of feed and fodder ingredients (including broad beans), gentle treatment and good housing conditions. Large-scale production is impossible. The Chianina requires skills, care and *impegno*, as farmers say (*impegno* literally means 'dedication', but in this context also refers to a high labour input per animal). There are additional features that make the animal unattractive for capital. One reason is that it is literally too slow. It takes 18 months to breed a calf to have enough weight (and quality) to be slaughtered. That is much longer than other breeds and would definitely slow down the accumulation of capital. Beyond that, the calves cannot be separated from the mother. In short: living nature (represented here by the Chianina) imposes its own rules on the agricultural labour process. Co-production has to respect these rules, and the more it does so, the more productive the outcomes.

The care given to the animals needs to be repeated in the other stages involved in bringing the animal to the table. Slaughtering (including transport to the slaughterhouse) needs to be done in a gentle way to avoid stress, which would irreversibly damage the quality of the meat. The butchers need to be knowledgeable and skilled. The animal needs to be hung for at least 14 days and to be carefully portioned to facilitate the ripening process. Consumers also need to be knowledgeable: they need to 'understand' Chianina meat in order to prepare it properly and

Figure 10.6 The Chianina (male and female) (Marleen Felius [1996]; reproduced with permission).

appreciate it's qualities. Bistecca fiorentina (made from Chianina beef) is one of the masterpieces of Italian cuisine.

In short: the production, processing, distribution and consumption of Chianina beef are, to borrow an expression from Marc Wegerif (2017), a symbiotic whole. The different stages (and the different locations in time and space that come with them) have to be neatly coordinated (in his thesis, Hielke van der Meulen [2000] gives a fascinating overview of such coordination). In each stage the combination of craft and care is decisive and needs to be optimized in order to deliver the best possible quality. The market for Chianina beef is a very 'long-standing' territorial, or nested, market involving very short circuits and jointly shared normative frameworks that link producers, butchers and consumers.

The re-emergence of craft and care in subsequent stages can be noted in several other chains as well. The production of the right kind of milk for making Parmigiano-Reggiano cheese is a craft on its own, implying a relatively high labour input on the farm (see Table 5.2). The transportation of this milk from the farm to the cheese factories (*caseifici* in Italian) and the subsequent processing of the milk into cheese are equally delicate operations. The milk should not be transported much more than 20 kilometres or else it will be degraded. The processing cannot be industrialized but requires a *casaro* (cheesemaker: see Figure 10.7), a profession

Figure 10.7 A *casaro* (cheesemaker) preparing Parmigiano-Reggiano cheese (photo by Consorzio).

Figure 10.8 Dutch farmhouse cheesemakers (photo by Hans Dijkstra).

that requires years of apprenticeship. It also requires a local network of farmers to deliver the milk. This network is needed to know exactly how, and under what conditions, the milk was produced and also for correcting the farmers – if the need arises. A nice detail, which extends the principle of co-production, is that in shops (including supermarkets) the cheese is grated in front of the consumers, giving added reassurance that they are being provided with genuine Parmesan cheese.

Finally, I want to discuss Dutch *boerenkaas* (farmers' cheese): cheese that is produced on-farm, only using raw milk. Raw milk (or *latte vivo* [living milk], as the Italians say) is not pasteurized (as is the case with all industrial dairy products). Working with raw milk involves co-production. It is about perceiving and understanding the particular characteristics of the milk (characteristics that change from day to day, from animal to animal, from one type of roughage to another, etc.) and translating this into the most appropriate treatment of (i.e., care for) the milk that is being processed into cheese. There is much to tell about *boerenkaas* (and similar cheeses from elsewhere, such as French Gruyere), but I will limit myself here to one telling aspect. *Boerenkaas* is sold at the farm gate (in farm shops, as discussed in Chapter 8), at daily and weekly markets and in specialized cheese shops. That is, it is sold by people who are knowledgeable about cheese. It is rarely distributed through large retailers. For one thing, knowledgeable sales agents are too expensive for supermarkets, but equally because the producers of *boerenkaas* are too small scale to be able to deliver the large quantities required by supermarket chains. Equally, most cheese-producing farmers do not want to become dependent upon large retail chains. And, finally, the product is too variable: it is not standardized and therefore not suitable for distribution through large retailers.

industrialized processes of agricultural production or food processing. To do so would effectively annihilate the very nature of the qualities that are embodied (i.e., 'materialized') in these food products.

A second feature shared by the food products discussed in Text Box 10.3, and many others like them, is that they enable co-production to *travel* to the spheres of processing, circulation and consumption. Chianina beef needs to be slaughtered, hung and proportioned in a particular, craft-like way, just as it assumes knowledgeable consumers who are able to prepare the meat in the required way. It requires consumers who 'know the manual'. If not prepared correctly, the meat will not reveal its taste and quality. The same applies to the other products mentioned. Parmesan cheese, for instance, is grated in front of the consumer to guarantee that it truly is Parmesan cheese that is being acquired. A third feature I want to note here is that all of these particular products are part of strong culinary traditions within which they are deeply anchored (to the degree that they are part of regional identities). For consumers in Umbria and Tuscany, Chinana beef is *carne nostrano* (our meat).

In more general terms (and now I jump to the level of concepts), it can be argued that *boerenkaas, carne Chianina, Parmigiano-Reggiano* (and many similar products)[11] essentially represent *extended co-production*. They *extend* the encounter with living nature to the domains of processing, circulation and consumption. They bring specific sets of rules that shape these domains into an extension of co-production and, by doing so, they keep the logic of capital at a considerable distance.

Parallel to (and sometimes combined with) this particular extension, there is yet another mechanism for extending co-production to other spheres of both society and nature. This is the symbiotic inclusion of landscape management, the protection of biodiversity, the reduction of fossil fuel use, etc., within the process of agricultural production (see Chapter 6).

The different ways of extending co-production partly operate through (new) commodity circuits. The delivery of ecosystem services is paid for (often by the state as a 'public good' but sometimes by conscientious consumers), just as the production of the right kind of raw milk for conversion into Parmesan cheese and *boerenkaas* is remunerated through the market. But these particular commodity circuits are *not* controlled by capital, just as it is *not* all-purpose money that is circulating in these circuits: *this money renders good labour incomes* for those who produce excellent food and/or protect landscapes and nature. It definitely is *not* oriented at obtaining, in the first and foremost place, a higher than average return on capital.

All of this suggests that actively extending co-production is probably an important vehicle for change (change that does not necessarily require an a priori takeover of power, as John Holloway [2000] argued). It pushes back material dependency on capital[12] and helps to create new alliances (in this case with consumers) and thus strengthens the economies of opposition.

Can capital take over these (potentially threatening) activities, 'co-opt them' and defuse their transformative potential? I think not – and the key for this resides precisely in the *combination* of productive activities (such as raising and breeding Chianinas) with other activities such as landscape management, direct marketing, etc. (see Chapter 7). Taken in isolation, each single field of activity provides a low level of remuneration. Only when skilfully combined do they provide an adequate labour income. However, the making and reproduction of such combinations critically depends on craft, on the well-tuned and place-bound intertwinement of different forms of co-production, and this often requires patience. They cannot be guaranteed through control at a distance. Multifunctionality (see Chapter 7) critically depends on skilful farmers, and while it can render a good labour income, it definitely fails to deliver the returns that capital would expect.

Reading markets

Markets make products flow, from producers to consumers, and they make money flow the other way, from consumers to producers. Markets also make information, expectations, confidence and trust flow in different directions. Brands and manuals are, in this respect, important vehicles – but they are far from the only ones. Currently, markets are far removed from direct transactions between producers and consumers. Nearly all products are compound assemblies of many ingredients coming from different places and passing through different domains – most of which are linked through multiple transactions.

Products and services do not flow in random, undirected and/or chaotic ways. They flow through purposely organized channels that have points of inlet and points of outlet. These channels can be said to form socio-material infrastructures which can be relatively simple or highly complex, localized or global, decentralized or governed from one clear *centre of command* (see Chapter 8) – the key is that they make products (and services) flow in particular and well-delineated ways that are in line with particular interests and perspectives, whilst marginalizing others.

Control over markets for food and agricultural products is a defining criterion of food empires. Food empires are extended, oligopolistic, globally operating networks that increasingly control the production, processing, distribution and even consumption of food. I have previously described food empires as three-tiered phenomena, shaped like a pyramid. At the lowest level there is the socio-material infrastructure, which consists of factories, farms, logistical systems, shops, etc. The in-between level is the level of movements: raw materials moving from farms to factories, food from factories to shops and then to customers, packing materials following yet other routes, whilst technological artefacts are moved to the places where they are needed, etc. The top level is the one of control. The movements of products and services are directed and controlled from the top (the *cupola*) of food empires, and it is here that the value created with these moves is appropriated. The ingredients needed are preferably obtained from impoverished spaces of production and the assembled food directed to rich spaces of consumption. Thus, food empires make ingredients and food travel long distances – both in time and in space.

Control over the movements of food and agriculture often resides in the ownership of the elements that make up the socio-material infrastructure. This coincides with the way in which most food empires are constructed:

through ongoing takeovers that aim to create and solidify their market power. Yet, at the same time, these takeovers represent one of the Achilles' heels of food empires. The acquisition of other enterprises is mostly funded with credit – thus, food empires often show very high levels of indebtedness. Text Box 10.4 gives an illustration of a particular food empire and the extent and connectedness of its global interests.

Text Box 10.4　De Heus, a Dutch food empire

De Heus company is a typical food empire. It is based, and has its origins, in the Netherlands. It first developed as a company importing ingredients for animal feed and now operates globally. One of the countries where it is quickly expanding is Myanmar, where it has a firm grip on imports of maize for cattle feed. Domestic maize only accounts for an estimated 35% to 40% of the country's total feed demand, which amounted to 3 million MT in 2020. De Heus is a leading supplier in Myanmar and also indirectly controls much domestic maize production. It also controls growing parts of the meat market, with three production locations and a new slaughterhouse. In addition, it owns the Grandparent (GP) Stock Breeding farm in Kayah State, where piglets are raised to be sold to commercial breeders for fattening, after which the animals are slaughtered and processed in other units that also belong to the infrastructure controlled by De Heus.

Figure 10.10 shows the arrival of 271 GP piglets on Myanmar soil. They were bred by Topigs Norsvin (located in France) and flown in from Paris by Boeing 747. This is empire in a nutshell: bringing in piglets from Europe in order to establish a breeding system that delivers genetic material to farmers in Myanmar, who will feed the animals with animal feed delivered by De Heus, and to be slaughtered in a facility owned by the same company, after which it will be sold in the domestic market.

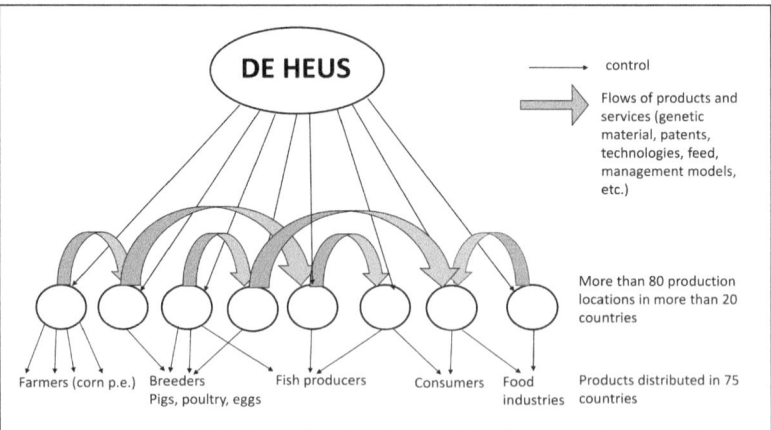

Figure 10.9 The three-tiered structure of the De Heus empire (author's own elaboration).

Figure 10.10 Pigs that fly (www.agroberichtenbuitenland.nl/actueel/nieuws; 2 September 2019).

The company has created similar webs of dependencies for fish products, eggs, etc. Control over genetic material, knowledge, technologies and long-distance trade patterns is decisive in these arrangements. In its branding, De Heus describes itself as 'powering progress'.

The imperial organization of markets nearly always implies considerable barriers for newcomers and often also leads to the exclusion of certain categories of consumers. The typical 'green deserts' (areas where it is almost impossible to buy fresh, unprocessed, food) typically located in poor, urban areas in the USA, the EU and elsewhere are just one expression of this. But the presence of such structural holes (see Methods Box #16) is not solely limited to social actors. It may very well also involve products: specific expressions of authenticity, locality, freshness and quality can easily be excluded. The same applies for specific innovations.

However, such structural holes can be bridged. Bridging the separation (the structural hole) between different value circuits has been, through the centuries, the typical activity of entrepreneurs, mostly occupying a marginal position: those who succeed in bridging these gaps make good money out of it, and some have gone on to create commercial empires.

The point I want to make here is that many of the structural holes inherent to the imperial ordering of food markets *point to the possibility of constructing new markets* that are nested in coalitions of farmers and urban consumer groups. Food empires and the obligatory passage points they represent can be *bypassed* (see Figure 10.11) through the construction of new, nested markets that build direct relations between producers and consumers – direct relations that are nested in

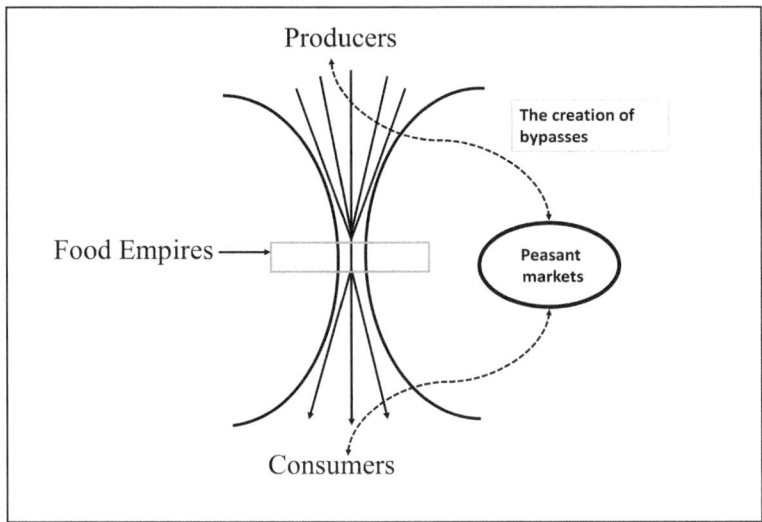

Figure 10.11 Empire-controlled markets and newly emerging peasant markets (author's own elaboration).

jointly shared definitions of both the brand and the manual of the exchanged products. These direct relations also create a new socio-material infrastructure in which proximity is central.[13]

Critically reading global food markets allows for the identification of particular mechanisms and features (that result in structural holes) that might further facilitate the construction of new, nested markets.

Global food markets are profit driven. They only move food is as far as moving it renders profit. Beyond that, they critically depend on the availability and use of credit. In this respect, the Brazilian Ecovida network (introduced in Chapter 8) represents some startling differences, two of which are of particular importance here. The first is that the circulation of food (from one area to another) is not profit driven but is oriented by the search for good labour incomes for the participating farmers. The second relates to having made credit redundant. In conventional markets, distribution is profit driven. In this sense, it is *trade*: a search for the lowest possible price for obtaining commodities and the highest possible price for selling them. In the case of Ecovida, products only go to another area (and the *feira* located there) if those products cannot be produced there (or at a given time). In such cases, the farmers of the supplying production area carefully assess a 'market price' based on the costs incurred, labour time dedicated and transport costs. Bridging ecological differences and the wish to increase labour incomes (by selling more products) without increasing the prices paid by consumers, rather than profit, are the ordering principles. If possible, the flow of food products from one area (A) to another (B) is balanced with another flow (representing more or less a similar value) that will go from B to A. If this is the case, no monetary exchange is needed.

Figure 10.12 Inside a Japanese farmers' market (photos by author).

Selling the products from A in B remunerates the farmers of B who sent their products to A. The value of the products from B sold in A remunerates the farmers in A. Thus, *tied sales*[14] allow for circulation without any dependency on monetary exchange.[15] Such mechanisms have considerably strengthened and expanded this peasant market, which now operates over large distances (see Figure 8.5).

Global food markets impose prices without making clear how they are arrived at. In this respect, the *mercati contadini* (peasant markets) of Rome show another interesting difference: they were linked to the central computer of the Ministry of Agriculture[16] and thus knew the average supermarket price of different products. This price was shown on huge boards together with the *prezzo amico* (the price to be paid by friends) of the peasant market. This *prezzo amico* was, and still is, 20% below supermarket prices.

Another difference is that global food markets are anonymous, while peasant markets showcase the producers: who can be seen, visited and spoken to directly by consumers. Figure 10.12 shows a farmers' market in Kyoto, Japan, with the portraits of the producers attached to one of the walls, nicely illustrating this feature.

In synthesis: a critical reading of the dominant markets can generate many insights that are very useful for the design of new, nested markets that offer a radically different manual.

Reading development

The problem with many contemporary studies of rural realities is that they are only loosely (and selectively) connected to *material* reality. This is due to a series of complex and partly interrelated causes. One of these is the use of case study methodology, which is currently very much *en vogue* in the social sciences.[17] Typically, case studies currently take only one, or at best two or three, cases on board (thus, they are far from the methodological principles outlined by Glaser and Strauss

[1967] in their classical book on the construction of grounded theory[18]). Generally, the cases that are selected are expected to 'confirm' the theory being expounded. Strictly speaking, such case studies can only render hypotheses – they are exploratory rather than explanatory. They test, at best, the *plausibility* of describing ('retelling') specific phenomena with specific, and mostly new, concepts. That is fine, but it is equally possible to retell them with *other* concepts. Farming realities (and, more generally, rural realities) are open to multiple interpretations: they can be made to fit within different, mutually contrasting, narratives. This is partly because these realities are actively structured according to different repertoires and logics and partly because there are different theoretical approaches. At the crossroads of the two there is an impressive range of possibilities to explore, study, theorize and represent rural and agricultural realities. The wide range of (often highly contrasting) case study approaches is further proof of such poly-interpretability.

The same applies for the much-used approach of cross-sectional analysis, which basically comes down to selecting two (or more) different data sets (sometimes based on stratified sampling) that are supposed to reflect the operation of a supposedly underlying structural feature (i.e., the independent variable). Yet, there is usually no real empirical check on the validity of the independent variable being the only and/or most important root cause of the variety among the dependent variables. A statistical association might merely be a spurious correlation.[19] Strictly speaking, cross-sectional analysis is only justified (and can only really reveal the root causes of certain empirical phenomena) if there is *also* an additional inquiry into, and thorough theoretical discussion of, the probability that certain variables do, indeed, operate as *explanans*; that is, that they really *explain* that what they are supposed to bring about. Just as with case studies, cross-sectional differences can be interpreted in different and mutually contrasting ways.

In short: the problem of contemporary rural studies is that they easily end up *floating* (and this applies as much to the social science side of rural studies as to the agronomic and technical sides).[20] There is no single, exclusive and well-tested relation between theory and empirical reality. The results and insights of rural studies may be true or untrue: most probably, they only *partly* reflect empirical realities (alongside other competing or conflicting theories). In this context, the 'right view' becomes a question of power.

To avoid *floating*, it is paramount to ground rural studies as much as possible in the analysis of time series that reflect the development of material constellations over longer periods of time. Focussing on *material* realities (fields, farms, cows, levels of employment, flows of products, etc.) reduces the probability of floating – simply because one cannot 'rethink' a field laying barren as being well cultivated or (un)employment being twice as high as it actually is. Materiality cannot be neglected, nor can it be easily represented as differing from what it actually is. At the same time, the artefacts that make up the material world are *socio-material* objects: they allow for a reading that goes backwards from these objects, via the way they have been constructed, towards the social actors who are/were involved in their co-production. Thus, we go from the material to well-grounded theories about the *social* side of the equation.

But rural studies should not just be about taking the material in all of its immediacies: they need to be based on *time series* that reveal and document how socio-material realities have been unfolded (moulded and remoulded) over longer periods of time. One reason for this is that it excludes the possibility of projecting *assumed* tendencies onto the empirical material.

Comparative analysis of the development of socio-material realities (not in terms of moving averages but through the diversity of actually existing empirically pathways)[21] is probably the most powerful methodological tool of the sociology of farming and should, I think, be a guiding example for rural studies in general.

A first example I want to present here relates to 'de-agrarianisation' (Bryceson, 1996), a process that refers to the disappearance of agriculture from large areas, which implies smallholders abandoning cropping their fields. Charley Shackleton and others (2019) assessed the validity of this theory (which tends to be supported by data describing the *average* proportion of regionally available land dedicated to crop production) in South Africa. They noted a 'variation in the extent and rate of decline and in the drivers of this change' (Shackleton et al., 2019:687). By taking landscape photographs from archives, they constructed 'photo pairs': photos of the same landscape, taken from the same angle at different moments, often covering a considerable time span. The first photo from each pair showed the situation in the 1930s or 1940s and the second presented the *same piece* of land in the second half of the 2010s (see Figure 10.13 for an example).[22] The differences between the initial and final year were subsequently categorized with values of −2 (more than 25% decline in cropland), −1 (between 5% and 25%), 0 (stable situation), +1 (expansion of crop land of between 5% and 25%) and +2 (more than 25% expansion). The outcomes of this inquiry showed that 'there is no simple abandonment but *a combination of deactivation and reactivation* of fields' (Shackleton et al., 2019:692, italics added). That is, the analysis highlighted that rural development is a differential process. In one-third of the photo pairs the decline was greater than 25% of the original area. One-fifth of the sites showed little change, while at 10% of the sites there was evidence of an increase in the area of cultivated fields. This result convincingly demonstrates that there is no single (and generic)

Figure 10.13 Paulshoek, Namaqualand, 1938 and 2005 (courtesy of Rick Rohde and Tim Hoffman).

process of 'de-agrarianization'. Instead, 'field cultivation clearly remains part of [...] livelihood strategies' in which 'home garden cultivation has intensified' as a result of 'rural people respond(ing) to opportunities and adversities at many different levels' (Shackleton et al., 2019:693).[23] Such insights allow for the formulation of policies that are far better attuned to local situations and prospects than current policies (aiming at food security) could ever do.

A second example I want to introduce briefly is derived from research I undertook with Eppo Bolhuis (see Ploeg, 1990). This research regarded the development of agricultural employment in an extended area (Luchadores) that was initially controlled by a large *hacienda* in Peru and later, after the land was invaded, by a workers' cooperative called Luchadores del 2 de Enero (Combatants of January the Second[24]; this cooperative is also discussed in Methods Box #17).

Figure 10.14 shows the development of productive employment over the years for the central area of Luchadores (i.e., for *one and the same* piece of land). Figure 10.14 is another *return to materiality*, as discussed in this chapter: it is about productive employment (people who were de facto integrated in the processes of production located here) and about people being remunerated by, and through, these processes of agricultural production. To put it bluntly: it is not about elegant modelling, nor about people living from social welfare, nor is it about dreaming. It is about hardship and the materiality of life in a poor area: standing barefooted in the fields and working the land in order to earn a daily wage for maintaining the family.

I will not detail here the specific periods considered in Figure 10.14 nor discuss the particularities of the associated levels of productive employment.[25] What is

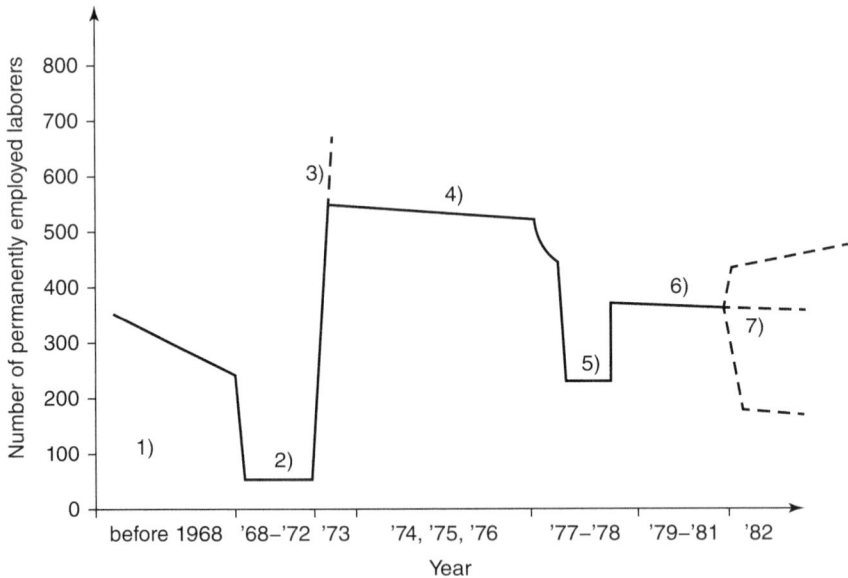

Figure 10.14 The development of productive employment in Luchadores (Peru) (Ploeg, 1990).

important here (from a theoretical point of view) is, in the first place, that the development of productive employment followed, almost like a compass needle, the changes in power balances. With the growing power of the labour union, employment levels went up (from points 2 to 3 in Figure 10.14). With the subsequent return of the hegemony of the landowner (and, later, the state and the military), employment levels went down (from points 3 to 5); point 7 indicates the insecurity reigning at the time the research was done. In addition, these changes in productive employment had a very strong impact on the cultivated area, cropping scheme, yields and, consequently, total production. The *diachronic* approach (i.e., taking into account a longer time series)[26] showed that labour-driven intensification was turned into a reality – even under difficult conditions. It also showed that such labour-driven intensification can collapse (at least partly) if and when the required conditions are demolished.

From a methodological point of view, the trends summarized in Figure 10.14 imply that the theoretical conclusions (summarized above) are firmly *rooted* in empirical reality: they are proven to be possible in *practice*. And that is the strength of the sociology of farming: showing that alternative trajectories, that go beyond the misery of the average trend, are far from a pipe dream and are very much attainable.

Notes

1. These material elements are partially composed by objects that belong to living nature (see Chapter 1). But that which applies to the co-production of 'man and living nature' generally also applies more specifically to the interaction of social actors and the material objects used in, and resulting from, agricultural production. The different elements mutually shape each other.
2. Stubbornness, endurance and distrust in new, unknown items and messages are likely all part of these skills. That is, the proverbial skills here are not limited to technical know-how but also include the social and the ways to deal with it.
3. All artefacts contain a manual. A code is a specific manual: it is the manual built into techniques and technologies.
4. This does not just apply to the micro level. At the macro-level it applies that 'grains make states' (Scott, 2017:128) just as mountains make for anarchism (Scott, 2009). In the same vein, pigs played an important role in the development of long-distance trade. As vividly discussed by Toynbee (1965), this was due to the fact that they could walk (i.e., 'transport themselves'). Annie Proulx (2002) made the same point for cows in the US Midwest, where driving cattle to the main fattening and slaughtering centres in the east was a way of getting rich – if you could find water and get past the Indians and rustlers. And it is not only crops and cattle: Hugh Campbell (2021) argued that *farms* have agency as well, a point of view that can also be encountered in, for example, *Stepmother Earth* by Theun de Vries (see Ploeg, 2021).
5. These data are derived from a study performed for and published by the National Council for Agricultural Research (NRLO), Report 94/1, 1994 (Ploeg et al., 1994).
6. β Shows both the strength and impact of the relation between two variables. When the first variable changes by $+s$ (with a unit identical to the standard deviation), the second variable changes by $\beta * s2$ (in which s2 is the standard deviation of the second variable. Betas can be calculated with multivariate techniques, and they are often used in path diagrams such as the one shown in Figure 10.1.

7. Implying that things can be produced (and/or consumed) in radically different ways.
8. See Meulen (2000), Dessein (2002), Black (2012), Viteri (2010), Hanser (2016), and Chikulo et al. (2020).
9. Capital here is understood here as the centralized control over production and consumption that allows for the appropriation of value and subsequent accumulation.
10. This subtitle is a wink to the 'uncaptured peasantry' of Goran Hyden (1980).
11. The problem is that while products such as *boerenkaas*, Parmesan cheese and Chianina beef are highly visible, there are many others that are hidden and whose values have not been realized or made explicit.
12. There is another strategic feature here: the food products that represent extended co-production. They are mainly (though not exclusively) marketed locally (even in the case of Parmesan cheese a large part of total production is consumed in the area of production). Thus, there is less need for long-distance trade and/or massive credits. Long-distance trade and the provisioning of credit have been, through the ages, favoured ways of accumulating capital (Braudel, 1992; viz. the historically lucrative trades in coffee, sugar and spices). In our times of globalized economies and the strong dependency of the real economy on the financial economy, long-distance trade and credit have become even more embedded as cornerstones in the accumulation of capital.
13. Thus, nested markets are not reducible to mere voluntarism – nor do they solely exist as an extra-economic luxury for middle- and high-class consumers. They are, instead, actively constructed responses to structural holes, which are a reason for the emergence and development of these nested (or peasant) markets.
14. Interestingly, the use of money is socially defined here. Again, it is not 'all-purpose' money.
15. The Dutch *landwinkels* discussed in Chapter 8 also function without any recourse to credit.
16. After pressure from the large retailers, the ministry had to stop providing these data, but the market found another way to obtain them.
17. This is, at least, partly understandable. It reflects the scarcity of resources for doing more extensive research, because case studies render quick results. It also reflects the pressure (especially in universities) to publish articles – preferably as many as possible. Case study methodology also allows the researcher to abstract from many empirically existing problems located 'outside' of the case.
18. This principle implies that the addition of extra cases should be continued until the last case adds nothing new. Then the database is *saturated*: it covers the large heterogeneity that can be found within a rural area. This principle avoids the limited, and therefore selective, empirical grounding of the theoretical projections being elaborated.
19. If there is a statistical correlation between, say, A and B (assuming that A causes B), there might very well be a third variable C that comes with both A and B. Then the association between A and B might be due to the presence of C. As such, the link between A and B is spurious.
20. Agronomic sciences increasingly focus on theoretical constructs (such as optimal yields, the lowest possible contamination per unit of end-product, etc.) that are imposed on empirical realities as normative parameters. This puts these theoretical constructs beyond the possibility of *falsification* by specific and awkward parts of highly diversified empirical realities that do not 'fit the script'. When 'mismatches' emerge, the falsifications are simply put aside as 'refrigerator anomalies' (deviations that are put to one side in order to be explored later, maybe).
21. This requires the use of constant samples (as illustrated in the empirical through-flow of farms obtained from Dutch agricultural census data over 26 years and how this varies from aggregated statistics that merely express the 'average', as explained in Chapter 5).

22. The photo at the left shows threshing of the wheat after harvest; the photo on the right shows an increase of the semi-toxic woody shrub *Galenia africana* after the field was abandoned.
23. See also Hebinck and Monde (2007).
24. This is the date in 1973 when the former hacienda was invaded by the labour union and the population of the surrounding villages.
25. The interested reader is referred to Ploeg (1990:207–233).
26. As opposed to the synchronic approach that is inherent to a cross-sectional analysis.

Bibliography

Benvenuti, B. (1982), 'De technologisch administratieve taakomgeving (TATE) van landbouwbedrijven', *Marquetalia* **5**, pp. 111–136.

Black, R. E. (2012), *Porta Palazzo: The Anthropology of an Italian Market*, University of Pennsylvania Press, Philadelphia.

Braudel, F. (1992), *The Wheels of Commerce: Civilization and Capitalism, 15th–18th Century*, Vol. 2, University of California Press, Berkeley.

Bryceson, D. F. (1996), 'Deagrarianization and rural employment in sub-Saharan Africa: A sectoral perspective', *World Development* **24** (1), pp. 97–111.

Büntzel, R., and F. Marí. (2016), *Gutes Essen – Arme Erzeuger: Wie die Agrarwirtschaft mit Standards die Nahrungsmärkte Beherrscht* [Good food – Poor producers: How the agricultural industry rules the food markets with standards]. Oekom, München, Germany.

Campbell, H. (2021), *Farming Inside Invisible Worlds. Modernist Agriculture and Its Consequences*, Bloomsbury Academic, Sydney, NSW, Australia.

Chikulo, S., P. Hebinck, and B. Kinsey. (2020), '"Mbare Musika is ours": An analysis of a fresh produce market in Zimbabwe', *African Affairs* **119** (476), pp. 311–337.

Delanda, M. (2006), *A New Philosophy of Society: Assemblage Theory and Social Complexity*, Continuum, London.

Dessein, J. (2002), *Het Stremmen en Stromen van de Markt*, Ph.D. thesis, Universiteit van Leuven, Leuven, Belgium.

Felius, M. (1996), *Rundvee, Rassen van de Wereld, een Systematische Encyclopedie*, Misset, Doetinchem, the Netherlands.

Glaser, B. G., and A. L. Strauss. (1967), *The Discovery of Grounded Theory: Strategies for Qualitative Research*, Sociology Press, Mill Valley, CA.

Groot, J. C. J., J. D. van der Ploeg, F. P. M. Verhoeven, and E. A. Lantinga. (2007), 'Interpretation of results from on-farm experiments: Manure-nitrogen recovery on grassland as affected by manure quality and application technique, 1, an agronomic analysis', *NJAS – Wageningen Journal of Life Sciences* **54** (3), pp. 235–254.

Groot, J. C. J., W. A. H. Rossing, E. A. Lantinga, and H. van Keulen. (2003), 'Exploring the potential for improved internal nutrient cycling in dairy farming systems using an eco-mathematical model', *NJAS – Wageningen Journal of Life Sciences* **51**, pp. 165–194.

Groot, J. C. J., W. A. H. Rossing, and E. A. Lantinga. (2006), 'Evolution of farm management, nitrogen efficiency and economic performance of dairy farms reducing external inputs', *Livestock Production Science* **100**, pp. 99–110.

Hanser, A. (2016), 'Street politics: Street vendors and urban governance in China', *The China Quarterly* **226**, pp. 363–382.

Haverkort, B. (2021), *Oude Wortels, Nieuwe Scheuten: Een Zoektocht in Wereldbeelden*, Vol. 42, African Studies Centre, Leiden University, Leiden, the Netherlands. https://hdl.handle.net/1887/3249100.

Hebinck, P., and N. Monde. (2007), Production of crops in arable fields and home gardens, in Livelihoods and Landscapes: The people of Guquka and Koloni and their resources, edited by P. Hebinck and P. C. Lent, 181–221, Brill Academic, Leiden/Boston.

Holloway, J. (2000), *Cambiar el Mundo sin Tomar el Poder: El Significado de la Revolución Hoy*, El Viejo Too, Madrid.

Hyden, G. (1980), *Beyond Ujamaa in Tanzania: Underdevelopment and an Uncaptured Peasantry*, University of California Press, Berkeley.

Kimanthi, H. (2019), *Peasant Maize Cultivation as an Assemblage: An Analysis of Socio-cultural Dynamics of Maize Cultivation in Western Kenya*, Ph.D. thesis, Wageningen University, Wageningen, the Netherlands.

Latour, B. (2005), *Reassembling the Social: An Introduction to Actor-Network-Theory*, Oxford University Press, Oxford.

Long, N. (2016), 'Actors, interfaces and development practice: Revisiting theoretical and ethnographic perspectives', in *Key Ideas in Contemporary Agrarian and Development Studies*, edited by J. Ye, Vol. 2, 85–98, Social Sciences Press, Beijing.

Meulen, H. van der. (2000), *Circuits in de Landbouwvoedselketen: Verscheidenheid en Samenhang in de Productie en Vermarkting van Rundvlees in Midden-Italie*, Ph.D. thesis, Wageningen University, Wageningen, the Netherlands.

Ploeg, J. D. van der. (1990), *Labour, Markets and Agricultural Production*, Westview Press, Boulder, CO.

Ploeg, J. D. van der. (2021), 'Stiefmoeder Aarde [Stepmother Earth] by Theun de Vries (1936)', *Journal of Peasant Studies* **48** (5), pp. 1124–1139. https://doi.org/10.1080/030 66150.2020.1774701.

Ploeg, J. D. van der, H. Renting, and J. Roex. (1994), *Meerdere Vergelijkingen en Veel Onbekenden*, NRLO, Den Haag, the Netherlands.

Proulx, A. E. (2002), *That Old Ace in the Hole*, Scribner, New York.

Reboul, C. (1984), 'Barrages contre le développement? Les grands aménagements hydrauliques de la vallée du fleuve Sénégal', *Revue Tiers Monde* **25** (100), pp. 749–760.

Scott, J. (2009), *The Art of Not Being Governed: An Anarchist History of Upland Southeast Asia*, Yale University Press, New York and London.

Scott, J. (2017), *Against the Grain: A Deep History of the Earliest States*, Yale University Press, New Haven, CT.

Sennett, R. (2008), *The Craftsman*, Yale University Press, New Haven, CT.

Shackleton, C. M., P. J. Mograbi, S. Drimie, D. Fay, P. Hebinck, M. T. Hoffman, K. Maciejewski, et al. (2019), 'Deactivation of field cultivation in communal areas of South Africa: Patterns, drivers and socio-economic and ecological consequences', *Land Use Policy* **82**, pp. 686–699.

Staudenmaier, J. M. (1985), *Technology's Storytellers*, MIT Press, Cambridge, MA.

Thiemann, L. (2022), *The Third Class: Artisans of the World, Unite?*, Ph.D. thesis, Institute of Social Studies (ISS), Erasmus University, The Hague/Rotterdam, the Netherlands.

Toynbee, A. J. (1965), *Hannibal's Legacy: The Hannibalic War's Effects on Roman Life*, 2 Volumes, Oxford University Press, London.

Viteri, M. L. (2010), *Fresh Fruit and Vegetables: A World of Multiple Interactions. The Case of the Buenos Aires Central Wholesale Market*, Ph.D. thesis, Wageningen University, Wageningen, the Netherlands.

Wegerif, M. C. A. (2017), *Feeding Dar es Salaam: A Symbiotic Food System Perspective*, Ph.D. thesis, Wageningen University, Wageningen, the Netherlands.

Index

Note: page numbers in italic type refer to Figures; those on bold type refer to Tables.

Page numbers followed by 'n' and another number refer to notes.